AF539715

# Climate Change and Sustainable Agriculture

nipa
Browse Subject
eBooks / eChapters / Articles
eBooks
eBooks / eChapters / Articles
Browse, Search, Read & Buy...
Subject Catalogues
eChapters
Publishers
Print Books
Forthcoming
eBooks
New Titles
scan for catalogue
NIPA
GENX
ONLINE RESOURCES
Publishing Books and Journals
Ebooks and e articles, Current Affairs
Reasoning, Logic and Aptitude
Competitive Examination Preparation
Language Learning and Development Programme
Document Quality Checker and Improvement Tool
Effective Public Speaking, Presentation and Interpersonal Skills
Online Programmes for Professional Development
Personality Development and Human Values
PAY USING
UPI
PayPal

# Climate Change and Sustainable Agriculture

*Edited by*

**P. Suresh Kumar**
Senior Scientist (Fruit Science)
ICAR-National Research Centre for Banana
Thogamalai Road, Trichy-620102, Tamil Nadu

**Manish Kanwat**
Senior Scientist & Programme Coordinator
Krishi Vigyan Kendra, Anjaw, ICAR RC for
NEH Region AP Centre, Basar-791101, Arunachal Pradesh

**P.D. Meena**
Principal Scientist (Pathology)
ICAR-Directorate of Rapeseed Mustard Research
Sewar, Bharatpur-321303, Rajasthan

**Vinod Kumar**
Senior Scientist (Computer Application in Agriculture)
ICAR-Directorate of Rapeseed Mustard Research
Sewar, Bharatpur-321303, Rajasthan

**Rajesh A. Alone**
Scientist (Agroforestry)
ICAR RC for NEH Region AP Centre
Basar-791101, Arunachal Pradesh

**NEW INDIA PUBLISHING AGENCY**
New Delhi – 110 034

**NEW INDIA PUBLISHING AGENCY**
101, Vikas Surya Plaza, CU Block, LSC Market
Pitam Pura, New Delhi 110 034, India
Phone: +91 (11)27 34 17 17 Fax: +91(11) 27 34 16 16
Email: info@nipabooks.com
Web: www.nipabooks.com

Feedback at feedbacks@nipabooks.com



ISBN No. 978-93-85516-72-6



Composed, Designed & Printed in India

## About the Editors

**Dr. P. Suresh Kumar** is a Senior scientist (Fruit Science), presently working at ICAR- National Research Centre for Banana, Trichy, Tamil Nadu. He obtained his PhD at IARI, New Delhi. He has been awarded JRF and SRF for his M.Sc and PhD degree from ICAR and IARI respectively. He has been awarded young scientist awards by Samgra Vikas Welfare Society, Lucknow and Bioved Research Institute of Agriculture & Technology, Allahabad, and also been awarded K.U. Patel Memorial award for his best paper in Indian Food Packer. He has visited USA under NAIP sponsored training programme and underwent advanced training on Biomolecules separation from plants at Michigan State University, East Lansing. During his tenure as Scientist at ICAR RC NEH Region, Arunachal Pradesh, he has handled CSS Technology Mission for Horticulture Development in NE states, NAIP Projects on livelihood improvement via establishment of low cost polyhouse, DBT project for tissue culture production of Khasi mandarin; five different NHB projects for holistic development of Horticulture in Arunachal Pradesh. He also worked at ICAR-CTCRI, Trivandrum and ICAR-NIASM, Baramati with special focus on developing mitigation strategies for edaphic and drought stresses in fruits and vegetables. Currently, he is interested in bio-prospecting and separation of bio-active compounds from horticultural crops and developing functional foods thus alleviate hidden hungers like vitamins, protein and essential bio-regulating compounds. He is an editorial member of several scientific societies and referee to reputed food and Horticultural journals. He has published more than 80 research papers of national and international repute, 25 popular articles and 6 review articles. He edited souvenir, written many book chapters and published five book, eight technical bulletins.

**Dr. Manish Kanwat** is currently Senior Scientist & Head at KVK Anjaw, ICAR Research Complex for NEH Region Arunachal Pradesh. He obtained his M.Sc. (Agril. Extension) at Institute of Agricultural Sciences, Banaras Hindu University (U.P.) and Ph.D. (Agril. Extension) at MPUAT, Udaipur (Rajasthan). He has been awarded Junior Research Fellowship and Senior Research Fellowship for his PG and Ph.D. from ICAR and UGC respectively. He has been selected for the CZECH Government Scholarship - 2006-07 for Doctoral programme from Mendel University of Agriculture and Forestry in Brno Czech Republic by Ministry of Human Resource Development, Department Secondary & Higher Education, Government of India, New Delhi. He has good experience of handling of various institutional projects like NAIP, NICRA, TSP and external funded projects from NABARD respectively. Natural Resource management, Medicinal & Aromatic Plants, Indigenous Traditional Knowledge in Agriculture, Communication, Participatory Rural Appraisal based research and Information and Communication Technologies (ICTs) etc. are the key areas of his research. He published many research papers in reputed National and International journals, book chapters, folders, popular articles in National magazine and the leading Newspapers. He is life members of several scientific societies. He has been bestowed with the Best KVK Professional Award by the Society Extension Education.

**Dr. P.D. Meena** obtained Ph D from University of Rajasthan and M Sc (Plant Pathology) from Rajasthan Agricultural University, Bikaner, India. Presently, working as Principal Scientist (Plant Pathology) at ICAR-Directorate of Rapeseed-Mustard Research, Bharatpur, Rajasthan since his selection as ICAR-ARS Scientist from 15 Nov 1998. His research interest includes resistance, epidemiology and management of rapeseed-mustard diseases. Published more than 70 research papers in national and international journals and authored 5 books. He is recipient of the Best Scientist Award of ICAR-DRMR, Dr. PR Kumar Outstanding Scientist Award, Fellow of Plant Protection Association of India, Fellow Society of Mycology and Plant Pathology, Fellow Indian Society of Oilseed Research. He has also been visited Rothamsted Research, UK.

**Dr. Vinod Kumar** obtained Ph D (Computer Science) from Institute of Information & Computer Science, Agra and MCA from MMM Engineering College, Gorakhpur, UP, India. Presently, he is working as Sr. Scientist (Computer Applications in Agriculture) at ICAR-Directorate of Rapeseed-Mustard Research, Bharatpur, Rajasthan. His research interest includes image data management and the development of online application for agricultural research information management using open source technology. Published more than 30 research papers in national and international journals which includes IEEE publications also. He has presented the paper in IEEE conference at China. Associate editor of the Journal Indian Research Journal of Extension Education, and Managing Editor of the Journal of Oilseed Brassica. Recipient of ICAR-DRMR Best

Scientist Award, Young Scientist Award, Best IT Professional Award, Dr. PR Kumar Outstanding Scientist Award and FSMR from different professional societies.

**Dr. Rajesh A. Alone** is a Scientist (Agroforestry) at ICAR Research Complex for NEH Region, Arunachal Pradesh Centre, Basar. He obtained his M.Sc. (Forestry) degree from Forest Research Institute, Deemed University, Dehra Dun and Ph.D. degree in Forestry from Indira Gandhi Krishi Vishwa Vidyalaya, Raipur (Chhattisgarh). He has published many research papers, technical bulletins, popular articles and review articles in journals of national and international repute. He is an active member of many scientific societies. Presently, he is working on the quantification of the carbon sequestration potential of the various Agroforestry systems in Arunachal Pradesh.

रानी लक्ष्मी बाई केन्द्रीय कृषि विश्वविद्यालय
ग्वालियर रोड, झांसी 284 003 (उत्तर प्रदेश) भारत
**Rani Lakshmi Bai Central Agricultural University**
Gwalior Road, Jhansi 284 003 (U.P.) India

डॉ अरविन्द कुमार
कुलपति
**Dr Arvind Kumar**
Vice-Chancellor

Phone : 0510-2730777
Telefax : 0510-2730555
E-mail : vcrlbcau@gmail.com
Website : www.rlbcau.ac.in

# Foreword

Climate change has direct impact on agricultural production, because of the climate-dependent nature of agricultural systems. This impact is particularly significant in developing countries where agriculture constitute employment and income sources for the majority of the population. This at a time when our population is still rising steadily. Developing countries like ours would be the worst hit. Already, for several years, the pattern of monsoon in India has become unpredictable, uncertain and erratic. Scientists attribute these changes to climate change. Agriculture production in many countries, including India will be impacted by climate variability. It is estimated that greater loss is expected in *Rabi* as compared to *Kharif* crops. By 2020 in some Asian countries, yield from rainfed agriculture could be reduced by up to 50 percent. The worst affected would be millions of small and marginal farmers and people who are already the most vulnerable. As these changes take place, agriculture is getting affected adversely, and the threat of decrease in food production is becoming very real. Climate Changeis likely to lead to some irreversible impact on biodiversity. Approximately 20 to 30 per cent of the species assessed so far are likely to face the threat of near extinction.

This book "**Climate Change and Sustainable Agriculture**" is an excellent compilation of research on climate change, aimed at promoting understanding of the most cost-effective and sustainable indigenous climate change adaptation practices to promote sustainable agriculture.

The spirited group of editors from different national ICAR Institute, Drs, Suresh Kumar, Manish Kanwat, P.D. Meena, Vinod Kumar and Rajesh A. Alone have invited 23 chapters encompassed broadly four major thematic areas as the key subjects to provide a broader perspective of climate change. I personally believe this publication will go a long way to highlight many existing and emerging concerns about climate change and their mitigation strategies and practices and shall serve as guide for researchers, teachers, extension personnel and students.

**Arvind Kumar**
Vice-Chancellor
Rani Lakshmi Bai Central Agricultural University
Jhansi, Uttar Pradesh

# Preface

Climate change is perhaps the biggest challenge facing the world today, and the very existence of man depends on how effectively this challenge is tackled. All governments have come together on a common forum to devise means to cope with this phenomenon, which threatens to play havoc with the lives of people across the globe. Our health, agriculture, habitation, everything depends on how effectively we are able to tackle this problem.

Climate change is one of the most complex challenges that humankind has to face in the next decades. As the change process seems to be irreversible, it became urgent to develop sound adaptation processes to the current and future shifts in the climate system. In particular, it is likely that the biggest impacts of changes will be on agricultural and food systems over the next few decades. There is significant concern about the impacts of climate change and its variability on agricultural production worldwide. Climatic changes and increasing climatic variability are likely to aggravate the problem of future food security by exerting pressure on agriculture. However, there are lot of uncertainties about the assessment of impact, adaptation and mitigation of climate change in agriculture. It is largely because the methodology followed for such assessments is not standardized and sometimes it is inaccurate and imprecise.

Within this framework, it is crucial to identify information and communication systems that the farmers need in order to cope with the new conditions. This is particularly true for poor smallholder farmers, as in India where the majority of small and marginal farmers do not have access to the scientific and technological advances that support agricultural decision-making because of the lack of reliable communication networks. With regard to agronomic research, one of the major challenges will be to study how to fill the information needs of policy makers, and how to report and communicate research results in an effective way for supporting the adaptation of food systems to climate change.

This book to comprehensively present the standard methodologies for studying the impacts of climate change on agriculture, measuring and developing inventories of greenhouse gas emission, and analyzing the vulnerabilities and mitigation options. The book describes the methodology in a simple and lucid way so that a researcher can adopt it in field studies. Individual chapters are dedicated to subjects such as quantification of climate change impacts on crops in controlled and field conditions, impacts of climate change on water resources, soil fertility, erosion and carbon sequestration, insects, pests, weeds, microbes and diseases; greenhouse gas emission assessment, assessment of regional vulnerability to climate change, selection of crop. The book converted its 23 chapters into four sections. The first section deals with recent mitigation strategies developed by the scientists to reduce the effect of climate changes and promote the Sustainable Agriculture by different ways viz., Role of Marker in Development of Climate Resilient Varieties. Section 2 deals with the conservation agriculture and mitigation strategies, Biofertilizers helps in development of sustainable agricultural practices. Section three deals with impact of climate change on pest scenario across the crops and has given insight on possible avoidance of pests. Final sections deals with the agroforestry and how they can minimise their affect of climate change. Finally section 4$^{th}$ deals with Impact of climate change crop and livestock sectors and strategies for mitigation and adaptation. Role of ICT and extension agencies on framing policies and effective dissemination of technologies were discussed.

We believe the book has given bird view sight on the such macro topic which governs the present day agriculture. We appreciate the efforts of all our authors to bring this book a highly informative with the content on variety of aspects. We welcome suggestion from all the readers and stakeholders for further improvement in future editions.

**Editors**

# Acknowledgement

The editors wish to express their heartful thanks with reverential honour and profound gratitude to Dr. T. Mahapatra, Secretary (DARE) and Director General (ICAR), Govt. of India, Dr. S.V. Ngachan, Director, ICAR RC for NEH Region, Dr. Dheeraj Kumar Singh, Director DRMR, Sewar, Dr. P.S. Minhas, Director, NIASM, Baramati and Dr S. Uma, Director, ICAR-NRCB, Tiruchirapalli for their support and encouragement for editing the book on convergence mode. We are highly indebted to Dr. A.K. Singh DDG (Agril. Extension) & DDG (Horti.), ICAR and Dr. N.S. Rathore DDG (Education) for their valuable suggestion, encouragement and writing the foreword. The editors would like to thanks Dr. B.C. Dekha Director ICAR-ATARI, Zone-III, ICAR RC for NEH Meghalaya, Dr. A.K. Tripathi, ATIC In-charge & Nodal Officer KVK ICAR RC for NEH Meghalaya and Dr. R. Bhagawati, Joint Director, ICAR A.P. Centre, Basar for their immense help and support in every stage of carrying out the activities successfully. The editors deeply value the painstaking efforts made by each contributor of this book. Needless to say the editors acknowledge the (NIPA) New India Publishing Agency for giving better format and cover page to the book.

**Editors**

# Contents

# Section 1

## Climate Change and System Approaches

# 1

# Abiotic Stresses Forced Challenges in Agriculture Under Future Climate

**P. Suresh Kumar, S.K. Bal, Yogeshwar Singh, V.K. Choudhary, A. Sangeetha and M. Kanwat**

## Introduction

Much of the world's agricultural biodiversity is found in environments marginal for agricultural production. It is in such environments where management of high levels of diversity can become a central part of the livelihood management strategies of farmers and the survival of their communities. Abiotic stress can be termed as the negative impact of environmental factors on the organisms in a specific situation. It is a natural phenomenon that occurs in multiples and interdependent, and its impact varies across the sectors of agriculture. Abiotic stress is the primary cause of crop loss worldwide, reducing average yields for most major crop plants by more than 50%. Drought and salinity are becoming particularly widespread in many regions, and may cause serious salinization of more than 50% of all arable lands by the year 2050 (Wang 2003). Environmental stresses such as erratic and insufficient rainfall, extreme temperatures, salinity, alkalinity, aluminium toxicity, acidity, stoniness and others limit yield and productivity of many cultivated crop plants. Not only are such problems serious today, it seems they are inevitably worsening.

Climatic variability is the biggest challenging factor that affects agriculture in India and elsewhere. Drought, salinity, extreme temperatures and oxidative stress are interconnected and affect the water relations of a plant on the cellular as well as whole plant level causing specific as well as unspecific reactions (Beck *et al.*, 2007). This leads to a series of morphological, physiological, biochemical and molecular changes that adversely affect plant growth and productivity (Wang *et al,*. 2003). Forecasting climate change is imperfect, complex, important, and often controversial. While disputes remain, the consensus foresees accelerating increases in average annual temperature and changes in precipitation coupled with increasingly erratic intra-annual weather patterns. Stemming from these two primary dimensions of climate change (higher averages and more volatility) are melting glaciers and ice caps, rising sea levels and more frequent and more severe extreme weather events. Some of these changes will likely be shared globally – most places will get hotter – but other changes will vary geographically. For agriculturally important agro-ecological zones, higher level forecasting of daily weather extremes (frosts, the intensity and form of precipitation, extreme temperature, etc.) is crucial but even more demanding.

Organized research and innovation have been central to agricultural policy for nearly two centuries, often with the goal of increasing output per unit of land, water, labour or other input. More recently, reducing the negative environmental spill over effects of agriculture as joined improving crop yields and other simple productivity indicators as a research pursuit. With a growing global population, with especially rapid population growth in some of the poorest places, with improved diets for the poor an imperative, and with evident local environmental impacts, agricultural innovation has never been more important. Climate issues add to this already challenging agenda. In this paper, we tried to link the impact of future climate on influencing the various abiotic stressors, and the policy decision to be taken to avoid, adopt and mitigate those climatic vagaries.

## Impact of climate change

Climate has obvious and direct effects on agricultural production. The effects of agriculture on GHG emissions are also large. Agriculture is a major part of the global economy and uses substantial fossil fuel for farm inputs and equipment. Animal agriculture also releases substantial GHGs in the form of nitrogen and methane. Furthermore, and probably more importantly, land clearing and preparation releases carbon from the living biomass that is removed from the land. The 2010 World Development Report draws on analysis of the Intergovernmental Panel on Climate Change (IPCC, 2007) to calculate that agriculture directly accounts for 14 percent of global GHG emissions in $CO_2$ equivalents and indirectly accounts for an additional 17 percent of emissions when land use and conversion for crops and pasture are included (World Bank, 2009). Regional disparities around the global average impact are substantial. India and Africa are projected to see reductions of agricultural output by 30% or more (Cline, 2007). Given that agriculture's share in global GDP is about 4 percent, these figures suggest that agriculture is highly GHG intensive. The climate implications of agricultural production and practices have broadened the agricultural agenda over recent years to include responses to climate issues, and the climate change agenda has similarly subsumed agricultural production as both a contributor to climate change and, through adjustment in practices, a potential mitigating force.

Innovation constraints in many developing countries stem from deeper problems in the agricultural sector. More generally, developing countries are vulnerable to climate change because they depend heavily on agriculture, they tend to be relatively warm already, they lack infrastructure to respond well to increased variability, and they lack capital to invest in innovative adaptations. Concerns about mitigating and adapting to climate change are now renewing the impetus for investments in agricultural research and are emerging as additional innovation priorities. In the coming decades, the development and effective diffusion of new agricultural technologies will largely shape how and how well farmers mitigate and adapt to climate change. This adaptation and mitigation potential is nowhere more pronounced than in developing countries where agricultural productivity remains low; poverty, vulnerability and food insecurity remain high; and the direct effects of climate change are expected to be especially harsh. Creating the necessary agricultural technologies and harnessing them to enable developing countries to adapt their agricultural systems to changing climate will require innovations in policy and institutions as well. In this context, institutions and policies are important at multiple scales. Impediments to the development, diffusion and use of relevant technologies can surface at several levels – from the inception and innovation stages to the transfer of technologies and the access to agricultural innovations by vulnerable smallholders in developing countries.

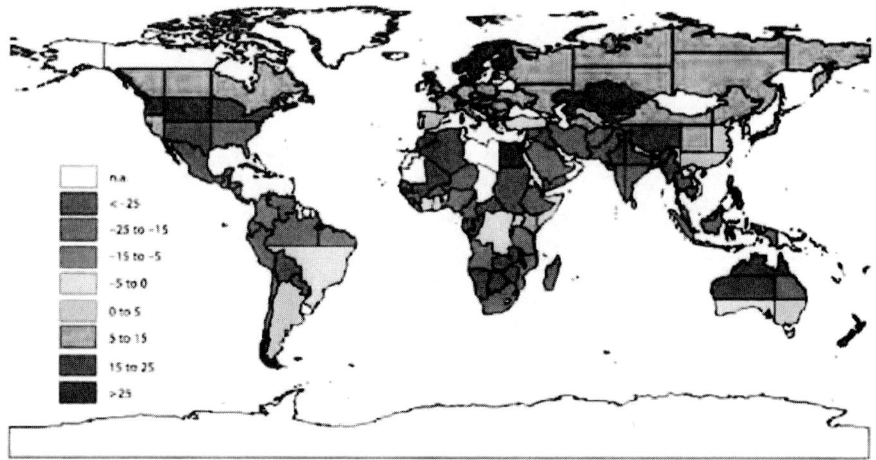


**Figure 1.** Projected impact of climate change on 2080 agricultural production assuming a 15% carbon fertilization benefit (*Source*: Cline 2007).

## The $CO_2$ fertilization effect

$CO_2$ is an essential plant 'nutrient', in addition to light, suitable temperature, water and chemical elements such as N, P and K, and it is currently in short supply. Higher concentrations of atmospheric $CO_2$ due to increased use of fossil fuels, deforestation and biomass burning, can have a positive influence on photosynthesis (Fig. 2) under optimal growing conditions of light, temperature, nutrient and moisture supply. Biomass production can increase, especially of plants with $C_3$ photo-synthetic metabolism above and even more below ground.

Plants are classified as $C_3$, $C_4$ or CAM according to the products formed in the initial phases of photosynthesis. $C_3$ plants: cotton, rice, wheat, barley, soybeans, sunflower, potatoes, most leguminous and woody plants, most horticultural crops and many weeds; $C_4$ plants: maize, sorghum, sugarcane, millets, halophytes (i.e., salt-tolerant plants) and many tall tropical grasses, pasture, forage and weed species; CAM plants (Crassulacean Acid Metabolism, an optional $C_3$ or $C_4$ pathway of photosynthesis, depending on conditions): cassava, pineapple, opuntia, onions and castor. $C_3$ species respond more to increased $CO_2$; $C_4$ species respond better than $C_3$ plants to higher temperature and their water-use efficiency increases more than for $C_3$ plants. There are some indications that enhancements can decline over time ('down-regulation').

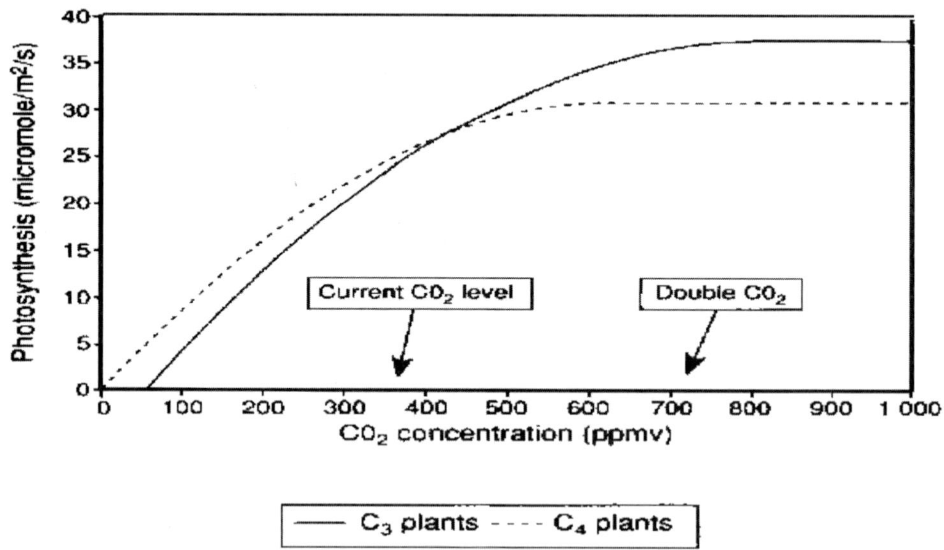


**Figure 2.** Schematic effect of $CO_2$ concentrations on $C_3$ and $C_4$ plants. The main mechanism of $CO_2$ fertilization is that it depresses photo-respiration, more so in $C_3$ than in $C_4$ plants

A total of 10 to 20% of the approximate doubling of crop productivity over the past 100 years could be due to this effect and forest growth or regrowth may have been stimulated as well. Further productivity increases may occur in the coming century, in the order of 30% or more where plant nutrients and moisture are adequate. Higher $CO_2$ values would also mitigate the plant growth damage caused by pollutants such as $NO_x$ and $SO_2$ because of smaller stomatal openings. Higher percentages of starch in grasses improves their feeding quality, implying less need for feed mixes when silaging.

### Effect on quality

According to the IPCC's "The importance of climate change impacts on grain and forage quality emerges from new research. For rice, the amylose content of the grain—a major determinant of cooking quality—is increased under elevated $CO_2$. Cooked rice grain from plants grown in high-$CO_2$ environments would be firmer than that from today's plants. However, concentrations of iron and zinc, which are important for human nutrition, would be lower. Moreover, the protein content of the grain decreases under combined increases of temperature and $CO_2$. Studies using FACE have shown that increases in $CO_2$ lead to decreased concentrations of micronutrients in crop plants. This may have knock-on effects on other parts of ecosystems as herbivores will need to eat more food to gain the same amount of protein. Studies have shown that higher $CO_2$ levels lead to reduced plant uptake of nitrogen (and a smaller number showing the same for trace elements such as zinc) resulting in crops with lower nutritional value. This would primarily impact on populations in poorer countries less able to compensate by eating more food, more varied diets, or possibly taking supplements. Reduced nitrogen content in grazing plants has also been shown to reduce animal productivity in sheep, which depend on microbes in their gut to digest plants, which in turn depend on nitrogen intake.

## Major Abiotic Stressors Influence the Crops Growth and Production

Abiotic stresses occurring during production can either be the primary cause (direct) for disorders that exhibit themselves during postharvest handling and storage practices or they can influence the susceptibility of a fruit or vegetable to postharvest conditions that cause abiotic stresses resulting in disorders (indirect). It is important to characterize the relationship between pre harvest abiotic stresses occurring during production and postharvest abiotic stresses that the fruit or vegetable is exposed after harvest and during storage and distribution, since the solution to these different problems will be best resolved by focusing on pre harvest or postharvest abiotic stress amelioration, respectively.

## Drought

Aberrant monsoon lead to moisture deficit which may affect the agricultural scenarios. Water stress refers to the situation where cells and tissues are less than truly turgid. It occurs whenever the loss of water in transpiration exceeds the rate of absorption. When plants are unable to absorb enough water to replace that lost by transpiration, water stress develops in the plant system. The results may be wilting, reduction in photosynthesis, disturbances in physiological processes, cessation of growth, or even death of the plant or plant parts. In fact water stress practically affects every aspect of plant growth, modifying the anatomy, morphology, physiology, and biochemistry which are related with decrease in turgor, water potential and osmotic potential. The occurrence of drought conditions during production of fruit and vegetable crops is becoming more frequent with climate change patterns (Whitmore 2000; Kumar *et al.,* 2013). The existing literature provides some insight which may lead to better understanding and perhaps also encourage future research. Water stress during the production phase of some fruits and vegetables may affect their physiology and morphology in such a manner as to influence susceptibility to weight loss in storage. There have been both positive effects reported for field water deficits (stress) in tree fruits and root vegetables. Timing of a water stress event can also be very important in determining response to postharvest abiotic stress response. Therefore it is critical to avoid water stress until the fruit has reached maximum size in order to minimize incidence of chilling-induced injury in storage. Water stress, particularly at the tuber forming stage, can also lead to a higher susceptibility of potatoes to postharvest development of black spot disorder.

## Temperature extremes

Susceptibility to high (heat injury inducing) or low (chill injury inducing) temperatures is known to be reduced by prior exposure of the sensitive fruit or vegetable to low ambient temperatures. However, if the pre harvest temperature leads to chilling induced injury in the field, then susceptibility to postharvest chilling injury can be increased. Therefore, the level of the pre harvest temperature extreme will be a determinant as to if the exposure will have positive or negative effects on postharvest stress sensitivity. Extreme high temperatures can occur in the field and apple fruit exposed to direct sunlight can reach in excess of 40 °C. High temperatures during the late season (leading up to harvest) can enhance susceptibility of apples to superficial scald which develops in storage.

## High temperature

Extreme high temperatures can occur in the field and apple fruit exposed to direct sunlight can reach in excess of 40 °C. High temperatures during the late season (leading up to harvest) can enhance susceptibility of apples to superficial scald which develops in storage. Sun scald of grapes, soft nose of mango, sun burn in papaya and citrus are mostly related to high temperature. Fruit cracking in pomegranate and litchi is associated with temperature aberrations coupled with fluctuation in moisture.

## Low temperature

When plant tissue temperatures fall below critical values, sensitive perennial crops such as grapevines and tender tree fruits can suffer irreversible cold injury, causing malfunction or death of plant cells.

## Frost/ temperature inversions

In clear sky and minimum wind, radiational cooling forces temperature inversions to occur where air temperatures 10-20 m above the field are at least 2 - 3°C warmer than at 1 m above ground. For formation of inversions, wind speeds must be less than7 km/h; for inversions to break up, wind speeds mustbe greater than 7 km/h. This is because the wind mixes the air above the field with the air at ground level. This is like a house furnace fan mixing air throughout a house to keep air temperatures more uniform. When that fan is off, air stratification can occur with colder, heavier air settling in the basement, while warmer, lighter air rises and remains upstairs.

## Cold wave

The prolonged cold spell causes large-scale damage to agricultural and horticultural crops. The damage to fruit plantations is so severe that many a times farmers up root their well-established orchards. Crop yields were lower by 10 to 40 per cent in wheat, 25 per cent in gram, 50 to 70 per cent in mustard and 60 to 95 per cent in *aonla* compared to a normal year. The orchards of mango, litchi, guava, ber and kinnow in Punjab were hit resulting in reduced size or poor quality of the fruit. The plant mortality in mango varieties was highest in Dashehari, followed by Amrapali and least in Langra with 18 to 48 per cent flowering damage in parts of Jammu, Punjab, Haryana, Chandigarh and Himachal Pradesh. The mortality rate in papaya ranged from 40 to 83 per cent in the Shiwalik region, plains of Uttar Pradesh, Bihar and the northeast.

## Heat waves

Three or more consecutive days of unusually high maximum and minimum temperatures for the location and time of year. A heatwave is a run of unusually hot days but what constitutes a heat wave will differ from region to region given the normal temperatures for that area. Inland areas can expect to be normally hotter than areas closer to the coast. Heat-wave intensity is classified as low, severe or extreme based on the actual temperatures and the previous 30 days' temperatures.

Fruit is less likely to set if heat waves occur during blossoming. There are no precise threshold temperatures or periods for reduced pollination because it is a complex interaction between bee activity and the availability of viable pollen and receptive stigmas. Sunburned fruit is caused by combined high air temperature and solar radiation. Sunburn typically occurs in the afternoon on the western side of trees and is often worse where cool, cloudy weather is suddenly followed by a change to hot, sunny conditions. This problem is becoming more common with the increased frequency of heatwaves. An early sign of fruit sunburn is white, tan or yellow skin. Severe sunburn results in dark brown to black sunken patches on the fruit surface. Browning associated with sunburn is likely when the fruit surface reaches 46–49°C. Skin necrosis occurs at 52°C. Air temperature provides a reasonable estimate of sunburn risk. It is normally 10 to 18°C lower than fruit surface temperature. Cherry doubling is caused by temperatures above 35°C after harvest when buds for the following year's crop are initiated. It is unclear how many days at this temperature are needed but heat waves almost certainly increase the risk. In grapes, hot and dry summers typically result in early harvests of mature fruit and abundant robust wines. Heat waves often associate with lower-yield harvests.

## Hailstorm

Hailstorms are the result of four atmospheric factors which are characterised as strong convective instability creating strong updrafts, abundant moisture at low levels feeding into the updrafts, strong wind shear aloft, usually veering with height, enhancing updrafts and some dynamical mechanisms that can assist the release of instability such as air flow over mountain ridges. The typical mechanism of hail formation in a thunderstorm is that when the ground is heated during the day by the sun, the air close to it gets heated as well. Hot air, being less dense and lighter than cold air, rises and cools. As it cools, its capacity for holding moisture decreases. When the rising, warm air has cooled so much that it cannot retain all of its moisture, water vapour condenses, forming puffy-looking clouds. The condensing moisture releases heat of its own into the surrounding air, causing the air to rise faster and give up even more moisture.

The development of hailstones typically occurs 5 to 7 km/hr above the earth's surface. As a thunderstorm moves along, it may deposit its hail in a long narrow band (often several kilometers wide and about 5-10 kilometers long) known as a hail-streak or hail-swath. Its size and shape depend on how fast the storm is moving and how strong the updrafts are inside the storm. A typical hail-streak is about 1.5 km wide and 8 km in length. However, these may vary from a few acres to large belts, about 16 km wide and 160 km long. Most storms that produce hail generate one or two hail-streaks during their lifetime. The volume of hail reaching the ground falls at a speed of about 40 m/sec, and is usually less than 10 per cent of the volume of rain produced by a thunderstorm.

## Frost

Frost occurs when water vapour in the air freezes upon contact with an object that has a surface temperature below 0°C. It is commonly seen as ice crystals on tree branches, grass stems, and car windscreens.

***Radiant frost*:** It occurs at night in cold temperatures when heat radiates into the atmosphere and leaves the ground and ambient air cooler. It most commonly occurs under clear skies with little or no wind. Clear skies encourage heat radiation, and windless nights encourage cold air to settle which increases the frost risk, particularly in low lying areas. Drought years are conducive to radiant frost due to the lack of cloud cover and the increased difference between day and night time temperatures.

***Advection frost:*** Forms when a mass of very cold air moves over an area, replacing the warmer air in that area. It is not affected by cloud cover. It occurs more often, possibly due to the southward movement of the subtropical ridge of high pressure zones. There are more southern highs bringing air masses from further south than in the past. This air is very cold and leads to widespread frost.

When plant tissues freeze, ice crystals form and rupture cell walls. The resulting dead tissue leads to the familiar water-soaked and scorched appearance of frosted plants. Late heavy frosts during spring can kill fruitlets at any stage of development. The severity of the damage varies with the phenological stage and type of fruit tree affected. The duration of exposure to frost and the rate of tissue defrosting also influence the amount of damage. Developing fruit can withstand relatively low temperatures early in the season, but become more susceptible to frost damage (GRDC 2014). Fruitlets which are killed, shed early in the season. Fruit can also suffer sub-lethal damage which distorts growth during maturation and makes it unmarketable. This damage is often expressed as banded areas of russetted, corky growth or constricted bands leading to misshapen fruit. Severe, late frosts can also kill the buds responsible for the following year's crop and have an ongoing effect on orchard productivity.

## Light

It would be considered logical to assume that effects of exposure to high light are difficult to dissociate from effects exposure to high temperatures. However, research has shown that low light in the preharvest interval reduced susceptibility of apples to developing superficial scald in cold storage and, in contrast, high ambient temperatures resulted in increased susceptibility. In the case of ambient low light, when lettuce is grown under low light which is sub-optimal for photosynthetic activity, shelf life of fresh cut lettuce (i.e. lettuce subjected to mechanical stress) is much shorter than lettuce produced under optimal light conditions. Tomato size is smaller when the crop is grown under ambient low light levels, such as in the early spring season in northern latitudes and since surface area to volume ratio is greater in smaller fruits, susceptibility to postharvest desiccation stress would increase. Low light also results in lower levels of ascorbate in many greenhouse-grown fruits and vegetables, which would render them less fit to deal with postharvest stresses since ascorbate contents are generaly directly proportional to relative levels or stress tolerance.

## Salinity

The coastal system, made up of waters from the sea and other estuaries, is highly coupled. As a result, local meteorological processes,such as precipitation, evaporation, wind events, and direct inflows of fresh-water through stream flow and runoff, strongly influence the salinity, water quality, and circulation of coastal waters. Long-term changes in sea surface temperature, sea-level rise, hurricane severity and frequency, and other more recently discovered phenomena, such asa rise in ocean acidification, are expected to occur as a result of natural and anthropogenic global climate variability. Annual wet season/dry season pattern causes noticeable changes in the regional sea surface salinity, most pronounced in the coastal zone near the river mouths and along the onshore edges.

Tomatoes grown under high salinity will produce smaller fruit with higher soluble solids. Smaller fruit will have higher surface area to volume ratios, hence greater susceptibility to postharvest water loss (i.e. desiccation stress). Firmness declines in tomatoes grown under 3 and 6 dS $m^{-1}$ salinity levels were increased by 50 to 130%, respectively, at two weeks holding at 20 °C compared with control fruit.

## Plant nutrition

Climate change impacts on soil fertility and crop nutrition in developing countries leads to the opening of Pandora's Box. 963 million people on our planet are underfed or malnourished including a 10% increase over the last 3 years due to rising food costs (FAO 2009). Demand for food continues to increase while per capita arable land area dedicated to crop production continues to shrink because of population growth, urbanization and soil degradation. From the supply side, this imbalance is largely driven by edaphic constraints that result from inherently low soil fertility and/or soil degradation from unsustainable farming practices. Under tropical and sub-tropical climate, wide spread deficiency of N and P occurs. Acidification of soil leads to K, Ca, Mg deficiency and toxicity of Al and Mn.

**Table 1.** Climate interactions with plant nutrition

| Process | Climate variable | Nutrient affected |
|---|---|---|
| Transpiration driven mass flow | $CO_2$, RH, Rain, Temp | $NO_3$, $SO_4$, Ca, Mg, Si |
| Root growth and architecture | $CO_2$, drought, temp | P, K |
| Root symbiosis | $CO_2$, drought, temp | N, P, Zn |
| Root exudates | $CO_2$ | Al tox, metal uptake, P |
| Tissue dilution | $CO_2$ | Stress responses, nutrient cycling, nutritive value |
| Growth responses | $CO_2$, drought, temp | N, P: "no clear pattern" |

Drought highly impacts the plant nutrition. Water dependent diffusion and mass flow of nutrients to the roots slows with increasing soil moisture deficit. Impaired root growth decreases the capture of less mobile nutrients such as phosphorus. Drought inhibits N-fixation in legume crops and disrupts N cycling by soil bacteria. Conversely, high intensity rainfall events can be a major source of erosion leading to loss of nutrient (N) rich top soil and surface broadcast fertilizer. Waterlogged soils become hypoxic which slows the $O_2$ dependent active transport of nutrients, lead to change in soil redox status resulted to Mn, Fe and Al toxicity. Waterlogging also leads to N losses through denitrification when nitrate is used as alternative electron acceptor in the absence of oxygen. High temperatures that result in extreme vapor pressure deficits, may reduce mass flow of nutrients by triggering stomatal closure.

Arguably, rising global $CO_2$ increases yield and decreases water use by crops, and this is often presented as one positive of atmospheric change with the expense of quality. $CO_2$ levels will likely reduce the levels of zinc, iron, and protein in wheat, rice, peas, and soybeans. In addition to wheat, rice, peas, and soybeans, which all use a form of photosynthesis known as $C_3$, Myers and his colleagues studied corn and sorghum, which use $C_4$ photosynthesis, a faster kind. They found relatively little effect of $CO_2$ enrichment on the nutritional value of the $C_4$ crops.

Calcium is also been suggested as a putative signalling molecule involved in the development of cross tolerance to abiotic stresses. Therefore the role of preharvest calcium nutrition is postharvest stress resistance may be complex, and dependent on whether the fruit or vegetable is also exposed to environmental abiotic stresses. In carrots, deficiency in potassium is associated with greater weight loss (desiccation stress) in storage. Improved potassium nutrition has also been shown to reduce susceptibility of potatoes to internal bruising in response to mechanical stresses imposed during postharvest handling. Relatively high preharvest nitrogen is often associated with poor postharvest quality of many fruits and vegetables. In regards to affecting susceptibility to postharvest stress, applying higher than recommended levels of pre harvest nitrogen for a specific crop have been linked to storage discoloration susceptibility in both cabbage and potato. In contrast nitrogen deficiency or lower than recommended nitrogen application rates will most often results in increased vitamin C content in many fruits and vegetables. Vitamin C content has been tightly linked with storage life potential which is likely a consequence of the importance of this antioxidant nutrient in forestalling oxidative injury that leads to quality losses in storage.

## Agricultural Technologies for Climate Change Mitigation & Adaptation

The core challenge of climate change adaptation and mitigation in agriculture is to produce

(i) more food, (ii) more efficiently, (iii) under more volatile production conditions, and (iv) with net reductions in GHG emissions from food production and marketing. Producers will grapple with these growing demands and shifting incentives amidst more volatile production conditions. Agricultural technologies will play a central role in enabling producers to meet these core challenges. Because agriculture is inseparably linked to climate and feedback runs in both directions, most agricultural technologies have direct or indirect climate linkages. Most new technologies change the use of farm inputs, often in ways that alter the impact of weather on production and of production on carbon emissions. While most agricultural technologies therefore have climate implications, there are a handful of current and emerging technologies with particular relevance to developing country agriculture and climate change.

### New Traits, Varieties & Crops

Increasing agricultural productivity requires technological advances in crop yields. In contrast to developed counties, which have seen dramatic yield gains in the past century through investments in agricultural innovation and operate close to the technological frontier, much of developing country agriculture is far from this frontier. Profitable adaptation and farmer adoption of suitable varieties and crops could spark substantial yield gains. These productivity gains could confer a substantial mitigation benefit in the form of foregone land conversion or even reversion of some sensitive lands to grass or forests. Since land use changes, including deforestation and conversion to agricultural production account for 17% of global $CO_2$ emissions (World Bank, 2009), productivity gains represent a significant mitigation mechanism in agriculture. New varieties and traits can also lead to less intensive use of other inputs such as fertilizers and pesticides and the associated equipment.

In addition to increasing productivity generally, several new varieties and traits offer farmers greater flexibility in adapting toclimate change, including traits that confer tolerance to drought and heat, tolerance to salinity (e.g., due to rising sea levels in coastal areas), and early maturation in order to shorten the growing season and reduce farmers' exposure to risk of extreme weather events. These promising new traits and varieties, which are mostly still in development,can emerge from traditional breeding techniques that leverage existing varieties that are well suited to vagaries of the local production environment as well as from more advanced biotechnology techniques such as marker assisted selection and genetic modification.

In many places, new traits and varieties for the crops, farmers have traditionally cultivated will confer sufficient scope for adaptation. In other places, shifting to a totally different mix of crops will be required to cope with dramatic changes in rainfall or temperature, and cropping systems will fundamentally change as a result. Even if adaptation does not imply an entirely new mix of crops, many producers will benefit from new crops and varieties as they diversify their production portfolios as a means of stabilizing their revenue or local production of basic foods in the face of more volatile conditions. These diversification benefits will be important because many households and many regions will continue to produce their own food even decades from now, when transportation, communication and financial infrastructure has penetrated many areas that are currently poor and remote.

Climate change will also lead to new pest and disease pressures. The nuances of temperature changes e.g. higher low temperatures and fewer freezes - could shorten dormant periods, speed pest and disease growth and change the dynamics of these populations and their resistance. Crops, varieties and traits that are resistant to pests and diseases will improve producers' ability to adapt to climate change. To the extent that these varieties reduce the need for pesticides, they also reduce carbon emissions by decreasing pesticide demand as well as the number of in-field applications. Breakthroughs in nitrogen use efficiency could substantially mitigate emissions in agriculture. The mitigation potential of new crops and varieties extends to direct carbon sequestration and perhaps to second generation biofuel crops. There are several second generation biofuel crops (i.e., beyond sugar cane and maize) that appear promising as fuel sources (e.g., miscanthus,). The damage to fruit trees was relatively more in low-lying areas where cold air settled and remained for a longer time on ground. But temperate fruits such as apple, peach, plum and cherry gave higher yield due to extended chilling. One of the best way to reduce the impact of the cold wave is to go in for proper selection of fruit species and varieties as per the site conditions. Apart from that, wind breaks or shelter belts should be provided. There should be frequent irrigation, smoking, covering young fruit plants with thatches or plastic sheets.

For the development of traits and varieties that help to mitigate and adapt to climate change, agricultural biotechnology stands out as an especially promising set of tools. While it remains controversial in some policy arenas and public fora, agricultural biotechnology has produced dramatic improvements in yield and reductions in production costs and input use intensity. Many of the promising traits and varieties discussed above owe their existence to biotechnology, including genetically modified crops with pest resistance (Bt) and herbicide tolerance (Roundup Ready) and conventionally bred varieties that benefit from breeding tools such as marker selection and tissue culture.

## Adoption of Soil and Moisture Conservation Techniques

Some of the basic methods to collect and conserve water are; harvesting of rain water, harvesting of water from snow melting, development of catchment area and storing runoff water for recycling, check dams and construction of waterways. Contour cultivation, contour trip cropping, mixed cropping, tillage, mulching, zero tillage, are some of the agronomical measures for the *in-situ* soil moisture conservation. Mechanical measures like contour bunding, graded bunding, bench terracing, vertical mulching etc. also need to be followed for effective soil and moisture conservation in dry lands. Another technology for efficient utilization of runoff is water harvesting recycling. Rainwater harvesting includes collecting runoff water into dug out ponds or tanks in small depressions, gullies and into storage dams of earth or masonry structures. Rain water harvesting is possible in areas having rainfall as little as 500 to 800 mm. Depending on the rainfall and soil characteristics, 10-50 % of the runoff can be collected in farm pond. Surface run off thus collected in a farm pond can be used to provide protective irrigation in the period of prolonged dry spell. Trench planting (0.5-1.0 m) is recommended for ber, *aonla* and custard apple to conserve moisture. The trenches collect rain water along with silt and organic matter and thus promote tree growth. Planting in trenches is common for pineapple under dry conditions. contouring, contour bunds, sloping area land technology (SALT), contour drains, graded bunding, terraces, broad bed furrow system for raising crop and conservation of water, contour Furrow, stabilisation structures, curved land ploughing for intensive land preparation; soil conservation and water retention through use of vegetation, live check of bamboo pieces and loose boulders are some of the other agronomical measures used to conserve moisture.

## Water Management & Irrigation

In the midst of increasing urban and environmental demands on water, agriculture must improve water use efficiency generally. Adding climate change to this mix only intensifies the demands on water use in agriculture. With hotter temperatures and changing precipitation patterns, controlling water supplies and improving irrigation access and efficiency will become increasingly important. Climate changes will burden currently irrigated areas and may even outstrip current irrigation capacity due to general water shortages, but farmers with no access to irrigation are clearly most vulnerable to precipitation volatility.

Across the Middle East and North Africa, Central Asia and Southern Africa, water availability is projected to decline dramatically with climate change and population growth in the next several decades. It is no exaggeration that the future of agriculture in these regions hinges primarily on improving the efficiency of existing irrigation systems and, where profitable, extending irrigation infrastructure. Drip irrigation systems are important on farmers' fields, but inefficiencies in delivery (e.g., canal construction and maintenance) are often more glaring than field-level inefficiencies in application (e.g., flood versus drip irrigation).

In places with limited access to irrigation, well timed 'deficit irrigation' can make a substantial difference in productivity. With dwindling water supplies, such deficit irrigation techniques will become increasingly important. In non-irrigated areas, water conservation and waterharvesting techniques may be farmers' only alternative to abandoning cultivation agriculture all together. Drip irrigation has proved its superiority over other conventional method of irrigation, in horticulture due to precise and direct application of water in root zone. A considerable saving in water, increased growth, development and yields of fruits and vegetables and control of weeds, saving in labour under drip irrigation are the added advantages. Drip irrigation can be adopted in fruit crops and also to all vegetable crops including closed spaced crops like onions and beans. The saving in water is to the tune of 30-50 % depending on the crop and season.

The following methods may be adopted under limited water conditions to save water:

(a) **Water Saving Irrigation Method**: Under limited water situations, water-saving irrigation methods like alternate furrow irrigation or widely spaced furrow irrigation and drip irrigation systems can be adopted. Studies conducted on methods of irrigation in capsicum, tomato, okra and cauliflower indicated that adopting alternate-furrow irrigation and widely-spaced furrow irrigation saved 35 to 40 per cent of irrigation water without adversely affecting yield.

(b) **Mulching Practices in Vegetable Production** - The technique of covering the soil with natural crop residues or plastic films for soil and water conservation is called mulching (Suresh Kumar *et al.*, 2012). Mulching can be practiced in fruits and vegetable crops using crop residues and other organic material available in the farm. Recently plastic mulches have come into use due to the inherent advantages of efficient moisture conservation, weed suppression and maintenance of soil structure. Wide variety of vegetables can be successfully grown using mulches. In addition to soil and water conservation , improved yield and quality, suppression of weed growth , mulches can improve the use efficiency of applied fertilizer nutrients and also use of reflective mulches are likely to minimize the incidence of virus diseases. For vegetable production generally polyethylene mulch film of 30 micron thick and 1 to 1.2 m width is used. Generally raised bed with drip irrigation system is followed while laying the mulch film.

Depending upon situation and availability of water, this technology can be used for fruits and vegetable crops. The cost of initial establishment is lower compared to drip system. Further in summer the sprinkling of water helps in reducing the microclimate temperature and increasing the humidity, thereby improving the growth and yield of the crop. The water saved is to the tune of 20 to 30 per cent.

## Other Production Inputs

Improvements in crop yields per unit of land are crucial as an alternative to extensive conversion of grassland and forestland to crops. Therefore practices or technologies with potential to increase the intensity of land use can yield mitigation benefits. This may even include application of additional fertilizer or pesticide inputs, where the "first round" GHG implication may not look favorable. There are, however, other amendments such as biochar, a charcoal soil amendment, that may offer both improved soil fertility and serve as a carbon sink (Lehmann, *et al.*, 2006). Similarly, herbicides and other inputs that reduce competition from weeds can improve productivity and thereby serve to mitigate GHG emissions associated with bringing additional land under cultivation. Furthermore, since potential cropland in different regions has very different capacities to sequester carbon, shifting crops to the land with the least negative carbon implication may have net GHG benefits. This may mean farming dry regions under irrigation which allows use of land that otherwise would not contribute to mitigation.

**Application of foliar nutrition**: The foliar application of nutrients during water stress conditions helps in the better growth by quick absorption of nutrients. The spraying of K and Ca induces drought tolerance in vegetable crops. Spraying of micronutrients and secondary nutrients improves crop yields and quality. Anti-transpiration coatings have been shown to be effective for maintaining quality through control of water loss. Anti-transpirants are chemicals which when sprayed on plants form a film which increases the diffusion resistance of water from stomata and thus reduces transpiration losses of water. Several chemicals have been successfully used like Acropyl in grapes, polycot in banana and kaolinite (3-8%) in different fruit plants. The energy input can be reduced by increasing plant reflectivity by using effective chemicals like zinc oxide, kaolinite, chalk etc alone or in combination with other anti-transpirants. These chemicals are used to reduce temperature on plant parts. Film forming compounds like wiltpruf, mobileaf, clear spray, vapour gard and folicoat can be used to reduce transpiration and water loss. Water loss can also be restricted by applying chemicals which facilitate stomatal closure. Some of these are; phenyl mercuric acetate (PMA), Decenyl succinic acid (DSA), Atrazine and Sodium azide.

## Production Management & Practices

Production techniques may be as important as production technologies in climate change adaptation and mitigation. One such technique stands out in particular: ***conservation or reduced tillage agriculture***. This technique aims to build up organic matter in soils and create a healthy soil ecosystem by not tilling the soil before each planting. Seeds are planted using seed drills that insert seeds to a precise depth without otherwise disturbing the soil structure. By increasing the organic matter in soils, conservation agriculture improves the moisture capacity of the soil and thereby increases water use efficiency. The practice also reduces carbon emissions by reducing tilling, although it also requires more sophisticated pest and disease control because the system is not 're-booted' at each planting. An array of other production management practices and technologies could similarly improve farmers' mitigation and adaptation to climate change, including equipment and information that enables more precise application of inputs, especially fertilizer. The key challenge is to assure that such practices do not reduce yields so that the demand for additional land offsets the benefits from on field sequestration.

The basic principles of conservation agriculture are soil cover, particularly through retention of crop residues on the soil surface as mulch or cover crops; a minimum level of soil disturbance, e.g., reduced or zero tillage practices; and sensible, profitable crop rotations. Soil moisture can be conserved through mulching either with black polythene or locally available mulches, growing cover crops or inter-culturing in the orchards to check soil erosion and runoff rain water. If area around tree basin remains covered with mulches, like grass mulch (10-12 kg/ basin) and black polythene mulch (100 micron) throughout the growth period, it helps in conserving the soil moisture, saving water for more critical stages during summer and reduced the weed population by 60% with grass mulch and 100% with black polythene mulch. The polythene mulch maintained 29% more soil moisture compared to un-mulched trees on soil available water content basis. Surface-applied mulches provide several benefits to crop production by controlling evaporation from the soil surface, heat energy and nutrient status in soil, buffering drastic changes in soil temperature. Soil temperature, especially at the surface layer is reported to be important for the translocation of photo-assimilates. Straw mulch ameliorates environmental stresses and improves the product quality and safety.

**Wind breaks, hedges and intercropping:** To overcome the adverse effect of high temperature and dry winds, tall growing trees need to be planted all along the boundary of the farm. Windbreaks help to deflect and to filter through the wind current thereby reducing the velocity of wind resulting lower displacement of wind around tree and cause reduction in transpiration and evaporation. The orientation of windbreak should more or less be at right angles to the prevailing winds. It is believed that the windbreak is effective for a distance equivalent to 3 to 4 times of tree height. *Acacia tortalis, Cassia siamea* and *Prosopis julifera* are some of the useful tree species suitable as wind break under less water conditions. Inter cropping of vegetable crops of the area can be practiced in orchards during summer months. Maize/ Sorghum can be grown all along the border of the plot to mitigate the effect of desiccating winds. Crop rotation, cover crops, strip cropping, mulching (Suresh Kumar *et al.* 2011), contour hedge row intercropping system are used to conserve moisture.

## Use of Plant Growth Regulators

Foliar sprays of 50 mMIAA, $GA_3$, or benzyl amino purine (BAP) partially counteracted the effect of water deficit on photosynthesis and transpiration. Growth regulator applications can also potentially enhance stress resistance, particularly for fruits and vegetables which are prone to show accelerated ripening or senescence in response to drought stress. Accelerated ripening or senescence is most often mediated by ethylene production in response to the stress. As consequence anti-ethylene products such as aminovinylglycine (AVG) and 1-methylcyclopropene (1-MCP) could be used to mitigate drought stress. While foliar cytokinin (CK) application can prevent ABA-induced photosynthetic limitation, the effects can be transient and of little consequence in the long term. Paclobutrazole (10 mg/ lit) is used to avoid moisture stress in mango. Accelerated ripening or senescence is most often mediated by ethylene production in response to the stress. As consequence anti-ethylene products such as aminovinylglycine (AVG) and 1-methylcyclopropene (1-MCP) could be used to mitigate drought stress. Other growth hormones, such as methyl jasmonate (which promotes leaf senescence), can enhance chilling resistance in avocado, grapefruit, bell peppers and zucchini squash. Abscisic acid has been demonstrated to reduce chilling-induced injury in some crops (Wang 1993).

## Beneficial microbes

Microbial activity in soils is stimulated by the release of carbon-rich material in the form of root border cells, and/or the selective exudation of specific sugars, carboxylic acids, or amino acids that encourage the development of cultivar-specific, plant-beneficial, microbial communities. Some of these microorganisms benefiting from the availability of fresh energy are also capable of enhancing plant growth and development, which are generally called as Plant growth promoting (PGP) rhizosphere microorganisms. Plant growth-promoting rhizobacteria (PGPR) and fungi (mycorrhizae) can facilitate plant growth directly by facilitating the uptake of nutrients from the environment, by influencing phytohormone production (e.g. auxin, cytokinin, or giberallin), and/or by enzymatic lowering of plant ethylene levels. In addition to facilitating the growth of plant, these microorganisms can protect plants from the deleterious effects of flooding and drought.

The beneficial plant-microbe interactions in the rhizosphere are the primary determinants of plant health and soil fertility. Rhizobacteria include mycorrhization helper bacteria (MHB) and plant growth promoting rhizobacteria (PGPR), which assist AMF to colonize the plant roots. Synergistic positive interactions have been reported between AMF and plant growth-promoting bacteria (PGPB) such as nitrogen fixers, fluorescent *pseudomonad's* and sporulating *bacilli*. The most common bacteria in the mycorrhizosphere are *Pseudomonas*. Some extremely mycorrhizal dependent plants, including grapes, citrus, melons, oaks and pines may quite literally starve to death in soils that lack this helpful fungi. These benefits of mycorrhizal symbioses in vegetables, fruits and tree species, both agronomically by increased growth and yield as well as ecologically by improved fitness, indicate that mycorrhizal plants are often more competitive and better able to tolerate environmental stresses.

## Hail control mechanisms

Artificial hail control is an important measure in disaster prevention and mitigation. Cloud seeding for hail suppression is based on the cloud microphysical concept in which seeding is postulated to reduce hail severity. The natural and artificial ice crystals compete for the available super-cooled liquid cloud water within the storm. Hence, the hailstones that are formed within the seeded cloud volumes will be smaller and produce less damage if they should survive the fall to the surface. If sufficient nuclei are introduced into the new growth region of the storm, then the hailstones will be small enough to melt completely before reaching the ground. Another concept is to create shock waves which can prevent the formation and growth of hail by melting altogether. Shockwaves are produced using hail guns/cannons. The super-cooled water situated on the external layer of hailstone is transformed from liquid state to solid state. Therefore the hail nuclei are not able to melt anymore and remain at a small size which thus minimise the damage when they hit the ground. Nowadays, acetylene or butane gas is used to generate hail disruptive shockwaves this allows the emission of a more powerful shockwaves with higher continues to 1 frequency. It is be reiterated that hail control mechanisms cannot eliminate hail completely but the cloud seeding can be beneficial.

Protective screens termed as anti-hail nets above the crop can be appropriately utilised especially for high value crops. The hail climatology, microclimatic effects of the cover, its durability properties and installation cost are useful in evaluation of the merit of such screens. These anti-hail nets are not effective against strong hail storms. Tree shelterbelts can markedly reduce hail damage in their immediate vicinity since hails are usually associated with strong winds. Some hail is intercepted directly by the trees protecting crops immediately downwind. The trees also create a change in the air flow so that the area in the lee of them is particularly sheltered with hail deflected laterally. Wind speeds will also be less in the lee of the shelter so the total hail kinetic energy, which results both from the vertical fall speed of the hail and the wind speed will be less.

## Use of mechanization in agriculture during extreme weather events

Wind machines can raise air temperatures around plants by about a third to half the temperature inversion difference. One protection method is to use wind machines in orchards of high value crops.Wind machines are tall, fixed-in-place, engine-driven fans that pull warm air down from at least 15 m above ground during strong temperature inversions, blowing it down and out, pushing away and replacing cold air near target crops. This raises air temperatures around cold-sensitive perennial crops such as grapes.

Protective screens termed as anti-hail nets above the crop can be appropriately utilised especially for high value crops. The hail climatology, microclimatic effects of the cover, its durability properties and installation cost are useful in evaluation of the merit of such screens. These anti-hail nets are not effective against strong hail storms. Tree shelterbelts can markedly reduce hail damage in their immediate vicinity since hails are usually associated with strong winds. Some hail is intercepted directly by the trees protecting crops immediately downwind. The trees also create a change in the air flow so that the area in the lee of them is particularly sheltered with hail deflected laterally. Wind speeds will also be less in the lee of the shelter so the total hail kinetic energy, which results both from the vertical fall speed of the hail and the wind speed will be less. Netting provides some protection from frost damage in light to moderate frost events. Nets hold heat radiating from the ground and raise the temperature in the orchard. Nets also block much of the cold air flow associated with advection frosts. Frost protection is really only effective for radiant frosts when winds are light or calm, and in low-lying, frost-prone areas. Protection efforts need to focus on the most frost-prone blocks and in response to forecasts of the critical temperatures needed for substantial crop damage. Protection has three basic objectives.

1. Reduce heat loss from the surface
2. Stir the air and break up the temperature inversion layer
3. Add heat to maintain the temperature above the danger level.

## Marketing & Supply Chains

Post harvest GHG emissions per unit of consumption mainly depend on efficiencies of transport (rail versus road, ocean shipping versus land shipping, and large loads versus small loads) rather than distance traveled. Improvements in transportation efficiency are therefore as important to reducing agriculture's GHG emissions as they are to other sectors of the global economy.

Post harvest losses represent one of the single greatest sources of inefficiencies in food production worldwide and therefore one of the best opportunities for effectively improving crop productivity. These losses - which are due to poorly timed or executed harvesting, exposure to rain, humidity and heat, contamination by microorganisms, and a host of other sources of damage and deterioration - often get far less attention than they deserve. Half or more of the total harvest of some crops can be lost post harvest. Investments in improved harvesting, processing, storage, distribution, and logistics technology and necessary training investments can pay off as well as improved crop yields in terms of gains to consumers and the climate. As climates become hotter and precipitation more erratic, the potential for postharvest losses may increase and thus improved transport and storage become even more important.

### Innovative and improved postharvest management strategies

Simple modifications to postharvest handling systems can sometimes result in significant reduction in stress exposure and consequently result in storage and/or shelf life extension. One of the most successful strategies is the application of plastic film packaging or wraps to prevent desiccation, resulting in significant improvements and shelf life and quality of many fruits and vegetables. In many cases, modified atmosphere packaging is considered to largely control humidity around product and thus prevent moisture loss of fresh-cut and whole fruits and vegetables. Also, anti-transpiration coatings have been shown to be effective for maintaining quality through control of water loss (Baldwin 2003). In regards to maintaining water content on the retail shelf, the application of misting systems can 'recharge' the vegetable and in so doing maintain quality over longer durations at less than ideal storage temperatures.

## Future Perspectives

- The best policy and institutional responses will enhance information flows, incentives and flexibility
- Policies and institutions that promote economic development and reduce poverty will often improve agricultural adaptation and may also pave the way for more effective climate change mitigation through agriculture.
- Existing technology options must be made more available and accessible without overlooking complementary capacity and investments

- Adaptation and mitigation in agriculture will require local responses, but effective policy responses must also reflect global impacts and inter-linkages.

## Conclusion

Agriculture has unique role in development. It is our primary source of food, has significant potential for mitigation of global GHG emissions, and is particularly sensitive to climate change. Almost certainly, climate change will be severe in most developing countries and will directly and, in some cases, dramatically hurt agricultural production in these countries. Yet, development cannot be taken for granted and the dual burden of climate change adaptation and mitigation may make economic transformation more difficult. As climate change affects input availability, especially water in many places, input use efficiency must increase with these productivity demands. Carbon emission polices may simultaneously encourage or force producers to recognize GHG emissions as an important and costly "input" in production processes and open new opportunities and incentives for on farm GHG mitigation. It is a fool's errand to attempt to fully catalogue in any comprehensive way agricultural technologies with potential for climate change mitigation and adaptation over the next seven decades. If history is any guide, the most important such technologies have yet to be developed or even conceived.

Innovations in agriculture have always been important and will be even more vital in the context of climate change. Thoughtful policy responses that encourage the development and diffusion of appropriate agricultural technologies will be crucial to enabling an effective technological response. A careful balance of institutional change and wise investments is required to deal with both the demands of climate change and the demands of improving lives of the poor. As we consider implications of and responses to climate change, continuing concerns for improvements in nutrition, food security, food safety, local environments, and rural communities must not be neglected. Agricultural development efforts cannot be diverted even while recognizing the importance of climate change and the interaction between climate and other agricultural issues. Given the reliance of the poor on agriculture and the sensitivity of agriculture to climate change, impending climate changes will almost certainly hit (currently) developing countries and vulnerable populations within these countries hardest. While this reality seems to make the development process more complex, it should also stimulate greater urgency in addressing rural poverty and vulnerability. These twin imperatives of climate change – greater complexity and greater urgency – are important to keep in mind when formulating policies and institutions aimed at improving climate change adaptation and mitigation in agriculture.

## References

Baldwin EA (2003). Coating and Other Supplemental Treatments to Maintain Vegetable Quality. In: Bartz JA, Brecht JK (ed) Postharvest Physiology and Pathology of Vegetables, 2nd Edn. Marcel Dekker, Inc, New York, p 413-456.

Beck EH, Fetitig S, Knake C et al (2007a). Specific and unspecific responses of plants to cold and drought stress. J Bio Sci 32(3): 501-510.

Cline, W.R. (2007).Global Warming and Agriculture : Impact Estimates by Country. Washington, DC: Center for Global Development : Peterson Institute for International Economics.

Goyary D (2009). Transgenic Crops, and their scope for abiotic stress environment of high altitude: Biochemical and Physiological Perspectives. DRDO Science Spectrum, March 2009, p 195-201.

GRDC (2014). Frost risk on the rise despite warmer climate. Ground Cover supplement: Turning up the heat on frost in cereals. Issue 109 March-April 2014. http://grdc.com.au/Media- Centre/Ground-Cover-Supplements/GCS109

IPCC (2013). Summary for Policymakers. In: Climate Change 2013: The Physical Science Basis. Contribution of Working Group I to the Fifth Assessment Report of the Intergovernmental Panel on Climate Change [Stocker, T.F., D. Qin, G.-K. Plattner, M. Tignor, S. K. Allen, J. Boschung, A. Nauels, Y. Xia, V. Bex and P.M. Midgley (eds.)]. Cambridge University Press, Cambridge, United Kingdom and New York, NY, USA.

IPCC. 2007. Climate Change (2007). Synthesis Report. Contributions of Working Groups I, II and III to the Fourth Assessment Report of the Intergovernmental Panel on Climate Change. Geneva: IPCC.

Kumar PS, MinhasPS, Govindasamy, V, ChoudharyRL (2014). Influence of Moisture Stress on Growth, Development, Physiological Process and Quality of Fruits and Vegetables and Its Management Strategies.**In.**(Eds. R.K. Gaur and Pradeep Sharma) Approaches to Plant Stress and their Management, Springer publications., New Delhi, pp.125-148.

Lehmann, J., J. Gaunt, and M. Rondon. (2006). "Bio-Char Sequestration in Terrestrial Ecosystems–a Review." Mitigation and adaptation strategies for global change 11(2):395-419.

Suresh Kumar P, Choudhary VK, Bhagawati R (2012) Influence of mulching and irrigation level on water-use efficiency, plant growth and quality of strawberry (Frageria x ananassa). Ind J Agric Sci 82(2):127-133.

Wang CY (1993) Approaches to Reducing Chilling Injury of Fruits and Vegetables.Hortl Rev 15:63-95.

Whitmore JS (2000) Drought Management on Farmland. Kluwer Academic Publishers, Dordrecht, The Netherlands.

World Bank. 2009 "Development and Climate Change." World Bank.

❑❑❑

# 2

# Climate Change Impact on Agriculture Adaptation Strategies1

**M.L. Dotaniya, V.D. Meena, Manju Lata and B.L. Meena**

## Introduction

Climate change is a big phenomena, it is having a huge impact on agriculture from ancient to present time. It has received wide attention and debate as the public concern has increased about unforeseen change in climate. The impact of climate change on Indian agriculture will be significant since more than 60% of Indian population depends on agriculture for their livelihood with the contribution of agriculture to GDP is approximately 20 percent. In recent decades, human induced changes in climate has the focus of scientific and social attention. The most imminent of this is increased concentration of Greenhouse Gases (GHGs) namely, carbon dioxide ($CO_2$), methane ($CH_4$) and nitrous oxide ($N_2O$) in the atmosphere (Figure 1). The Concentration of $CO_2$, $CH_4$ and $N_2O$ have increased markedly by 30%, 145% and 15%, respectively as a result of human activity since the industrial revolutions (1750) (IPCC, 2007). The $CO_2$, $CH_4$ and $N_2O$ concentration in the atmosphere were 280 ppm, 715 ppb and 270 ppb in 1750 AD. In 2005, these values have become 379 ppm, 1774 ppb and 319 ppb, respectively (IPCC, 2007, Kundu *et al.*, 2013b). The increase in concentration of these gases in the atmosphere was higher in recent years *i.e.* 70% increase of GHGs between 1970 and 2004 has been reported. The global increase in $CO_2$ concentration is primarily due to the consumption of fossil fuel and land use change, while those of $CH_4$ and $N_2O$ are primarily due to agriculture. For Indian region the IPCC has projected 0.5 - 1.2°C rise in temperature by 2020, 0.88-3.16°C by 2050 and 1.56-5.44°C by 2080 depending on the scenario of future development (IPCC, 2007). It is projected that by the end of the 21$^{st}$ century rainfall over India will increase by 15-40%, and the mean annual temperature by 3-6°C (NATCOM, 2004). Agricultural productivity is particularly vulnerable to the disruption of global climate changes.

Increasing concentration of the gases such as atmospheric $CO_2$, $CH_4$, nitrous oxide ($N_2O$), and sulphur dioxide ($SO_2$) have direct impact on crops, and on the other hand these are also critical in increasing air temperature (Table 1). Climate change effect governs by the global warming potential of GHGs and it causes disturbance in climate and weather activities. The United Nations Framework Convention on Climate Change (UNFCCC) identifies two responses to climate change: mitigation of climate change by reducing greenhouse-gas emissions and enhancing sinks, and adaptation to the impacts of climate change. Most industrialized countries have committed themselves, as signatories to the UNFCCC and the Kyoto Protocol, to adopting national policies and taking corresponding measures on the mitigation of climate change and to reducing their overall greenhouse-gas emissions (United Nations, 1997).

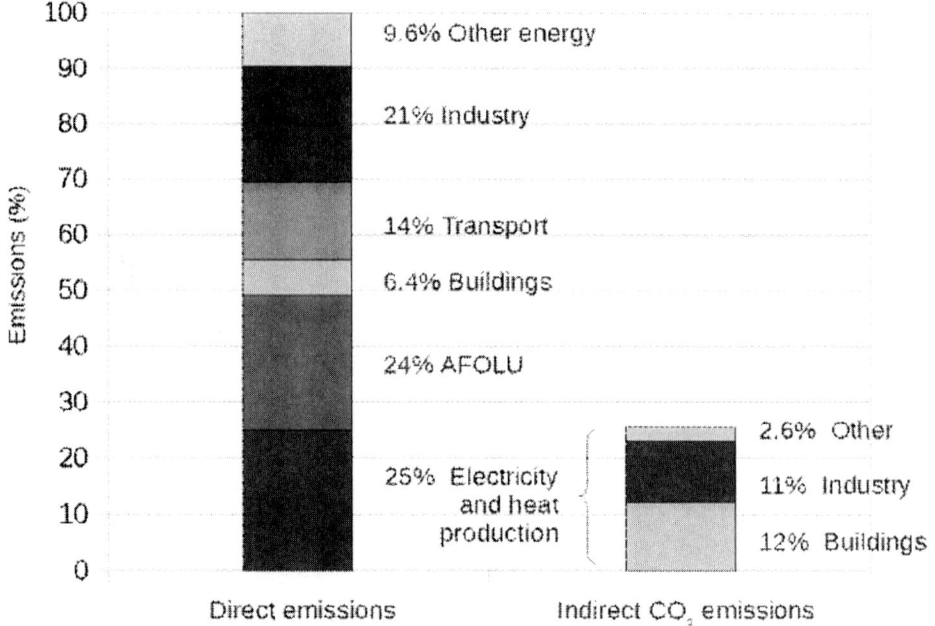


*AFOLU" stands for "agriculture, forestry, and other land use"
**Figure 1.** Greenhouse gas emissions by sector, in the year 2010.

**Table 1.** Rise in greenhouse gas concentration in atmosphere as influenced by anthropogenic activities (Sharma and Bazaz, 2012).

| Major GHGs | $CO_2$ | $CH_4$ | $N_2O$ | CFC-12 |
|---|---|---|---|---|
| Pre-industrial atmospheric concentration | 280 ppmv | 0.70 ppmv | 280 ppbv | 0 |
| Current concentration | 387 ppmv | 1.72 ppmv | 320 ppbv | 533 pptv |
| Current annual increase (%) | 0.5 % (1.5–1.8 ppmv) | 0.8 % (0.013 ppmv) | 0.25 % (0.75 ppbv) | 4 % (18–20 pptv) |
| Atmospheric life time (years) | 50–200 | 12–17 | 150 | 102 |
| Global warming potential relative to $CO_2$ | 1 | 25 | 298 | 10,600 |

## Impact of Climate Change on Agriculture

Climate change is not a one day phenomena, it is an accumulated long period activities. It is mainly consisted elevated $CO_2$ concentration and temperature across the world. Warming will be more pronounced over land areas with the maximum increase over northern India (Kundu *et al.*, 2013a). It is more important when agricultural growth and development are primarily governed by the environmental conditions of the soil and climate. While in the sub-tropics where scares rainfall and higher temperature induced drought and desertification will reduce the vegetation cover and carbon sequestration potential of the soil. Climate change mediated soil organic matter dynamics is a complex process that influenced by many factors and processes will differently interplay on a long term basis is unpredictable. Carbon dioxide is essential for the photosynthesis and its optimum concentration required for plant growth (Kumar *et al.*, 2012). Carbon dioxide take parts in fixes $CO_2$ process, RuBisCo, also fixes oxygen in the process of photorespiration. Rising $CO_2$ concentration in the atmosphere can have both positive and negative consequences. Increased $CO_2$ is expected to have positive physiological effects by increasing the rate of photosynthesis. This is known as 'carbon dioxide fertilization in low $CO_2$ areas. Leaf nitrogen concentrations in plant tissues typically decrease under elevated $CO_2$, with nitrogen per unit leaf mass decreasing on average by 13%. This decrease in tissue nitrogen is likely due to several factors: dilution of nitrogen from increased carbohydrate concentrations; decreased uptake of minerals from the soil, as stomatal conductance decreases and plants take up less water; and decreases in the rate of assimilation of nitrate into organic compounds. The effects of an increase in carbon dioxide would be higher on $C_3$ crops (such as wheat) than on $C_4$ crops (such as maize), because the former is more susceptible to carbon dioxide shortage. The consumption of atmospheric $CO_2$ occurs mainly through carbon sequestration; improved crop production increased the residue return and better nutrient management increase C input in soil. In humid tropics and temperate region, climate change induced increase in rainfall will have a positive influence on C storage in the soil. This increased plant growth is also reflected in the harvestable yield of crops, with wheat, rice and soybean all showing increases in yield of 12–14% under elevated $CO_2$ (Seneweera *et al.*, 1996).

The significance of climate change impacts on grain and forage quality emerges from new research agenda across the globe. For rice, the amylose content of the grain-a major determinant of cooking quality is increased under elevated $CO_2$ (Conroy *et al.*, 1994). Moreover, the protein content of the grain decreases under combined increases of temperature and $CO_2$ (Ziska *et al.*, 1997). Lator on, Loladze (2002) reported in Free-Air Carbon dioxide Enrichment (FACE) experiment that increases in $CO_2$ lead to decreased concentrations of micronutrients in crop plants. Rising $CO_2$ also helps to improve plant water use efficiency. However, still it is not very conclusive that how important this $CO_2$-enhanced water use efficiency might be in counterbalancing warming-induced desiccation because higher $CO_2$ also leads to higher plant biomass, and therefore greater transpiration surface (Morgan *et al.*, 2004; Leakey, 2009). The increasing global temperature due to interactive effect of atmospheric activities, disturb the ecological system of nature. The imbalance of natural cycles like water, carbon, nitrogen etc. The warmer atmospheric temperatures observed over the past decades are expected to lead to a more vigorous hydrological cycle, including more extreme rainfall events. Erosion and soil degradation is more likely to occur. It reduced the soil fertility, and promote to poor quality of land. Addition of C from atmosphere to agricultural field, it affected the soil C:N ratio, and affected the nutrient dynamics in soil system. Increasing global temperature it affect the weeds growth, some exotic weed emerged fast in compared to crop plants. Emerged weeds are more resistance to herbicidal application. Global warming would cause an increase in rainfall in some areas, which would lead to an increase of atmospheric humidity and the duration of the wet seasons. Combined with higher temperatures, these could favor the development of fungal diseases. Similarly, because of higher temperatures and humidity, there could be an increased pressure from insects and disease vectors (Karuppaiah and Sujayanad, 2012).

## Principle of Soil Carbon Sequestration to Minimize Climate Change Effect

World soils constitute the third largest global C pool comprising of two distinct components: (i) soil organic C (SOC) estimated at 1550 Pg, and (ii) soil inorganic C (SIC) pool estimated at 950 Pg, both to 1-m depth. Other pools include the oceanic (38,400 Pg), geologic/fossil fuel (4500 Pg), biotic (620 Pg), and atmospheric (750 Pg). Thus, the soil C pool of 2500 Pg is 3.3 times the atmospheric pool and 4.0 times the biotic pool. The SOC pool is at a dynamic equilibrium under a specific land use and management system (Lal, 2004). At equilibrium, the $C_{input}$ into a system equals $C_{output}$. Upon conversion to another land use and management, depletion of SOC pool occurs if $C_{input} < C_{output}$, and sequestration if $C_{input} > C_{output}$ (Eq. 1 to Eq. 3).

Steady state . . . . . . . . . . . . $C_{input} = C_{output}$ . . . . . . . . . . . . . . Eq. 1

Depletion . . . . . . . . . . . . . $C_{input} < C_{output}$ . . . . . . . . . . . . . . . Eq. 2

Sequestration . . . . . . . . . . $C_{input} > C_{output}$ . . . . . . . . . . . . . . . Eq. 3

However, soils of the poorly managed ecosystems have lost 50 to 75% of the original SOC pool. The magnitude of SOC depletion is higher in soils prone to erosion and those managed by low-input or extractive farming practices. The loss of SOC pool is also high in soils of coarse texture and those with a high initial pool (Lal, 2004a).

Climate policy is being expanded to consider a wide range of options aimed at sequestering carbon in vegetation, oceans and geological formations, at reducing the emissions of non-$CO_2$ greenhouse gases, and at reducing the vulnerability of sectors and communities to the impacts of climate change by means of adaptation. Consequently, the Third Assessment Report (TAR) provided a more balanced treatment of adaptation and mitigation (Figure 2).

The TAR demonstrated that the level of climate change impacts, and whether or not this level is dangerous, is determined by both adaptation and mitigation efforts (Smith *et al.,* 2001). Adaptation can be seen as direct damage prevention, while mitigation would be indirect damage prevention (Verheyen, 2005). However, only recently policy makers have expressed an interest in exploring interrelationships between adaptation and mitigation. Recognising the dual need for adaptation and mitigation, as well as the need to explore trade-offs and synergies between the two responses, they are faced with an array of questions (Clark *et al.,* 2004; Figure 3). How much adaptation and mitigation would be optimal, when, and in which combination? Who would decide, and based on what criteria?

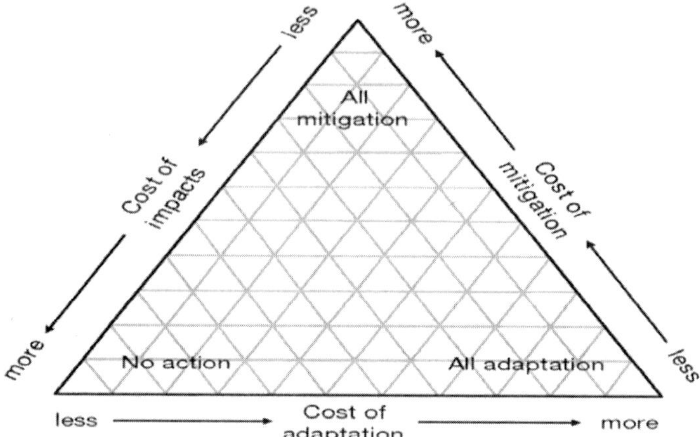


**Figure 2.** A schematic overview of inter-relationships between adaptation, mitigation and impacts, based on Holdridge's life-zone classification scheme (Holdridge, 1967).

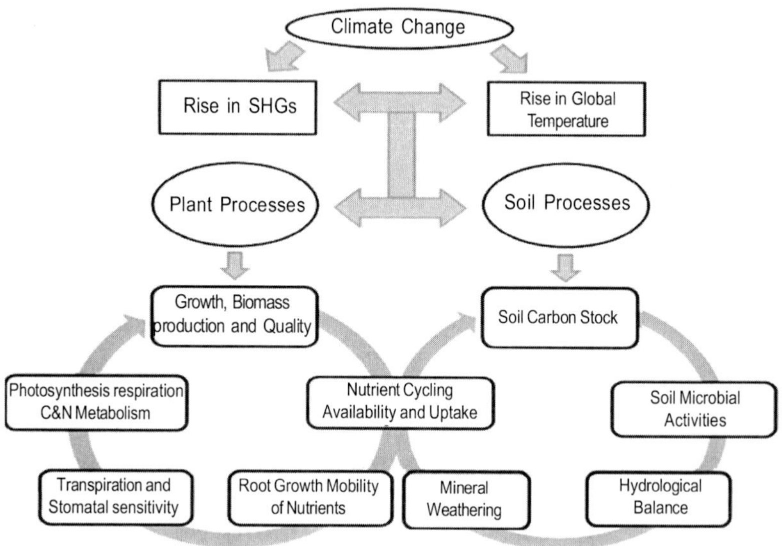


**Figure 3.** Interactive impact of climate change on plant and soil processes (Chakraborty *et al.,* 2013)

## Adaptation strategies

Any perturbation in agriculture can considerably affect the food system and thus increases vulnerability of a large fraction of the resources poor population. Thus, we can understand the effect of climate change on agricultural crops and combat strategies by different sections and different categories of producers in India. Some of strategies directly involved in minimizing global climate and others are indirectly involved. These strategies having more valuable when it is practiced in long period.

1. **Agricultural biodiversity**

   Agricultural biodiversity is the basis of our agricultural food chain, developed and safeguarded by farmers, livestock breeders, forest workers, fishermen and indigenous peoples throughout the world. The use of agricultural biodiversity (as opposed to non diverse production methods) can contribute to food security and livelihood security. It increases resilience to changing environmental conditions and stresses. Genetically diverse populations and species rich ecosystem have greater potential to adapt to climate change. Selection of crops and cultivars with tolerance to abiotic stress *i.e.* high temperature, drought, flooding, high salt content in soil, pest and disease resistance. Mixed farming has given good results in some countries to overcome the climate change effect on agricultural crops (Watson, 2010).

2. **Land use and management system**

   This is also one of the parameters which play a vital role in combating adverse effects of climate change in agricultural crop production. Application of modern technologies and new land management techniques, and water use efficiency related techniques are key to long term adaptation strategies (Lenka, 2013). Predictions or estimates of likely future adaptations are an essential element of climate change impact and vulnerability assessment. The degree to which a future climate change risk is dangerous depends greatly on the likelihood and effectiveness of adaptations in that system. Reilly and Schimmelpfennig (1999) defined the following major classes of adaptation :

   - Seasonal changes and sowing dates
   - Different varieties or species
   - Water supply and irrigation system
   - Promotion of agro-forestry
   - Management and sustainable use of agricultural inputs

3. **Emphasis on localization**

   Most of the Indian population living in rural areas, which is more important to our researcher and policy maker to replacing globalization with localization. Effect of climate change is not similar at every place, so the technologies may be machinery, agricultural inputs or management strategies should be on local need. Farming system should aim at maximizing plant biomass production from locally available diversified resources. The direction and magnitude of the climate change impacts will depend on the specific cropping system as well as regional conditions, but the adaptation measures should be flexible, so that can adjust as per the needs. Some of farmers in Himachal Pradesh, growing mango tree in lower altitude, but earlier it was not possible due to extreme cold weather.

4. **Fertilizer / Nutrient management**

   Optimal nutrition and most favorable soil tillage greatly affect water circulation within the plant, which is a highly effective method of combating drought. Under poor nutrient condition plants have to absorb more amount of water to full fill nutrient requirement. In low moisture conditions, this has a negative effect on plant growth and crop yield. Thus applying the right amount and at the right time can address the problem of environmental stress. Application of silicon as a fertilizer also helping to alleviation of negative stress of climate change, it's increasing plant resistance to drought stress. One of the alternative methods involving the silicon fertilization for alleviation of negative stress of climate change (Meena *et al.*, 2013). Unfortunately the silica that occurs in soil is in an unavailable polymerized form and for its absorption by plants it has to be depolymerized and rendered soluble by means of biological or chemical reactions in the soil.

5. **Use of waste water for irrigation**

   Most of the cities generating huge amount of waste water, which can be used as a vegetable production in peri urban areas. Wastewater is mainly comprised of water (99.9%) together with relatively small concentrations of suspended and dissolved organic and inorganic solids. Among the organic substances present in sewage are carbohydrates, lignin, fats, soaps, synthetic detergents, proteins and their decomposition products, as well as various natural and synthetic organic chemicals from the process industries. A number of potentially toxic elements such as arsenic, cadmium, chromium, copper, lead, mercury are also present (Dotaniya *et al.*, 2014a). In major Indian cities, the absence of a suitable network of sewage results pollution of urban environment. With adequate treatment, urban waste water, with an acceptable threshold quality, can be used as a source of water for irrigation (Dotaniya *et al.*, 2014b). Even in the rural areas, reuse of kitchen wastewater for micro irrigation should be encouraged. It reduces the global climate effect in drought prone areas across the world.

6. **Use drought management techniques**

   Drought limit the growth of crop plant and affected the productivity adversely. It is a climatic anomaly, characterized by deficient supply of moisture resulting either from sub-normal rainfall, erratic rainfall distribution, higher water need or a combination of all the three factors. About two thirds of the geographic area of India receives low rainfall (less than 1000 mm), which is also characterized by uneven and erratic distributions. Improvement in water management techniques will enhance the water use efficiency and improved the water for crop production and household use. Majority of fresh water is utilizing by agriculture and industrial sectors. Some of the activities given below can help to overcome the climate change effect.

   - Avoidance and tolerance mechanism
   - Develop water logged and water stress tolerance crop varieties
   - Proper water management
   - Soil moisture conservation, enhance ground water recharge
   - Rain water harvesting
   - Collective action by societies and NGOs

7. **Agricultural insurance**

   It is directly not involve in reducing climate change effect in agricultural sector. But it offers a valuable contribution to adaptation. Different insurance schemes given more risk bearing capacity to combat climate change effectively. Management of risk in agriculture is one of the major concerns of the decision makers and policy planner, which is indirectly more important parameters in any climate change related policies. Crop insurance not only stabilizes the farm income but also helps the farmers to initiate production activity after a bad agriculture year. Some of the agricultural insurance schemes are as follows.

   - National Agricultural Insurance Scheme (NAIS) 1999
   - Pilot crop insurance scheme 1979-1984
   - Farm income insurance
   - Livestock insurance
   - Weather based crop insurance/rainfall insurance

## Conclusion

Change in climate in India is predicted to be more severe than in other regions and are expected to have significant impact. No single set of adaptive policy recommendations can be universally appropriate at every region. Right strategies used in the right place to overcome the effect of climate change. The focus any adaptive strategies should be to empower our farmers which is possible only if the option is viable, feasible and profitable; otherwise government should provide incentives for adoption and insurance against failure. With this we can sustain social, economic and institutional pressure which is generated by global climate change.

## References

Chakraborty K., Bhaduri D., Uprety D. C. and Patra A. K. (2013). Differential Response of Plant and Soil Processes Under Climate Change: A Mini-review on Recent Understandings. Proc. Natl. Acad. Sci., India, Sect. B Biol. Sci. DOI 10.1007/s40011-013-0221-7.

Clark, W.C., Crutzen P.J. and Schellnhuber H.J. (2004). Science for global sustainability: toward a new paradigm. Earth System Analysis for Sustainability, H.-J. Schellnhuber, P.J. Crutzen, W.C. Clark, M. Claussen and H. Held, Eds., MIT Press, Cambridge, Massachusetts, 1-28.

Dotaniya M. L., Das H. and Meena V. D. (2014a). Assessment of chromium efficacy on germination, root elongation, and coleoptile growth of wheat (*Triticum aestivum* L.) at different growth periods. Environ. Monit. Assess. 186:2957-2963. DOI 10.1007/s10661-013-3593-5.

Dotaniya, M. L., Saha J. K., Meena, V. D., Rajendiran, S., Coumar, M. V., Kundu, S. and Rao, A. Subba (2014b). Impact of tannery effluent irrigation on heavy metal build up in soil and ground water in Kanpur. Agrotechnol 2(4):77.

Holdridge, L. R. (1967). Life Zone Ecology. Tropical Science Centre, San Jose, Costa Rica.

IPCC (2001). Climate Change (2001). Synthesis Report. A Contribution of Working Groups I, II, III to the Third Assessment Report of the Intergovernmental Panel on Climate Change, R.T. Watson and the Core Team, Eds., Cambridge University Press, Cambridge and New York, 398 pp.

IPCC (2007). The physical science basis. Summary for Policymakers. Intergovernmental Panel on Climate Change.

Karuppaiah, V. and Sujayanad, G.K. (2012). Impact of climate change on population dynamics of insect pests. World J. Agric. Sci. 8 (3): 240-246.

Klein, R.J.T., Schipper E.L. and Dessai S., (2005). Integrating mitigation and adaptation into climate and development policy: three research questions. Environ. Sci. Policy 8: 579-588.

Kumar, M., Swarup, A., Patra, A.K., Chandrakala, J.U., Manjaiah, K.M. (2012). Effects of elevated atmospheric $CO_2$ and temperature on phosphorus efficiency of wheat (*Triticum aestivum* L.) grown in an Inceptisol of subtropical India. Plant Soil Environ. 58(5):230–235.

Kundu, S., Dotaniya, M. L. and Lenka, S. (2013a). Carbon sequestration in Indian agriculture. *In:* Climate change and natural resources management (Ed. S. Lenka, N. K. Lenka, S. Kundu and A. Subba Rao) New India Publishing Agency. ISBN 978-93-81450-67-3. pp. 269-289.

Kundu, S., Rajendiran, S. and Coumar, M.V. (2013b). Elevated atmospheric $CO_2$- Its indirect effect on soil process. *In:* Climate change and natural resources management (Ed. S. Lenka, N. K. Lenka, S. Kundu and A. Subba Rao) New India Publishing Agency. ISBN 978-93-81450-67-3. pp. 75-89.

Lal, R. (2004). Soil carbon sequestration in India. Climatic Change 65: 277-296.

Lal, R. (2004a). Soil carbon sequestration impacts on global climate change and food security. Sci. 204: 1623-1627.

Leakey, A.D.B. (2009). Rising atmospheric carbon dioxide concentration and the future of C4 crops for food and fuel. Proc. Roy. Soc. London B Bio. 276:2333–2343.

Lenka, S. (2013). Adaptation strategies in agriculture in context of climate change. *In:* Climate change and natural resources management (Ed. S. Lenka, N. K. Lenka, S. Kundu and A. Subba Rao) New India Publishing Agency. ISBN 978-93-81450-67-3. pp. 159-175.

Loladze, I. (2002). Rising atmospheric $CO_2$ and human nutrition: toward globally imbalanced plant stoichiometry?. Trends Ecol. Evolution 17 (10): 457. Doi:10.1016/S0169-5347(02)02587-9.

Meena, V. D., Dotaniya, M. L., Rajendiran, S., Coumar, M. V., Kundu, S. and Rao, A. Subba (2013). A case for silicon fertilization to improve crop yields in tropical soils. Proc. Natl. Acad. Sci., India Sec. B: Biol. Sci. 84(3): 505-518. DOI 10.1007/s40011-013-0270-y.

Morgan, J.A., Pataki D.E. et al (2004). Water relations in grassland and desert ecosystems exposed to elevated atmospheric $CO_2$. Oecologia 140:11–25.

NATCOM (2004). India's Initial National Communication to United Nations Framework Convention on Climate Change. Ministry of Environment and Forests, Government of India, pp 392.

Seneweera, S., Blakeney, A., Milham, P., Barsa, A.S., Barlow, E.W.R. and Conroy, J. (1996). Influence of rising atmospheric $CO_2$ and phosphorus nutrition on the grain yield and quality of rice (*Oryza sativa* cv. Jarrah). Cereal Chem. 73(2):239–243.

Sharma, S.K., Bazaz, A.B. (2012). Sustainable management of biodiversity in the context of climate change-issues, challenges and response. Proc. Natl. Acad. Sci. India Sect. B Biol. Sci. 82: 251–260.

Smith, J.B., Schellnhuber H.J., Mirza M., Fankhauser S., Leemans R., Lin Erda, Ogallo L., Pittock B., Richels R., Rosenzweig C., Safriel U., Tol R.S.J., Weyant J. and Yohe G. (2001). Vulnerability to climate change and reasons for concern: a synthesis. Climate Change 2001: Impacts, Adaptation, and Vulnerability. Contribution of Working Group II to the Third Assessment Report of the Intergovernmental Panel on Climate Change, J.J. McCarthy, O.F. Canziani, N.A. Leary, D.J. Dokken and K.S. White, Eds., Cambridge University Press, Cambridge, 914-967.

United Nations (1997). Kyoto Protocol to the United Nations Framework Convention on Climate Change. http://unfccc.int/resource/docs/convkp/kpeng.pdf.

Verheyen, R. (2005). Climate Change Damage and International Law: Prevention, Duties and State Responsibility. Martinus Nijhoff, Netherlands, 418 pp.

Watson, R. (2010). Global biodiversity: Indicators of recent declines. Sci. 328: 1164-1168.

❑❑❑

# 3

# Hailstorm Occurrence and Its Management Strategies with the Changing Climate

**S.K. Bal, P. Suresh Kumar, Yogeshwar Singh, D.D. Nangare, B.P. Fand, Sunayan Saha and P.S. Minhas**

## Introduction

Food production systems in India are becoming increasingly vulnerable to climate variability and change which are largely characterized by altered frequency timing and magnitude of precipitation and temperature that trigger a series of other devastating events. High rates of snow and glacial melting, frequent floods and droughts, heat waves, hail and increased incidence of pests and diseases are already causing widespread damage and losses to agricultural sector in tropical and sub-tropical countries including India. Overwhelming direct damage to crops and livestock within few minutes can be caused by extreme weather events such as hail, wind-storm, or heavy frost. While occurrence, losses and post disaster management have been discussed extensively for natural calamities such as excess rainfall, drought and flood, little attention has been given to hail storm.

Hail is a solid, frozen form of precipitation that causes damage to properties and growing crops worth rupees hundreds of crores every year in this country and the world. It is formed when water droplets moving upward through clouds, encountering temperatures below freezing, causing the water to supercool and freeze on contact with condensation nuclei like dust particles. Hail usually forms over a relatively small area and leaves within a few minutes. But there have been instances when hail events persist in the same area for tens of minutes to one hour, leaving several inches of ice on the ground.

Hail storm often occur under hot, humid unstable weather conditions in the summer, although damaging storms have occurred in every month of the year. Generally, they occur in late afternoon or early evening when thunderstorms build up, but again, damaging hail storms have occurred in the night or early morning as well. The largest forms of hail stones can fall at rapid speeds higher than 160 $Kmh^{-1}$. Hail damages are determined by the size ranges and the number of hail stones that fall per unit area during a hail fall,wind force during the event and the property of the target. The extent of crop-hail damages also varies depending on the time of occurrence of hail duringthe growing season of a given crop. In the above context, this present paper has been prepared to give a comprehensive account on various aspects of hail and its management strategies.

## Science Behind Hail Formation

Hail is often associated with thunderstorm activity and changing weather fronts. This is formed in huge cumulonimbus clouds, commonly known as thunderheads. Hailstorms are the result of four atmospheric factors. These factors are as follows:

1. Strong convective instability creating strong updrafts
2. Abundant moisture at low levels feeding into the updrafts.
3. Strong wind shear aloft, usually veering with height, enhancing updrafts.
4. Some dynamical mechanisms that can assist the release of instability such as air flowover mountain ridges.

The typical mechanism of hail formation in a thunderstorm is described in Fig. 1. When the ground is heated during the day by the sun, the air close to the ground is heated as well. Hot air, being less dense and lighter than cold air, rises and cools. As it cools, its capacity for holding moisture decreases. When the rising, warm air has cooled so much that it cannot retain all of its moisture, water vapour condenses, forming puffy-looking clouds. The condensing moisture releases heat of its own into the surrounding air, causing the air to rise faster and give up even more moisture.

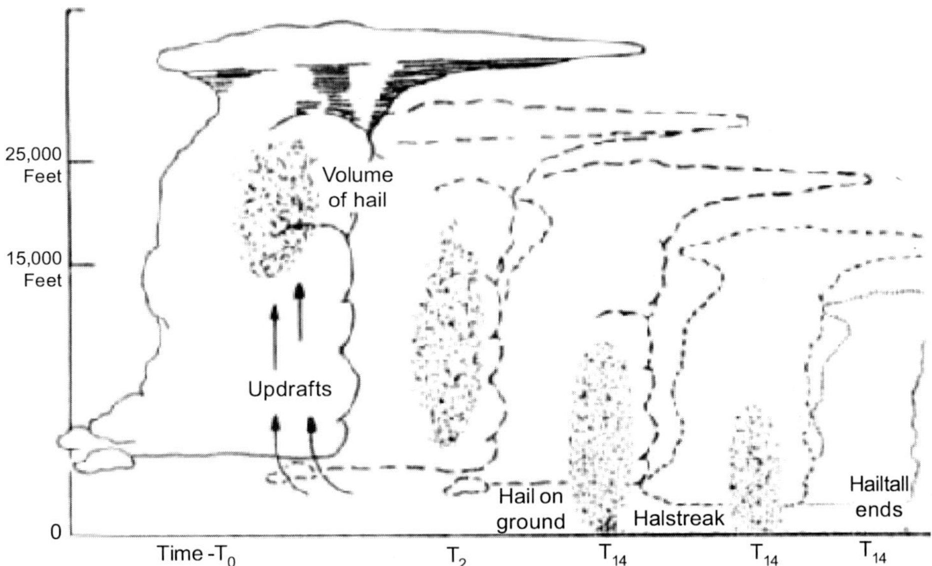


**Fig. 1.** A sequence showing the development of hail inside a thunderstorm, then its descent and arrival at the ground after 4 minutes (T4). Its deposition forms a path of hail labelled as a hailstreak, ending after 14 minutes (T14).(*Source:* Illinois State Water Survey, USA)

Cumulonimbus clouds contain vast amounts of energy in the form of updrafts and downdrafts. These vertical winds can reach speeds over 175 Km per hour. Hail grows in the storm cloud's main updraft, where most of the cloud is in the form of "super-cooled" water. This is water that remains liquid although its temperature is at or below 0°C (32°F). At temperatures above 0°C (-40°F), a super-cooled water droplet needs something on which to freeze, or it remains liquid. Ice crystals, frozen raindrops, dust, and salt from the ocean are also present in the cloud. On collision, super-cooled water will freeze onto any of these hosts, creating new hail stones or enlarging those that already exist. The faster the updraft on these balls of ice, the bigger they can grow. Cross sections of hailstones often reveal layers, much like those of an onion.

The growth of hail stones sufficiently large to reach the ground requires very strong updrafts, forces creating taller than usual thunderstorms. Strong updrafts support the hail stones aloft and allow hailstones to grow, often to 1 inch diameter or larger. If the falling hailstones enter another strong updraft, they can get carried aloft again in the moist air and grow even larger. This repetitive growth process is reflected in the structure of hailstones that often shows layers of ice around their embryo. When the hailstone becomes so heavy that the updraft can no longer support it or when there is dominance of downdraft over the updraft, it falls from the sky. The development of hailstones typically occurs 5 to 7Km above the earths surface. Themost of the hails come from thunderstorms; however, only about 60 percent of all thunderstorms ever generate hailstones aloft.

When a volume of hailstones descending from a storm reaches the surface, the stones often cover an area about 1-2Km in diameter at the earths surface. As the hailstorm moves over time, the falling hailstones produce an elongated area of hail that the scientists called as ***"hail-streak" or hail swaths***. Its size and shape depend on how fast the storm is moving and how strong the updrafts are inside the storm. A typical hail-streak is about 1.5 Km wide and 8 Km in length.However, these may vary from a few acres to large belts, about 16 Km wide and 160 Km long. Most storms that produce hail generate one or two hail-streaks during their lifetime. Some organized lines of thunderstorms produce many hail-streaks with hail covering hundreds of square Km as the storms move across the terrain. Infrequently a thunderstorm becomes a well-organized giant and lasts for three or more hours. These "supercell storms" generate very large hail-streaks.

The volume of hail reaching the ground falls at a speed of about 40 meters per second, and is usually less than 10 percent of the volume of rain produced by a thunderstorm. Hail produced by many thunderstorms never reaches the ground because it melts as it descends into warmer air near the ground, becoming raindrops. That is why thunderstorms in warmer climate zones seldom produce hail at the ground. On the other hand, hailstorms damages are more common and severe in the mid latitude and temperate nations owing to higher frequency and intensity.

Hailstones are usually layered (Fig 2), sometimes with sharp edges and range in size from pellets to golf balls or larger or sometimes get interesting shapes like pyramids (Fig 3). Hailstones are seldom perfectly circular and most have unusual shapes as a result of the different atmospheric conditions where they were formed. Most hailstones are of oblate shapes and some have knobs of ice radiating outwards.

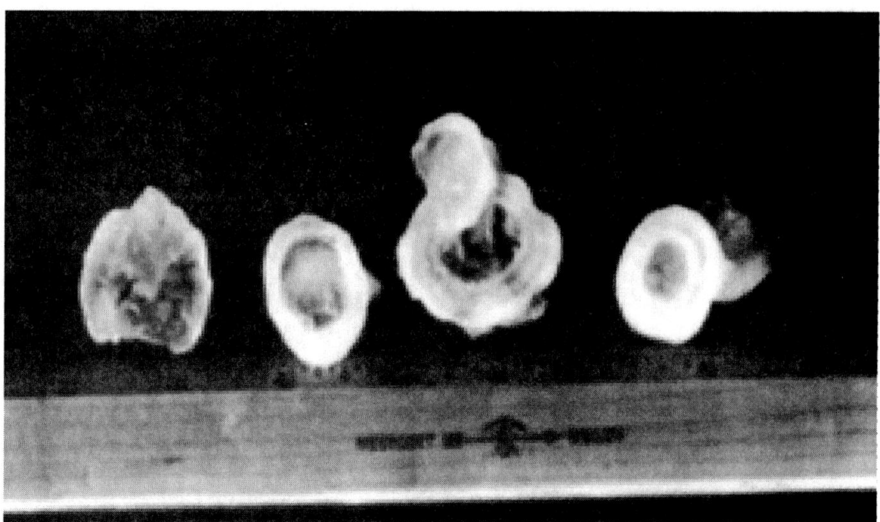

**Fig. 2** Inside layered structures of hailstones that have been cut in half (*Source:* Illinois State Water Survey, USA)

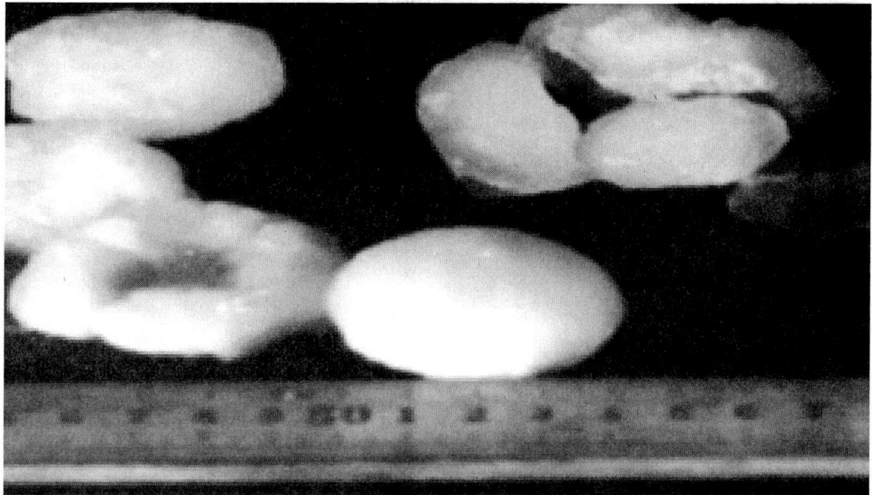

**Fig. 3.** Shape and size of hailstones (*Source:* Illinois State Water Survey, USA)

## Hail Characteristics and damage potential

Hail damages are a function of two conditions, a) hail characteristics and b) the characteristics of the target (property and crops). Hail characteristics that vary from storm to storm and cause damages of varying degrees include the size and number of hailstones that fall per unit area, shape of the hail and the strength of winds during a hail fall. The damage also varies according to the target. Some delicate-leaf crops such as tea and tobacco suffer damage from small hailstones, whereas other crops such as maize are not damaged unless hailstones are 3/4 inch or larger. The extent of crop-hail damages also varies during the growing season of a given crop. A specific type of hailstorm may not cause much damage early in the crops growing season but the same storm in mid-crop season can be very destructive.

## Hailstone size and damage potential

One of the important aspects of hail intensity is the size of the hailstones that occur. During most hail falls there is a range of stone sizes. The size of a hailstone matters considerablywhen accounting for damage. Studies have concluded that most property damage begins when hailstone diameters are 0.75 inch or greater. The larger the stones, typically the greater the property damage. However, most crop loss is not linearly related to stone sizes. For example, some crops at certain stages of development are most susceptible to damage from small 0.25-inch stones, and other crops (and growth stages) are susceptible only when stones exceed 0.5 inch in diameter. Hail mass is often the most critical factor causing crop damage(Table 1) provides a picture of the severity of damages by hails of different sizes borne out of hailstorms of different intensities.

**Table 1.** Hailstorm Intensity Scale

| Intensity | General description / Damage caused | Size of Hail (diameter in inches) |
|---|---|---|
| H0 | None | 0.2 to 0.4 (Pea size) |
| H1 | Makes holes in leaves | 0.2 to 0.8 (Marble size) |
| H2 | Strips leaves from plants | 0.2 to 1.2 (Penny size) |
| H3 | Breaks glass panels and can scrape paint | 0.4 to 1.8 (Nickel size) |
| H4 | Breaks windows, scrapes paint | 0.6 to 2.4 (Golf ball size) |
| H5 | Breaks some roof tiles, dents cars, strips bark | 0.8 to 3.0 (Tennis ball size) |
| H6 | Breaks many roof tiles, damages roofs | 1.2 to 3.9 (Baseball size) |

| | | |
|---|---|---|
| H7 | Shatters roofs, serious damage to cars | 1.8 to 4.9 (Grapefruit size) |
| H8 | Cracks concrete roofs, splits trees, injury to people | 2.4 to 5 (Softball size) |
| H9 | Marks concrete walls, kills people, and fells trees. | 3.2 to 5 (Softball size) |
| H10 | Damages brick homes, kills people | 4 to 7 (Volleyball size) |

*Source:* Illinois State Water Survey, USA

## Hail Frequency, Intensity and Impact: India Perspectives

Hailstorms occur in many parts of the world; however, its frequency and intensity are more in temperate and mid-latitude. USA, the worst hail affected nation in the world, has more systematic reports about atmospheric conditions that cause hail-producing systems. In case of India, thunderstorms occur at variable frequencies in different parts of the country. The hailstones were reportedly as big as oranges weighing up to 1 kg, and in some places the hails accumulated to heights as much as 2 feet.

Generally, before and after the monsoon season, thunderstorms with associated showers of rain and hail are the predominant weather phenomena. When sufficient moisture is not present in the air, as over north-west India, only a dust storm may result. Some violent thunderstorms are accompanied with hail, especially in northern and central India and occasionally in the interior of peninsular India.

The Western Disturbance coming from Afghanistan moves to the north of Maharashtra, bringing cold winds and at the same time the moisture incursion from the Bay of Bengal has got in warm winds. These activities are resulting in hailstorm in the northern parts of Maharashtra. There were two opposing systems easterlies bringing in moisture-laden winds, while northerlies bringing in dry and cold winds. The dynamic instability coupled with the sudden conflict of the easterlies and westerlies and availability of moisture has caused this hail storm phenomenon. Clouds measuring up to 22 km in height and sub-freezing temperatures have been observed during these events. Leaving the coastal Maharashtra, all divisions of the state and parts of Madhya Pradesh and north Andhra Pradesh witness this activity.

The frequency of hail storms is small in winter, but increases generally as the season advances to summer. During monsoon season, hailstorms are practically absent from the whole country. The number of days with hail is about 6-7 per year over Himachal Pradesh and its neighbourhood, but it decreases to 1 in 2 per years over the adjoining plains. Over Bengal, Bihar, Uttar Pradesh and Madhya Pradesh, hail storms occur on an average once a year. Hail storms are comparatively rare over the coastal tracts of the Peninsula. In central India, especially in February and March, hailstorms occur on 1 or 2 occasions in the year, and pose a potential hazard to *rabi* crops. The intensity and duration of the hail storm determines how much damage will be done.

Various parts of the country have recently experienced wide spread unseasonal rain and hailstorms.The freak weather event that battered the Country's six states Punjab, Uttar Pradesh, Rajasthan, Madhya Pradesh, Maharashtra and Andhra Pradesh for an unprecedented 20 days, from February 24 to March 14, has left millions of farmers in a similar state of shock. It occurred when farmers were getting ready to harvest rabi crops such as wheat, pulses, potato, sugarcane, maize, groundnut and mustard, and horticultural crops like grapes, papaya, mango, banana, onion and other vegetables.

## Types of hail damages

Two main types of hail damages in crops were observed. Interestingly, the damage occurred mostly on windward side of the stems or branches. The leeward sides of the plants were seen without or very less damage due to hailstones.

1. *Primary injuries:* heavy defoliation, shredding of leaf blades, breaking of branches and tender stems, lodging of plants, peeling of bark, stem lesions, cracking of fruits, heavy flower and fruit drop, death of farm animals and poultry birds etc.

2. *Secondary injuries:* dieback or wilting of damaged plant parts, loss of plant height, staining, bruises, discolouration of damaged parts like leaves, fruits affecting their quality, rotting of damaged fruits and or tender stems and branches due to fungal and bacterial infections.

Type of damages on various crops due to hailstorm

| Crop | Type of damage |
|---|---|
| Cotton | Plant stem lodging and breakage, deterioration of fibre quality |
| Sorghum | Plant stem lodging, ear head broken, grain shattering and secondary infection |
| Wheat | Lodging and ear head damage, secondary infection |
| Maize | Defoliation, lodging, stem bruises |
| Chickpea | Pod sheding |
| Grapes | Lodging of training structure, scars in bunches, berry drop, defoliation |
| Pomegranate | Defoliation, fruit and flower drop, fruit cracking, stem bruises |
| Sweet lime | Immature fruit dropping and lesions on mature fruits, secondary infection |
| Guava | Defoliation, fruit and flower drop, fruit cracking, stem bruises |
| Papaya | Petiole breaking and defoliation, fruit scars and drop |
| Mango | Flower bud and immature fruit drop |
| Fig | Stem bruises, fruit lesions, flowers fell off, leaves shattering |
| Water melon | Lesions on fruits, drying of leaves and rotting of fruits |
| Onion and Garlic | Leaves burning from terminal end, lesions on leaves and rotting |
| Tomato | Lesions on fruits, fruit drop and plant lodging |
| Brinjal | Lesions on fruits and leaves |
| Drum stick | Lesions on drum stick, bruises on stems, breakage of twigs |
| Lucerne | Lesions on leaves and secondary fungal infection |
| Fisheries | Migration, damage to fish nets |
| Poultry | Mortality and Fungal infections |

## Hail Preparedness: Forecasting and Suppression

In the last decade, there has been an increase in the phenomenon of Arctic air being pushed down leading to unusual regional climate anomalies across the Northern Hemisphere.Weather anomalies greatly increase the intensity and frequency of extreme events. There are strong indications that some types of extreme event, most notably heat waves and precipitation extremes, will greatly increase in a warming climate, and have already done so. Protection against violent hailstorms is possible by adopting artificial hail suppression methods. It is difficult to forestall a hailstorm, since its occurrence is sporadic and confined to very limited areas in a thunderstorm. While hail formation continues to elude scientists, sophisticated radar has been developed that can detect the presence of hail before it falls to the ground.

## Severe hail Climatology: Probability concept

These hail occurrence probability values are estimated from a 30-year period of actual hail events. Reports for each day are put onto a designated grid (for example 80 km x 80 km). If one or more reports occur in a grid box, that box is assigned the value "1" for the day. If no reports occur, it's a zero. The raw frequency for each day at each grid location is found for the period (number of "1" values divided by number of years) to get a raw annual cycle. The raw annual cycle at each point is smoothed in time, using a Gaussian filter with a standard deviation of specified period (eg. 15 days). The smoothed time series are then smoothed in space with a 2-D Gaussian filter with a standard deviation of certain period (eg. 120 km in each direction).

## Hailstorm forecasting using Weather Radar

Efforts to detect and measure hail have been underway for many years. Forecasters use Radar technology for detection of hail and have met with various degrees of success depending on conditions. RADAR is an acronym that stands for Radio Detection and Ranging. Early attempts using single wavelength radars were not successful as conventional Radars, could provide only rough subjective guidelines for identifying or inferring hail, turbulence, tornadoes etc. Dual wavelength radars make simultaneous use of two different wavelengths, hence offer better scope compared to single wavelength radars for detecting hail and distinguish it from rain showers. Later, polarimetric radars were tested to detect hail aloft.A weather radar consists of a parabolic dish (it looks like a satellite dish) encased in a protective dome and mounted on a tower of up to five stories tall. The radar itself does not delineate between rain and snow but use of algorithms do, based on atmospheric conditions. The exclusive conglomerate of weather radar data base plots the expected movement of significant storms over the next hour, and also gives access to storm details like hail and rotation.

## Hail Suppression

Artificial hail suppression is an important measure in disaster prevention and mitigation. With the development of atmospheric science and related science and techniques, the ability of hail cloud identification and hail suppression improves continuously. In recent years, with the rapid development of new-generation weather radar techniques, great progress has been made in the mechanisms of hail formation and new artificial hail suppression techniques, which provides the basis and foundation of scientific hail suppression operations. Based on the analysis of new-generation radar detection data, different radar parameters are concluded as evaluating indicesof hail suppression operations.

To a certain degree, the technical merit of hail suppression is raised. Meanwhile, problems are also noticed. For example, the accurate identification of hailclouds. People often take thunderclouds as hail clouds by mistake and cause economic waste. Especially, a complete, scientific operation scheme design in implementation is not formed. The casualness in determining operational parameters and the use of unreliable seeding methods reduce the effect of hail suppression.

## Identification of Hailclouds

For artificial hail suppression, the first step is to distinguish hail cloud from severe thunderstorms, then, decide whether or not to operate, further, choose proper technical scheme. The experience indicates that effective hail suppression is based on identifying hail clouds efficiently and accurately. According to the formation theories of hailcloud and the latest research results of hail suppression, the Doppler radar data are analysed statistically. The discriminant indices and models of spring, summer and autumn hail clouds are derived as follows: If the echo intensity of a strong convective cloud fits to any one criterion above step by step, the cloud can be identified to be a hailcloud which will be seeded. Specifically, determines the height of radar initial echo. If strong echo appears on the top of cloud or 45dBz strong echo top e" the height of l -30°C level, initially, it can be called hail cloud. After this, further identification can be proceed according to the variation of radar characteristic values with time.

## Classification of Hailcloud for Operational Decision-making

The hail clouds are mainly divided into three categories: single-cell hail clouds, multi-cell hailclouds and super-cell hailclouds. Radar characteristic parameters of echo intensity, echo top, >= 30d Bz echo centre height, intensive echo top and VIL in all types of hail clouds are analysed statistically to discriminant indices of hail cloud types, which are also called decision-making indices for hail suppression operations.

## Determination of Operational Scheme

Hail suppression operation scheme includes: elevation angle, azimuth angle, bomb consumption amount and operation mode. During practice, the technical scheme should be made by considering cannon or rocket types, hail cloud types and operational conceptual models. Based on artificial hail catalysis principles, as well as, the latest research results and Dalians local features, six different operation modes are designed. Therefore, the Doppler radar data is capable to discriminate between hail and severe thunderstorm clouds. Radar main characteristic parameters including echo and their variations with time are analysed to model discriminant index for identifying hail clouds and establishment of decision making index model.

## Post Hail Storm Management Options

As hail is the sudden event, and highly not possible to predict, farmers can take some precautions and corrective measures to avoid huge hail damage and cope up after hail.

### General recommendations to be followed

- For orchard crops, it is advised to remove all the broken branches and twigs due to hail damage.
- Remove fallen fruit to reduce disease and pests. Large wounds on trunks and branches should be covered with a water-based paint to avoid desiccation and disease infection.
- Field sanitation by collecting and destroying fallen fruits and flowers to avoid carryover of pests and diseases.
- Pulverisation of plant rhizosphere should be practiced to reduce the compactness.
- Application of 1.0 Bordeaux paste on wounded parts of plants like branches, stem etc. Alternatively spraying of copper oxychloride @ 2.5 g/liter or 1.0 bordeaux mixture can be done. This aids in rapid would healing by preventing secondary fungal and or bacterial infections.
- It is always advisable to have shelter belts and wind breaks around orchard to avoid heavy damage to the main crop.

- Harvest at 3/4$^{th}$ maturity is advisable, if advisory was already given for likely heavy rainfall/hail storm during the period.
- In frequent hail storm affecting areas farmers may use roof net practice in grapes to reduce hail impact and bird scaring.
- Use fruit thinning to selectively remove hail-damaged fruit and to improve yield and quality of remaining fruit.
- Contingency crops like green gram and black gram may be sown to take the advantage of available soil moisture before taking the next main crop.
- Farmers need not be panic if the damage is less severe, and can wait till the next sprout to come. Replacement of young saplings may be necessary if damage to the plants is very severe.
- Pruning may be necessary to retrain young trees and optimize new growth.
- Bud breaking chemicals and growth regulators may be applied to induce the vegetative growth in orchard crop along with balanced dose of NPK fertilisers.
- Proper drainage facilities are to be provided to avoid water stagnation and secondary infection.
- Near maturity bulb crops like onion and garlic may be harvested to avoid rotting.

## Contingent planning for secondary damage due to insect pests and diseases in hail storm affected crops

Consequent to the hailstorm, cloudy weather followed by intermittent drizzling rains is being continuously experienced by the crops during this period. Considering the increased humidity in atmosphere, persistence/retention of wet conditions on foliage/leaves and soil moisture, there is increased risk of incidence of insect pests and diseases in both crops standing in the field as well as in post-harvest produce. Hence, following measures are suggested for protection of pest and disease spread in hail storm affected orchards and field crops.

In fruit crops like pomegranate, grapes, guava, etc. heavy flower and fruit drop has occurred due to hail damage. The damaged flowers and fruits started rotting due to cold shocks and infection by bacterial and fungal pathogens like grey sooty mould. Hence, such fruits, flowers should promptly be collected and destroyed to avoid further spread of infection. This will help in preventing spread of insect pests like fruit borers that are hiding inside damaged fruits.

In case of severely damaged barks on branches and stems, become very susceptible to attack by insect pests like stem borers, bark borers, and infection by fungal diseases like blight, rot, wilts, etc. The damaged/ broken branches of plants should be pruned immediately. The spraying of 1% Bordeaux mixture (1 g Copper sulphate, 1 g lime, 100 l of water) should be carried out which will help in healing of wounds. Alternatively, spraying of copper oxychloride @ 2- 2.5 g/liter can be done.

Due to cloudy weather coupled with increased atmospheric humidity and wet foliage conditions, the plants suffered from hail damage are likely to be affected severely with sucking pests like thrips, jassids, mites, and diseases like powdery mildew, leaf blights, fruit and flower rot, etc. For the control of sucking pest attacking new foliage/growth, spraying of systemic insecticides/acaricides like Spinosad (0.3 ml/lit), fipronil (@ 1.5 ml/lit) is advised. For controlling powdery mildew, spraying of Hexaconazol (@ 1 ml/lit), propiconaole (1 g/lit) is useful. Spraying of Mancozeb (@ 2.5 g/lit) or Carbendazim (bavistin) (@ 1 g/lit) will help to prevent infection of blights and rot diseases.

Pomegranate could recover rapidly from hail damage in response to post disaster management that included application of bioregulators to promote the growth and pesticides to protect the injured pomegranate plants, which were almost totally defoliated, had broken twigs and had all its flowers and fruits withered. Copper oxychloride was sprayed the day after hailstorm to prevent secondary infection and then plants were treated with chemicals like cytozyme (100 ppm), vigour (0.1%), thiourea (0.02%), potassiumnitrate (2%), hydrogen cyanamide (0.02%), silixol (3ml/ lit), bio-formulation which can prevent fungal infection and stimulate the growth. This helped in total recovery of plant canopy within 27 days.

## Conclusion

Hail being a very short term and localized phenomena, its prediction well in advance so as to inform all stakeholders for adequate preventive measures is a major challenge for even the most technologically advanced and hail affected nation such as USA. India, being situated in the tropical and subtropical region, the frequency of hail damage is much less compared to the mid-latitude and temperate countries. Growers who are using hail-altering technology need to be alerted to hail development at any time of the day or season. Unfortunately even a short episode of hail can cause severe injury to fruit and trees which ultimately reduce the quality and lead to development of diseases like fire blight, cankers and fruit rots. Tough road lies ahead for India, says the IPCC report. It warns the country of severe food crisis due to extreme weather events, and estimates countrywide agricultural loss. This will severely affect the income of 10 per cent of the population with increasing weather extremities. So what can the changes in weather conditions be attributed to? Is this natural variability in climate, or climate change? In case of hailstorm, like drought, pest and disease attack, assessment is done by an Inter-Ministerial Group (IMG) which is headed by the Secretary, Ministry of Agriculture and Co-operation. Based on the recommendations of the IMG, a high level committee (HLC) approves the central assistance to be provided to the affected States from the national calamity release funds. Nevertheless, there is need for institutional arrangement in place in this country including forecasting agencies, disaster management committees, contingency plan and crop revival strategies from ICAR and SAUs to oversee all aspects of management of natural calamities.

❑❑❑

# 4

# Impact of Climate Change under Coastal Ecosystem and Adaptation Strategies

**B.L. Meena, R.L. Meena, Manish Kanwat, A. Kumar and M.L. Dotaniya**

## Introduction

Increasing evidence over the past few decades indicate that significant changes in climate are taking place worldwide and its effects are surreal. Major cause to climate change has been ascribed to the increased levels of greenhouse gases like carbon dioxide ($CO_2$), methane ($CH_4$), nitrous oxides ($N_2O$), chlorofluorocarbons (CFCs) beyond their natural levels due to the uncontrolled human activities such as burning of fossil fuels, increased use of refrigerants, and enhanced agricultural activities. Global average sea level has risen since 1961 at an average rate of 1.8 (1.3 to 2.3) mm/yr and since 1993 at 3.1 (2.4 to 3.8) mm/yr, with contributions from thermal expansion, melting glaciers and ice caps, and polar ice sheets. The climatic changes are resulting in erratic weather patterns and the impacts are already being manifested in many parts of the world and coastal agricultural systems are nearing crisis point. The IPCC has projected that during the 21$^{st}$ century the global surface temperature is likely to rise a further by 1.1 to 2.9 °C as per their lowest estimation model and by 2.4 to 6.4 °C as per their higher estimation model.

Increase in population and urbanization, the availability of arable land dwindled considerably from 0.48 ha in 1950 to 0.15 ha in 2005 and is likely to further reduce to 0.08 ha by 2020 (Mall *et al.*, 2006). Under climate change scenario, agriculture which feeds the entire human population is under great threat. More and more chemical fertilizers are now needed to produce per kg of food grain. Since the last few decades the average surface temperatures of the earth have risen appreciably. Water is becoming scarce in many regions. These phenomena have serious implications on bio-diversity as well as economics and livelihood patterns of the people of coastal ecosystem. This is particularly true in developing countries where a large portion of their population depends directly on agriculture for their livelihoods and their economies are closely linked to agriculture.

## Causes of Climate Change

The major cause of climate change is the Global warming. Global warming is the rise in the average temperature of the earth's atmosphere Earths temperature depends on the balance between leaving the planets system (www.epa.gov/climatechange/science/causes.html). When incoming energy from the sun is absorbed by the Earth system, it warms. When the suns energy is reflected back into space, Earth avoids warming. When energy is released back into space, Earth cools. The factors causing changes in Earths energy balance are:

1. Changes in the greenhouse effect, which affects the amount of heat retained by Earths atmosphere.
2. Variations in the suns energy reaching Earth.
3. Changes in the reflectivity of Earths atmosphere and surface.

Since the early 20$^{th}$ century, the earth's mean surface temperature has increased (Fig. 1) by about 0.8 °C (1.4 °F) of which about two-thirds of the increase occurred since 1980 (NASA 2010; NRC 2010). Recent climate changes, however, cannot be explained by natural causes alone such as changes in solar energy, volcanic eruptions, and natural changes in greenhouse gas (GHG) concentrations. Research indicates that natural causes are very unlikely to explain most observed warming, especially warming since the mid-20$^{th}$ century. Rather, human activities can very likely explain most of that warming (NRC, 2010). Scientists are certain that global warming is primarily caused by increasing concentrations of greenhouse gases ($CO_2$, $CH_4$ and $N_2O$) produced by human activities such as the burning of fossil fuels, deforestation, industrial production, agricultural activities (IPCC 2007; NRC 2010). The approximated contribution of different sectors of human activities to global warming is shown in Fig. 2. Carbon dioxide ($CO_2$) comes from the combustion of fossil fuels in cars, factories and electricity production and is most responsible for global warming. The content of $CO_2$ in the atmosphere is increasing rapidly in the recent years (Fig. 3). The major contributors of $CO_2$ are the developed countries and USA in particular (Fig. 4). The developed world's emissions had contributed most to the stock of greenhouse gas (GHGs) in the atmosphere and the per capita emissions *(i.e.* emissions per head of population) are still much low in developing countries. The emissions from the developing countries will now be increasing to meet their development needs. Besides $CO_2$ the other GHGs contributors include methane ($CH_4$) released from agricultural activities and from the digestive systems of grazing animals (ruminants), nitrous oxide ($N_2O$) from fertilizers used in agriculture, gases used for refrigeration (chlorofluorocarbons) and industrial processes and the loss of forests that would otherwise store $CO_2$.

The Intergovernmental Panel on Climate Change (IPCC) estimated that 31% of total emissions of greenhouse gases in 2004 came from agriculture and forestry. Different greenhouse gases have different heat trapping abilities. The other GHGs can trap more heat than $CO_2$. A molecule of methane ($CH_4$) produces more than 20 times the warming of a molecule of $CO_2$, Nitrous oxide ($N_2O$) is 300 times more powerful than $CO_2$. Other gases, such as chlorofluorocarbons (its use has been banned by many countries as it degrades the ozone layer) have heat-trapping potential thousands of times greater than $CO_2$. But because their concentrations are much lower than $CO_2$, none of these gases adds as much heat to the atmosphere as $CO_2$ does. The IPCC has also projected that during the 2151 century the global surface temperature is likely to rise a further by 1.1 to 2.9 °C as per their lowest estimation model and by 2.4 to 6.4 °C) as per their higher estimation model. The heat absorbed by the earth is not uniformly distributed throughout. The ocean is the main absorber which absorbs about 93% of the total heat absorbed by the earth.

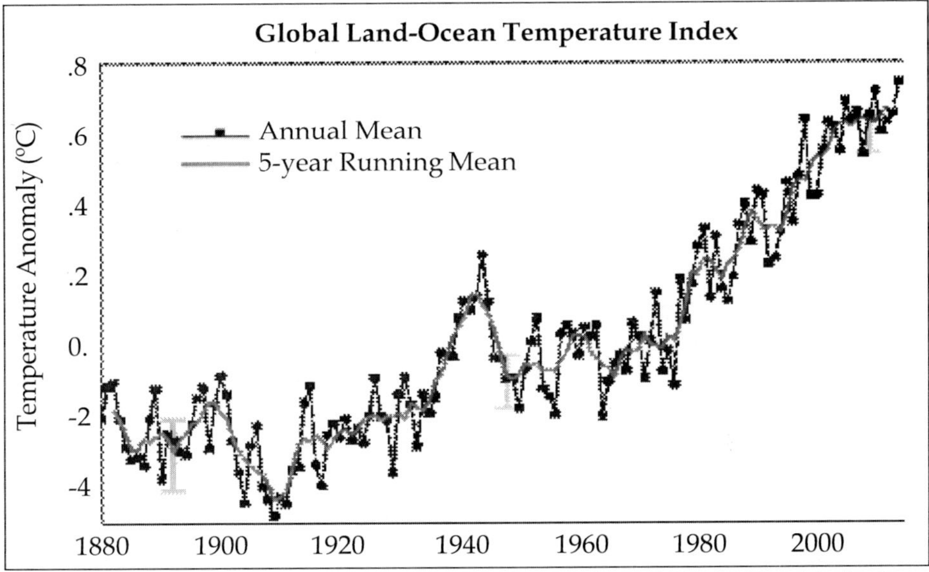


**Fig. 1.** Global mean land-ocean temperature changes from 1880-2000, with the base period 1951-1980. The green bars show uncertainty estimates.
*(Source:* NASA GISS)

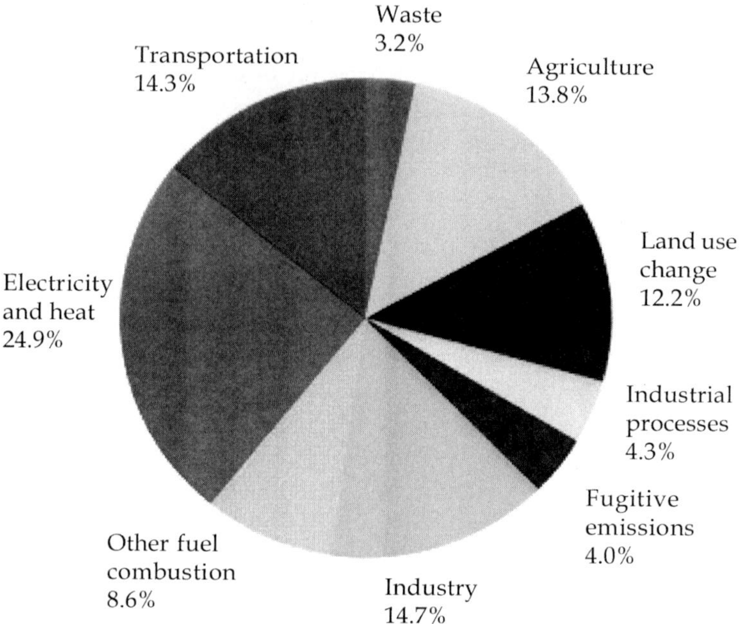


**Fig. 2.** Annual worldwide greenhouse gas emissions in 2005 by different sectors
*Source:* https://upload.wikimedia.org

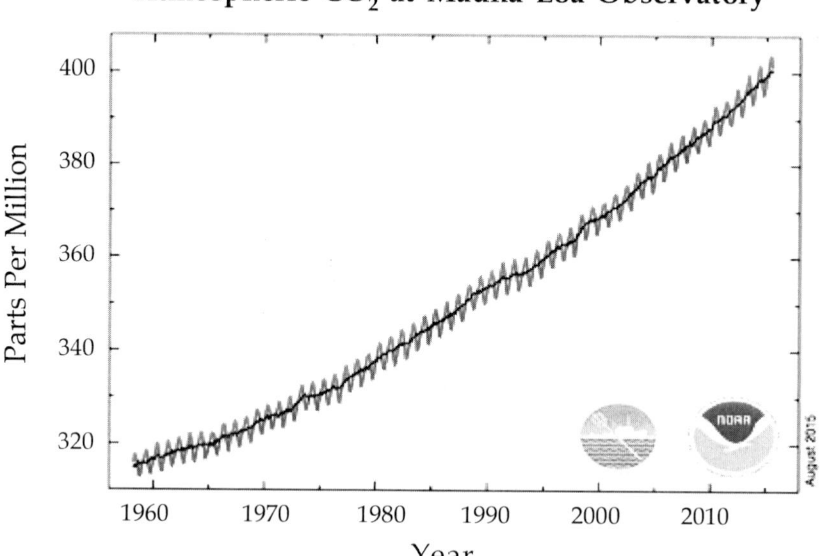


**Fig. 3.** Increase in atmospheric carbon dioxide over the year 1960-2010
(*Source:* http://www.esrl.noaa.gov)

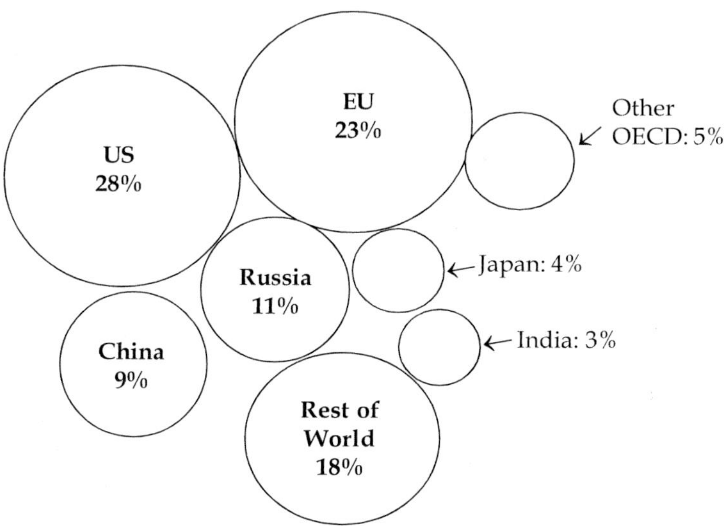


**Fig. 4.** Bubble diagram showing the share of global cumulative energy-related carbon dioxide emission for major emitters between 1890-2007.
(*Source:* https://www..wikimedia.org )

## Impact of climate change

Climate change may result in great threat to food security (due to decreasing crop, fish and meat yields) of human being and the loss of habitat due to sea level rise, particularly in the coastal region. In coastal regions agriculture and aquaculture co-exist as major livelihood of rural communities. The impacts of climate change on coastal ecosystem are briefly given below;

1. **Agriculture:** There may be considerable impact on microbes, pathogens and insects, which may result in more infestation of crops by disease and pests and deterioration of soil quality. Increase in temperature and soil-water salinity and decrease in fresh water availability would decrease crop yield drastically. The effect is expected to be very severe in major river flood plains of the world including Indo-Gangetic plain. Sea level rise will cause saline water intrusion through rivers and estuaries and many fertile lands will be 'inundated by saline water making them unfit for agriculture.

2. **Water resource:** The ground water quality will deteriorate by saline water intrusion. There may be changes in the water cycle and fresh water availability may be more scare in coastal areas. Global average sea level has risen since 1961 at an average rate of 1.8 (1.3 to 2.3)

mm/yr and since 1993 at 3.1 (2.4 to 3.8) mm/yr, with contributions from thermal expansion, melting glaciers and ice caps, and polar ice sheets (Rao, 2011).

3. **Natural ecosystems:** Mangroves, coral reefs, sea grass, marine life, *etc.* are vulnerable due to frequent changes in rainfall pattern, sea level rise and warmer sea temperature. Coastal wetlands including salt marshes and mangroves are likely to be very badly affected by sea-level rise especially where they are starved of sediment.
4. **Fisheries and aquaculture:** Increasing sea and river water temperatures are likely to affect fish breeding, their migration and production. Migration of different marine and inland species to favorable climate region
5. **Livelihood of people:** Sea level rise will make many coastal families homeless. Climate change may increase high tides levels, storms, floods, seismic sea waves (tsunami), erosion and more climatic hazards like, cyclone, *etc.* for which coastal areas will be exposed to increasing risks. This will threaten vital infrastructure, settlements and facilities that support the livelihood of coastal communities including the island communities. A new group of refugees will emerge called the environmental refugees. Some of the most vulnerable coastal areas in India are Sundarbans, Gujarat coast, Mumbai, South Kerala, South West Bengal, Lakshadweep Islands, Andaman Islands and many other small Islands (small Islands are especially vulnerable). In India, Vishakhapatnam beaches are severely eroding, Puri beaches are slowly disintegrating, and water has started entering close to solid ground in Goa beaches and would affect tourism and local coastal communities.

## Mitigations Strategies to Climate Change

Assess biophysical and socio-economic implications of mitigation of climate change before developing policy for their implementation. The IPCC defines mitigation as activities that reduce greenhouse gas (GHGs) emissions, or enhance the capacity of carbon sinks to absorb GHGs from the atmosphere. Climate change mitigation also includes acts to enhance natural sinks, such as reforestation, increased carbon sequestration in soil through appropriate management of agriculture, *etc.* Most countries are now members to the United Nations Framework Convention on Climate Change (UNFCCC), whose ultimate objective is to prevent dangerous anthropogenic climate change activities. The member countries of the UNFCCC (UNFCCC 2011) have agreed that urgent actions are to be taken to reduce the emission of GHGs and that future global warming should be limited to below 2.0 °C relative to the pre-industrial level. Many believe that a rise of 2.0 °C is the threshold beyond which impacts are likely to be severe, and dangerous to environmental systems. Agriculture, forestry and land use practices may have a major role to play in mitigation measures. According to IPCC (IPCC 2007), forestry accounted for 17% of greenhouse gas emissions in 2004. But if deforestation can be halted, reforestation initiated and existing forests are managed more sustainably by communities, forests could become part of the solution instead of part of the problem. Agriculture contributes about 14% of GHGs emission. But if soils can be better managed it can store more carbon. Agroforestry is also an underutilized mitigation option in agriculture; it can store more carbon in trees and in soil, while improving the soil quality. Many of the options that relate agriculture and natural resources management will have immediate development benefits towards, the productivity of the natural resource or system.

## Adaptations Strategies to Climate Change

Indian farmers have tremendous experience of coping with adverse climate and a large number of indigenous practices have been evolved over time. The local knowledge is much more important in adaptation, while mitigation require a national effort, new technologies and several policy initiatives. The different adaptation strategies may be adopted for coastal areas are summarized as follows.

***Crop And Cropping System Based Strategies:*** These are mainly centered on promoting the cultivation of crops and varieties that fit into the changed crop calendars and seasons, development of varieties with changed duration that can overwinter the transient effects of change, varieties for high temperature and heat stress tolerance, salinity, drought and sub-mergence tolerance and varieties which respond positively to high $CO_2$ (Challinor *et al.*, 2007). Farmers will like to have their crops/ varieties performing well in difficult environments, but also to produce very high yields when conditions are more favorable. Adjusting showing time of crops according the changing pattern of rainfall, soil-water salinity, humidity, temperature and other climatic variables may decrease the losses to crop production.

***Integrated pest and disease management (IPDM):*** Changing climate will change the pattern and intensity of attack of pests and diseases to crops and livestock. Temperature, humidity, rainfall and other weather parameters influence the spread of pests and diseases. Higher temperatures speed up the lifecycle of some pests and diseases, and their vectors thus; infestation by pests and diseases may rise. Anticipation of pest and disease outbreaks, and integrated pest and disease management (IPDM) is an urgent need in view of climate change.

*Resource conservation based technologies:* The major resource conservation based technologies in combating climate change are; *in situ* moisture conservation, rainwater harvesting and recycling, efficient use of irrigation water, conservation agriculture, energy efficiency in agriculture and use of poor quality water. Low quality water/ saline water can be managed and used in conjunction mixing with good quality water and more tolerant crops/ varieties are to be developed. Agriculture in developing countries mostly uses the traditional technologies for irrigation. Agriculture consumes about 70-90% of total use of water. Alternative improved technologies must be used to produce more food with less water to feed the growing population. The key approaches include: integrated watershed development; improving rainwater use efficiency through on-farm harvesting and recycling; contingency crop planning to minimize loss of production during drought/flood years. Watershed management is now considered an accepted strategy for development of rainfed agriculture. Watershed approach has many elements which help both in adaptation and mitigation. For example, the soil and water conservation works, farm ponds, check dams etc. moderate the runoff and minimize floods during high intensity rainfall. The small farm reservoirs (farm pond, furrows, channels, *etc.)* technology developed (Ravisankar *et al.,* 2008) by Central Agricultural Research Institute, Port Blair, India for harvesting rainwater in farm and its multiple uses turned to be very successful for coastal areas of India. Wastewater is rich in several plant nutrients but it may contain load of harmful bacteria and heavy metals. Research at Central Soil Salinity Research Institute, Kamal (CSSRI, 2013) has shown that waste water can be used successfully for irrigation of agro-forestry, non-edible crops and even for grain crops like wheat.

*Land shaping:* Land shaping means re-shaping the surface morphology of land for raising a portion of it in form of raised land ridges through excavation of soil by making ponds/ furrows/ channels, *etc.* to address the problems related to the soil drainage as well as on-farm harvesting of rainwater for multipurpose farm uses including irrigation, aquaculture, animal husbandry, poultry, *etc.* The technology was developed and tested by Central Soil Salinity Research Institute (Ambast *et al.,* 2011). It may be particularly important for low-lying coastal delta regions of Indo-Gangetic plain in India and other delta regions of major rivers in the world, which have poor drainage condition and are subjected to submergence following heavy rains, sea level rise, *etc.*

*Soil Management:* Appropriate soil management will be conserving/ improving soil quality and will reduce vulnerability of farming systems in the face of climate change. Organic manure is a critical input which improves soil health. Soil organic matter may be added by compost/ vermin-compost, azolla, crop residues, green manure, animal manure, intercropping, alley cropping ($N_2$ fixation), crop rotation with inclusion of legumes, agroforestry and growing cover crops especially the legume cover crops. More organic matter in the soil means, better microbiological health of soil, better availability of nutrients, less loss of nutrients, less carbon dioxide in the atmosphere (due to increased carbon sequestration in soil) and in the long run higher yield, better adaption and mitigation to climate change, and better environment. Judicious fertilizer application, a principal component of SSNM approach, thus has twofold benefit, i.e. reducing greenhouse gas emissions. The microorganisms in bio-fertilizers restore the soil's natural nutrient cycle, build soil organic matter and improve soil quality/ soil health. Through the use of biofertilizers, healthy plants can be grown, while improving the sustainability and environmental quality.

*Reforestation:* Reforestation, wherever possible, should be undertaken as an effective step towards climate change mitigation by reducing the vulnerability of the globe to climate change through fixation of atmospheric carbon load. Mangrove plants stabilize the coastal soils and protect it from erosion through its intricate root system. Mangroves not only provide effective protection to the coastal zone but also protect the country habitations against cyclones, sea surges, Tsunami waves, *etc.* Mangroves are important habitat for breeding and growth of fries of many fish species including shrimp. Massive mangrove reforestation programme should be undertaken in all the coastal areas.

*Agroforestry:* Agroforestry systems like agri-silvi-culture, silvipasture and agri-horticulture offer both adaptation and mitigation opportunities. Agroforestry systems buffer farmers against climate variability, and reduce atmospheric loads of greenhouse gases. Agroforestry can both sequester carbon and produce a range of economic, environmental, and socio-economic benefits (Jose, 2009). For example, trees in agroforestry systems improve soil fertility through control of erosion, maintenance of soil organic matter and physical properties, increased N accretion, extraction of nutrients from deep soil horizons, and promotion of more closed nutrient cycling. In rangeland areas, pasture improvement is essential to combat impeding changes through planned grazing processes, enclosures for recovery, or enrichment planting.

## Integrated Farming Systems (IFS)

Integrated farming systems approach also provides insurance against climate risks. Multiple-enterprise agriculture consisting of crop, livestock, poultry, fish farming and trees in a single unit of land will ensure protection against projected loss due to climate change and will also benefit from on farm resource use (Meena *et al.,* 2014). Several farming systems modules have been developed both for irrigated and rainfed conditions, which consists of one or more enterprises and the risk minimization potential of these modules is well documented (Behera *et al.,* 2008). The included breeds of livestock in IFS should have high feed efficiency so that they can produce high yield with less feed and can withstand multiple stresses. Climate change could open up new opportunities for fish farming as the sea encroaches on coastal lands. Research has shown that there are some strains in *Talapia fish* which have tolerance to saline environments too (Ravisankar *et al.,* 2008).

## Conclusion

The recent changes in global trends such as population growth, urbanization, increasing demands for water, over-exploitation of ecosystems, *etc.* will also have great adverse impact on climate change. If we really want meaningful mitigation we must re-look into the present trend in land use pattern and agricultural practices, energy generation and industrial activities within the global system as a whole. The effective measures for adaptations and mitigations to climate change require trans-boundary partnership programmes and regional and international cooperations as well as local measures. The impact of climate change is already being distinctly felt in fragile coastal ecosystems. The farming communities of coastal areas need intensive awareness training for appropriate management of soils, crops and trees to capitalize on their carbon storage potential. Policy incentives will play crucial role in adoption of climate needy technologies in agriculture too as in many other sectors.

## References

Ambast S.K., Ravisankar N. and Velmurugan A. (2011) Land shaping for crop diversification and enhancing agricultural productivity in degraded lands of A&N Islands. Journal of Soil Salinity and Water Quality, 3(2): 83-87.

Behera U.K., Yates C.M., Kebreab E., and France J. (2008). Farming systems methodology poor efficient resource management at the farm level: A review from an Indian perspective. Journal of Agricultural Science, 146: 493-505.

Challinor, A.J., Wheeler, T.R., Craufurd, P.Q., Ferro, C.A.T., Stephenson, D.B. (2007). Adaptation of crops to climate change through genotypic responses to mean and extreme temperature. Agricultural Ecosystems and Environment 119 (1-2): 190-204.

CSSRI. (2013). Annual Report of Central Soil Salinity Research Institute, Kamal 2012-2013, p. 52.

IPCC (2007) Intergovernmental Panel on Climate Change Synthesis Report, Fourth Assessment Report (AR4) SYR 2007.

Jose Shibu (2009). Agroforestry for ecosystem service and environmental benefits: An overview. Agroforestry System, 76: 1-10.

Mall R.K., Ranjeet Singh, Akhilesh Gupta, Srinivasan G and Rathore L.S. (2006). Impact of climate change on Indian Agriculture: A Review. *Climate Change* 78:445-478 (Springer Science, Business Media B.V.).

Meena B.L., Meena R.L., Ambast S.K. and Pandey M. (2014). Impact assessment of agriculture technological interventions in tsunami affected south Andaman - a case study. Bhartiya Krishi Anushandhan Patrika, 28(3): 141-148.

NASA (2010) 2009 ends Warmest Decade on Record. NASA Earth Observatory Image of the Day, 22 January 2010.

NRC (2010). *Advancing the Science of Climate Change*. National Research Council. The National Academies Press, Washington, DC, USA.

Rao,GSLHV Prasad( 2011). *Climate Change Adaptation strategies in Agriculture and Allied Sectors.* Scientific Publication (India), 336 p.

Ravisankar, N., Ambast, S.K. and Srivastava R.C. (2008). *Crop Diversification through Broad Bed and Furrow System*. Central Agricultural Research Institute, Port Blair, 220p.

UNFCCC (2011) *Status of Ratification of the Convention.* UNFCCC Secretariat Bonn Germany: UNFCCC.

□□□

# 5

# Advances of Biotechnology for Climate Resilient Agriculture

**Amit Sen, Kshitiz Kumar Shukla, Sanjay Singh, G. Tejovathi and Arifa Khatoon**

## Introduction

Agriculture plays a crucial role in ensuring food security, accounting for a significant share of India's Gross Domestic Product (GDP). Agriculture is not just about food production alone, since it is a complex multi-dimensional and multi-faceted sector which concerns the efficient use of the natural resources, productivity enhancement, and preservation of the ecosystems in a manner that can sustain the needs and improvements of human livelihoods (Devendra, 2012). It is also a most vulnerable sector to climate change which has been recognized as one of the most serious challenge for the world- its people, the environment and its economies. Increased incidence of abiotic stresses (drought, high & low temperature, salinity, submergence and oxidative stress) and biotic stresses (pest and diseases) have became major cause for stagnation of productivity in major crops (Grover *et al.*, 2011). Abiotic stresses are directly correlated with the change in climate while biotic stresses are indirectly correlated as the pattern and incidence of disease varies with the variation in climatic condition. Also, crop productivity rely on extent and type of stress in the particular area.

In the past years, several adaptation and mitigation strategies have been adopted to cope up with such changes in environmental conditions. Being sessile, plants are able to adapt or acclimate after constant exposition to environmental stress condition (Bremont *et al.*, 2013). This capability of plants could be refined through technological advances, especially biotechnological approaches. GM Crops can contribute positively by reducing $CO_2$ emission and mitigating the impact of climate change on food security.

Under climate change scenario, use of herbicide tolerant crops have reduced the application of chemicals, insecticides/pesticides, thereby reduced the contamination of soil. Soil carbon sequestration is an important part of mitigation strategy for climate change. Crops developed with agricultural biotechnology reduce the need for tillage or ploughing and allow farmers for no or zero tillage. As a result over time soil quality is enhanced and become more carbon enriched. Above practice also prevent loss of soil water, as soil moisture is trapped in un-tilled soil. Nitrogen use efficiency technology produces plants (*eg*. GM rice and canola) with yeilds equivalent to conventional varieties which require significantly less nitrogen fertilizer. This technology has the potential to reduce the amount of nitrogen fertilizer leading to less nitrogen pollution of ground and surface water (Anonymous, 2009).

All the above mentioned development has been possible through advances in plant molecular biology and biotechnology which have resulted in discovery of superior allele, study of tissue specific promoters, functional characterization of genes, gene pyramiding and generation of efficient methods for plant genetic transformation or plant genome manipulation.

## Biotechnological interventions for development of stress tolerant crop

The genetic fidelity of the regenerated plants is highly desirable for developing new improved plant varieties, especially under changing climate and limited arable land. Biotechnology has emerged as a viable option for developing genotypes that are able to resist pest & disease, drought, high temperature and salinity stresses. Allele mining using TILLING/EcoTILLING and next generation sequencing has become known for identifying and validating new genes responsible for particular trait and mining of favourable genes (Kumar *et al.*, 2010).

Molecular breeding approach (MB approach) and genetic engineering approach (GE approach) are key strategies where genetic enhancement for stress tolerance has led to crop improvement. Superior genes or alleles conferring stress resistance are identified in the same species (wild relative or germplasm collection) which can be transferred into elite genotypes through molecular breeding, while there is no barrier to transferring useful genes or alleles across different species from the animal or plant kingdom through GE (Varshney *et al.*, 2011). QTL mapping and marker assisted selection plays significant role in MB and GE approach for the generation of genetically modified crops.

Plant tissue culture techniques are also widely used since last few decades for the generation of stress tolerant plants. Micropropagation, protoplasts, embryo rescue, somatic embryogenasis, somaclonal variation increases the efficiency in obtaining tissue cultured variation, selection and multiplication of the desired genotypes (Brown and Thorpe, 1995). *In vitro* mutagenasis is another approach for generation of improved crop variety under climate resilience. Traditionally mutation is induced by physical and chemical methods in seeds and vegetatively propagated crops. The mutagen treatment breaks nuclear DNA and during the process of DNA repair mechanism, new mutations are induced randomly. Recently high energy beams have been used for mutation induction. Linear energy transfer through high energy beam induces higher biological effects such as lethality and chromosomal aberration as compare to commonly used physical mutagens. A range of mutants have been induced in several different ornamental plants as well as in maize, rice and wheat (Jain, 2010).

## Current scenario of crop biotechnology in the context of India

In the decade of 60s, green revolution dramatically increased the crop production in India. In this succession, the first transgenic crop, BT cotton with *cry 1Ac* and *cry 1Ab* gene was released in 2002 and it is the only transgenic crop under commercial cultivation in India. In the year 2014, 11.7 million hectare out of 181.5 million hectare of land is cultivated under BT cotton and India holds 4$^{th}$ position (Table-1). BT brinjal was another improved crop approved by Genetic Engineering Approval Committee (GEAC) and released for commercial production but this was blocked in 2010 by ministry of environment and forests (MoEF). The Indian variety of rice with golden event to ameliorate vitamin-A deficiency was developed by three public sector institutions (Tamil Nadu Agriculture University, ICAR-Directorate of Rice Research) and is yet to be released commercially. Presently, there are about eight GE traits in 17 crops at different stages of development by about 15 private and 17 public institutions in India (Rao and Annadana, 2014). Some of the GM crops under field trial with specific traits are given in Table 2.

**Table 1.** Top ten countries cultivating biotech crops (in million hectares) (James, 2014)

| Rank | Country | Area(million hectare) | Crops grown |
|---|---|---|---|
| 1. | USA | 73.1 | Maize, soybean, cotton, canola, sugar beet, alfalfa, papaya, squash |
| 2. | Brazil | 42.2 | Soybean, maize, cotton |
| 3. | Argentina | 24.3 | Soybean, maize, cotton |
| 4. | India | 11.6 | Cotton |
| 5. | Canada | 11.6 | Canola, maize, soybean, sugar beet |
| 6. | China | 3.9 | Cotton, papaya, poplar, tomato, sweet pepper |
| 7. | Paraguay | 3.9 | Soybean, maize, cotton |
| 8. | Pakistan | 2.9 | Cotton |
| 9. | South Africa | 2.7 | Maize, soybean, cotton |
| 10. | Uruguay | 1.6 | Soybean, maize |

*S-adenosylmethionine decarboxylase, pyruvate decarboxylase, alcohol dehydrogenase* gene, *codA, cor47* gene in rice, *bar, HVA1, PIN2* gene in wheat, *arabidopsis annexin, choline dehydrogenase* gene in mustard/rapeseed have been used as transgene in different Indian institutions for development of different abiotic stress tolerant crops (Sharma *et al.*, 2003).

**Table 2.** GM crops under field trial (upto 2013)

| S.No. | Crops | Organization | Traits/Gene |
|---|---|---|---|
| 1. | Brinjal | IARI, Sungro Seeds Ltd., MAHYCO, TNAU | Insect resistance/ *cry 1Aa, cry 1Abc, cry 1Ac* |
| 2. | Cabbage | Nunhems India Pvt. Ltd. | Insect resistance/ *cry 1Ba* and *cry 1Ca* |
| 3. | Cauliflower | Sungro Seeds Ltd., Nunhems India Pvt. Ltd. | Insect resistance/ *cry 1Ac, cry 1Ba* and *cry 1Ca* |
| 4. | Cotton | MAHYCO, Monsanto, Rasi, Nuziveedu, Ankur, JK Seed, CICR, UAS-Dharwad | Insect resistance, herbicide tolerance, *cry 1Ac* |
| 5. | Groundnut | ICRISAT | Virus resistance/chitinase gen |
| 6. | Maize | Monsanto | Shoot borer/ *cry 1Ab* gene |

| | | | |
|---|---|---|---|
| 7. | Chickpea | ICRISAT | Insect resistance/Pod borer, *cry 1Ac* |
| 8. | Mustard | UDSC | Hybrid seed, barnase/ barstar gene |
| 9. | Okra | MAHYCO, Beejo Sheetal | Borer *cry 1Ac, cry 2Ab* |
| 10. | Pigeonpea | ICRISAT, MAHYCO | Pod borer and fungal pathogen, *cry 1Ac* and chitinase |
| 11. | Potato | CPRI, NIPGR | *Ama 1* and *Rb* gene derived from *Solanum bulbocastanum* |
| 12. | Rice | MAHYCO, TNAU | *cry 1B- cry 1Aa* fusion gene, *cry 1Ac, cry 2Ab*, Rice chitinase (*chi 11*) or tobacco osmotin gene |
| 13. | Sorghum | NRCS | Insect resistance, shoot borer |
| 14. | Tomato | IARI, MAHYCO, NIPGR | Antisense relpicase gene of tomato leaf curl virus *cry 1Ac* |

## Constraints

Although GE crops and their products are acceptable, but at the same time it raises food and environmental safety questions, as well as economic and social issues, which must be addressed through scientific basis to farmers and consumers. In small-scale farming that predominates in India, the transition from R&D to innovation and commercialisation in agricultural biotechnology should also be addressed. In this connection, anti-tech activism which opposes GM crops is a major threat for India which is a biggest hurdle in field trial and release of crop variety, *eg*. BT brinjal. Unjustified delay in regulatory approval constraints the farmer to access the novel technologies (Rao and Annadana, 2014).

In the perspective of research and development, a major challenge is to fill up the gap between modern technology and conventional breeding programme. Integration of cultivation of GM crop with agricultural mitigation & adaptation strategies in accurate manner under climate change is highly required. For example, use of herbicide tolerant plant extensively may develop resistance toward resistance for insecticides and pesticides. Several reports are available on the identification or even validation of QTLs or markers for abiotic stress tolerance, their successful deployment in the development of a superior cultivar has had only limited success due to variation in nature and intensity of stress (Varshney *et al.*, 2011).

All products of modern technology have to be delivered in the form of seeds of improved cultivar. But small farmers do not have easy access to improved seeds in and in the short term, it is the failure of seed system which is a major limitation. Simplifying of biosafety regulation and appropriate awareness of technological advances is also needful for widespread cultivation of GM crops.

## Future prospects

Biotechnology approaches have potential to enhance crop production under different stress condition. Nevertheless, with current and fast emerging technologies such as RNAi, targeted gene replacement using zinc-finger nucleases, chromosome engineering, marker assisted recurrent selection (MARS) and genome wide selection (GWS), next generation sequencing (NGS) and nanobiotechnology, the future seems bright with respect to the development of designer crops with improved features that can use natural resources such as water, soil nutrients, atmospheric carbon and nitrogen with a far greater efficiency than ever before (Varshney *et al.*, 2011). Strengthening of public institutions and research programmes is also required to accelerate the ongoing and future research projects to fight with climate change.

## Conclusion

Environmental stresses which include both biotic and abiotic stress are the major forces that affect crop production in developing countries including India, thereby leading to food insecurity. Therefore, appropriate measures should be taken in management of environmental stress and sustainable agricultural production. One of the key approach is promotion of GM crops for cultivation with appropriate adaptation and mitigation strategy. However, environmental effect of transgenic crops should be addressed, monitored and compared with non-transgenic crops grown by conventional practices. Also, maintenance of conventional bred varieties for use as germplasm should be regularly done for each crop.

## References

Anonymous. (2009). Green Biotechnology and climate change. EuropaBio.

Bremont, J.F.J., Kessler, M.R., Liu, J.H., Gill, S.S. (2013). Plant stress and biotechnology. BioMed Research International, Editorial.

Brown, D.C.W., Thorpe, T.A. (1995). Crop improvement through tissue culture. World Journal of Microbiology & Biotechnology, 11: 409-415.

Devendra, C. (2012). Climate change threats and effects: Challenges for agriculture and food security. Academy of Science Malaysia.

Grover, M., Ali, S.K.Z., Sandhya, V., Rasul, A., Venkateswarlu, B. (2011). Role of microorganisms in adaptation of agriculture crops to abiotic stresses. World Journal of Microbiology and Biotechnology, 27: 1231-1240.

Jain, M. (2010). Mutagenasis in crop improvement under the climate change. Romanian Biotechnological Letters, 15 (2): 88-106.

James, C. (2014). Global status of biotech/GM crops. International service for the acquisition of agri-biotech applications, No. 49.

Kumar, R.G., Sakthivel, K., Sundaram, R.M., Neeraja, C.N., Balachandran, S.M., Shobha Rani, N., Viraktamath, B.C., Madhav, M.S. (2010). Allele mining in crops: Prospects and potentials. Biotechnology Advances, 28: 451-461.

Rao, C.K., Annadana, S. (2014). Genetically engineered crops in India- A gordian knot needing an alexandrian solution. Foundation for Biotechnology Awareness and Education, Bangalore.

Sharma, M., Charak, K.S., Ramanaiah, T.V. (2003). Agricultural biotechnology research in India: Status and policies. Current Science, 84 (3): 298-302.

Varsheny, R.K., Bansal, K.C., Aggarwal, P.K., Datta, S.K., Craufurd, P.Q. (2011). Agricultural biotechnology for crop improvement in a variable climate: hope or hype? Trends in Plant Science, 16 (7): 363-371.

❑❑❑

# 6

# Role of Marker Assisted Selection in Development of Climate Resilient Rice Varieties

**M. Girijarani and P.V. Satyanarayana**

## Introduction

Climate change has become an issue of global concern and has a direct large scale impact on land and marine biosphere. The expected and unlikely events like increase in temperature, melting of glaciers and rising of sea level, increase in magnitude and frequency of drought and floods, cyclones/tsunamis, desertification, heat and cold waves, snow fall are most threatening challenges that are being witnesses in the present era. Rain fall is likely to decline by 5-10 % over southern parts of India where as 10-20% increase likely over the other regions. There is probable decrease in number of rainy days over major parts of the country pointing at likely increase of extreme events of floods or drought. In India, salt affected soils currently estimated to be 6.73 M ha in different agroecological regions and the area is likely to be almost turn to 20Mha by 2050. Framing breeding strategies to develop climate resilient varieties is necessary to mitigate adverse effects of climate change.

Rice is one of the most important food crop of India and 2$^{nd}$ of the world. It feeds more than 50% of the world population. It is the staple food of most of the people of South-East Asia. Asia accounts for about 90% and 91% of world's rice area and production respectively. The large gap in the productivity of rainfed ecosystems compared with irrigated ecosystem is largely caused by abiotic stresses such as drought, floods and salted affected soils. Even productivity of irrigated ecosystem is also becoming uncertain due to adverse effects of climate change. Development of climate resilient rice varieties on fast track is important to mitigate ill effects of climate change. Marker assisted breeding is one of the important breeding strategies to incorporate targeted trait in popular varieties for the development of climate ready rice.

Most of the abiotic stresses such as drought, floods, lodging resistance to withstand under cyclones and salinity are controlled by polygenes with complexity of inheritance. Precise targeted trait selection in regular conventional breeding programmes is cumbersome and time consuming in evolving high yielding rice varieties even under adverse climatic conditions. Blending of molecular markers in regular breeding programmes will help in precise selection transmission of the complex traits like drought, floods, lodging resistance and salinity.

With the genomic era, molecular markers linked to complex traits such as floods, lodging resistance, salinity and drought are available for molecular breeding. QTLs (Quantitative Trait Loci) associated with these complex traits were detected, cloned and deployed in molecular breeding for the development of climate resilient rice varieties globally and at national level. Marker Assisted Back cross breeding involves incorporation of targeted trait in the back ground of widely adapted rice cultivars. It involves three major steps viz., fore ground selection, recombinant selection and back ground selection.

1. **Fore ground selection:** It is the selection for the targeted trait using gene/QTL linkedmarker associated with the trait. There is need to identify molecular markers linked to targeted trait based on parental polymorphism of donor parent and recipient parent. Selection for heterozygous allele in early generation up to BC1F1 generation and homozygous allele for donor in advanced generation is needed to fix the genomic region of the targeted trait in the recipient parent.
2. **Recombinant selection:** Selection exercised using flanking markers of the targeted gene/QTL to avoid linkage drag of unfavorable alleles. Selection of homozygous allele for recipient parent using recombinant markers will help in precise transfer of targeted trait with minimal linkage drag.
3. **Back ground selection:** At least 6 parental polymorphic makers covering all the chromosomes will be used to get maximum recovery of recipient parent. Selection for homozygous allele for recurring parent will be practiced.

Even though, back ground selection is exercised, sometimes maximum recovery of recurrent parent may not possible because we are using fraction markers of vast genome size. Now a days, with the availability of genome sequencing facilities SNP back ground selection are being practiced. Practical experience indicated that adaption of phenotypic selection after genotypic confirmation would help in realization of desirable phenotype.

On other hand, to recombine favourable alleles simple fore ground selection for the targeted trait in regular breeding would help in precise transfer of the trait in addition to other novel characteristics of the new genotype.

In this chapter, a brief progress of research on development of climate resilient rice varieties using marker assisted selection globally and scope of identification of novel genes will be discussed.

In this chapter, use of marker assisted selection for flood tolerance, lodging resistance, salinity and drought is emphasized.

## Flood tolerance

Flood-prone ecosystems are prevalent in South and Southeast Asia, and are characterized by periods of extreme flooding and drought. Yields are low and variable. Flooding occurs during the wet season from June to November, and rice varieties are chosen for their level of tolerance to submersion.

*Flooding during germination* (anaerobic germination; AG): a problem when direct seeding is practiced and heavy rains result in submergence before germination.

b) *Flash flood* (submergence): plants are completely submerged for up to 2 weeks. Submergence tolerance is required for this condition.

c) *Stagnant flooding* (medium deep or semi-deep): flooding occurs for a longer duration, more than 2 weeks and often several months, at depths up to 50 cm. Varieties tolerant of stagnant flooding conditions are required.

d) *Deeper stagnant flooding* (deepwater or floating rice): water depth increases throughout the season to depths above 50 cm and often a meter or more. Varieties with tall plant height or rapid internode elongation are required.

Type of flood decides formulation of strategies for the development of varieties. Varieties tolerant to flash floods adopt quiescence strategy and stagnant flooding with moderate shoot elongation under submergence. So far varieties developed for stagnant flooding does not have tolerance under flash floods and vice versa. But in real farming situation flash floods and stagnant flooding are taking upper hand under unpredicted cyclonic rains due to vagaries of climatic change. The major determinant of submergence tolerance is the *SUB1A* gene has been successfully introgressed into mega varieties swarana, IR64, sambamahsuri, Tadukan, CR1009 (*SUB1*), PSRC18 sub1, PSRC6 sub1, Chierang sub1 (Neeraja *et al.*, 2007, Septiningsih *et al.*, 2009). Ethylene responsive sub 1 gene isolated form FR13A is effective for flash flood tolerance and sub 1 version of Swarna gave 1-2 tons/ha average yield. Sub 1gene was introgressed into most of the popular variety like swarana tolerates flash floods but vulnerable to stagnant flooding (Reddy *et al.* 2010).

Combining sub1 with the genotypes showing moderate shoot elongation tolerates both flash floods and stagnant flooding. Identification of genotypes for both flash floods and stagnant flooding is under pipe line at IRRI ( Mackill et al., 2010) and sub1 versions with moderate shoot elongation was developed for Amara (MTU 1064) and Pushaymi ( MTU 1075) at Andhra Pradesh Rice Research institute, Maruteru ( Girija Rani *et al.*, 2014 and 2015). Introgression of sub1 is under way into popularrice varieties globally (Mackill *et al.*, 2012, Ismail *et al.*, 2013) and at national level (Singh *et al.*, 2013).

Recently rice farmers are shifting to direct seeding under puddled condition because of acute labour shortage. Breeding for submergence tolerance during germination (anaerobic germination) is becoming mandatory to get good seedling establishment under direct seeding conditions during monsoon in coastal areas. Varietal differences for anaerobic germination, AG have also been observed (Girija Rani *et al.*, 2013). Submergence tolerant varieties with the *SUB1*gene do not usually possess the AG trait, indicating that these two traits are independent. Major QTLs have been identified for anaerobic germination. Pelayo *et al.* 2014 validated major QTL AG1 identified from Khao Hlan On (KHO) for anerobic germination across nils of IR64 for anaerobic germination. Efforts to combine sub1, anerobic germination is under way at IRRI.

The chromosome 12 QTL is the major determinant of the rapid elongation response of deepwater varieties. It has been shown to consist of two ERF genes, named *SNORKEL1* and *SNORKEL2*, that are very similar in sequence to the *SUB1* genes (Hattori *et al.*, 2009).

## Lodging resistance

Rice genotypes differ widely in their lodging resistance, which is a complex phenotype determined by morphological (basal internode lengths and thickness, plant height and stem wall thickness) and biochemical components (silicon, cellulose, lignin). Ookawa *et al.* (2010) identified effective quantitative trait loci (QTL) for culm diameter, *STRONG CULM2* (*SCM2*) on chromosome 6 with flanking markers

RM20546-RM20562 and cloned QTL conferred lodging resistance in the near isogenic lines developed in cross between Habataki and the *japonica* variety Sasanishiki.

Yano *et al.* (2014) isolated an effective quantitative trait loci (QTL), *STRONG CULM3* (*SCM3*), confers culm strength and the causal gene of which is identical to rice *TEOSINTE BRANCHED1* (*OsTB1*), a gene previously reported to positively control strigolactone (SL) signaling. A near-isogenic line (NIL) carrying *SCM3* showed enhanced culm strength and increased spikelet number despite the expected decrease in tiller number, indicating that SL also has a positive role in enhancing culm strength and spikelet number. They produced a pyramiding line carrying *SCM3* and *SCM2*, another QTL encoding *APO1* involved in panicle development. The NIL-*SCM2*+*SCM3* showed a much stronger culm than NIL-*SCM2* and NIL-*SCM3* and an increased spikelet number caused by the additive effect of these QTLs.

## Salt tolerance

Soil salinity is another major problem in coastal areas and it is increasing by intensive cultivation due to secondary salinization of excessive chemical fertilizers and non judicious use of ground water. In inland areas, salt deposition is expected to increase as a consequence of increased evapotranspiration and water shortage with rising temperatures. Rice is suitable for reclaiming these soils because it thrives well under flooding and with high potential for genetic manipulation. Rice variety tolerant to salinity has to adopt different mechanisms to sequester excessive salts accumulated. Most of the salt tolerant varieties developed so far are with low yield potential having undesirable traits. Major QTL for salinity has been identified in saltol region confers salt tolerance is the skcl locus on chromosome 1 which maintains K+ homeostasis.

A major QTL was identified conferring salt tolerance on chromosome 1 and designated *Saltol* (Bonilla *et al.,* 2002). This QTL has been the target of marker assisted selection (Thomson *et al.,* 2010). Markers, RM8094, RM3412, RM493, RM10748 and RM10793 are most closely linked to saltol region were found to be effective. In India, efforts are under way for the introgression of sub1 and saltol into popular varieties separately. Identified QTLs for salinity tolerates at seedling stage. Ammar *et al.* (2009) detected QTLs for cl,Na- ratio and Na+ in the leaf at reproductive stage on chromosome 2, 3 and 8. Saltol versions of BR11 performing well at Bagladesh. Sarker *et al.,* 2014 introgression of saltol into BRRI dhan29. FL 478 developed from pokkali with saltol was used by several workers for the development of salt tolerant varieties globally and nationally by several workers (Singh *et al.* 2014).

## Drought tolerance

Drought is most complex characters controlled by poly genes. Though several QTLs for secondary traits such as root growth, leaf rolling etc detected, the use of direct measurement of yield under drought stress has shown more promising results. Bernier *et al.* (2007) detected a QTL on chromosome 12 in a large population from the cross of Vandana/Way Rarem that accounted for about 50% of the genetic variance, and was expressed consistently over 2 years. This QTL seems to be related to increased water uptake of plants under stress (Bernier *et al.* 2007). A chromosome 3 QTL had a large effect on drought tolerance in the cross between the tolerant variety Apo and the widely grown susceptible variety Swarna (Venuprasad *et al.*, 2009). Incorporation of DTY 1.1, 2.2,3.1 and 9.1 into popular rice varieties is under way by Singh *et al.*, 2014.

Aravind Kumar (2014) The study with over fifty mapping populations over the last ten years led to the identification of fourteen large effect QTLs, among which seven- *qDTY1.1, qDTY2.2, qDTY3.1, qDTY3.2, qDTY4.1, qDTY6.1, qDTY12.1* showed effectiveness in two or more genetic backgrounds as well as in diverse upland and lowland environments. Yield advantages of 0.5 t ha-1 in Vandana NIL with *qDTY12.1*; 0.8-1.0 t ha-1 in Anjali with *qDTY12.1* and *qDTY3.1*; 0.8 to 1.2 t ha-1 in IR64 and Sambha Mahsuri with *qDTY2.2, qDTY4.1*; 1.0-1.5 t ha-1 in Swarna and Swarna-Sub1 with *qDTY1.1, qDTY2.1, qDTY3.1* were achieved.

## Combining various biotic stresses for climate resilient varieties

In coastal areas, submergence and salinity will prevail together and at rainfed ecosystem both drought and floods come across in the same season. To mitigate these situations, combining submergence and saltol into one genotype (Siangliw *et al., 2014)* and drought and submergence into IR64 and TDK1 (Dixit *et al.,* 2014 and Singh *et al.*, 2014).

## Identification of novel sources for major abiotic stresses

There is need to search vast gene pool for exploitation of untapped genes tolerating major abiotic stresses to mitigate climate changed conditions. Venu Prasad *et al.* (2014) identified novel sources for major abiotic stresses. Several *O. glaberrima* accessions have good anaerobic germination capacity and, compared to standard *O. sativa* check (Khao Hlan on), they have better vigor and tolerance to higher temperatures and to Fe-toxic soil conditions. *O. glaberrima* was very suited to stagnant flood conditions and many accessions outperformed the standard *O. sativa* checks (IRRI119 and IRRI154) both in terms of biomass production and also yield. In flash flood condition, none of the tested *O. glaberrima* accessions showed better tolerance than the standard tolerant check (Swarna sub1). Many *O. glaberrima* accessions gave good yield under moderate levels of drought.

## Conclusion

Continuous research efforts resulted in identification of genes/ QTLs linked to major abiotic stresses and marker assisted selection is playing an important role in bringing climate resilient rice varieties for sustainable food security. The research should be focused on identification of untapped genes for the major abiotic stresses and pyramiding of these genes to get sustainable high yields even under adverse climatic conditions. Farmers and scientific community together could mitigate climate changed condition for global food security.

## References


Bonilla P, J Dvorak, DJ Mackill, K Deal, G Gregorio (2002). RFLP and SSLP mapping of salinity tolerance genes in chromosome 1 of rice (Oryza sativa L.) using recombinant inbred lines. Philipp. Agric. Sci. 85:68-76

Bernier J, A Kumar, V Ramaiah, D Spaner, G Atlin (2007). A large-effect QTL for grain yield under reproductive-stage drought stress in upland rice. Crop Sci. 47:507-518.

David J. Mackill, Abdelbagi M. Ismail, Alvaro M. Pamplona, Darlene L. Sanchez, Jerome J. Carandang and Endang M. Septiningsih.(2010). Stress Tolerant Rice Varieties for Adaptation to a Changing Climate. Crop, Environment & Bioinformatics 7:250-259.

Dixit, S., Singh, A and Kumar, A. (2014). Developing drought- and submergence-tolerant rice for south asia using marker-assisted backcrossing (MAB): IR64 SUB1 + DTY . Abstracts of 4th international rice congress, at Bangkok from 27th October to November, 2014.

Girija Rani, M., Satyanaryana, P.V., Suryanaryana, Y., Ramana Rao, P.V. (2014). Genetic improvement of amara rice variety for flood tolerance towards climate resilience. Abstracts of 4th international rice congress, at Bangkok from 27th October to November, 2014.

Girija Rani, M., Suryanaryana, Y., Satyanaryana, P.V., Ramana Rao, P.V. et al. (2015). Incorporation of sub1 into Pushyami( MTU 1075) through marker assisted breeding. Compendium of abstracts of Second international conference on Bioresources and stress Mangement held at PJTSAU and ANGRAU, Hyderabad from 7-10 , January, 2015.

Hattori Y, K Nagai, S Furukawa, XJ Song, R, Kawano, H Sakakibara, JZ Wu, T Matsumoto, A Yoshimura, H Kitano, M Matsuoka, H Mori, M Ashikari . (2009). The ethylene response factors SNORKEL1 and SNORKEL2 allow rice to adapt to deep water. Nature 460:1026-1031.

Ismail, A.M., , Singh, U.S., Singh, S,., Dar, M.B and Mackill, D.J. (2013). The contribution of flood tolerant rice varieties to food security in flood prone rainfed low land areas in Asia. Filed crops Research. http://dx.doi..org/10.1016/j.fert2013.0.1.007.

Kumar, A. (2014). Drought tolerance conferred through cross-talk of multiple loci in rice. Abstracts of 4th international rice congress, at Bangkok from 27th October to November, 2014.

Mackill, D.J.., Ismail., A.M., Singh, US., Labios, Ri.V and Paris , T.R. (2012). Development of sub1 and rapid adaption of submergence tolerant sub1 rice cultivars. Advances in Agronomy 115:299-352.

Neeraja CN, R Maghirang-Rodriguez, A Pamplona, S Heuer, BCY Collard, EM Septiningsih, G Vergara, D Sanchez, K Xu, AM Ismail, DJ Mackill (2007). A marker-assisted backcross approach for developing submergence-tolerant rice cultivars. Theor. Appl. Genet. 115:767-776.

Ookawa, T., Hobo, T., Yano, M., Murate, K., Ando, T., Miure, H., Asno, K., Ochiai, Y., Ikeda, M., Nishitani, R., Ebistani, T., Ozaki, I., Angeles, E.R., Hirasana, T and Matsuoka, M. (2010). New approach for rice improvement using a pleiotropic QTL gene for lodging resistance and yield. Nature Communications. http://nature.com.

Pelayo, G.M., Kretzschmar, T.L., Gabunada, L., Tamisin, I. ., Ismail, A , Mackill, D. and Septiningsih E. (2014). Gene validation of AG1, a major qtl for tolerance to anaerobic germination for direct-seeded systems. Abstracts of 4th international rice congress, at Bangkok from 27th October to November, 2014.

Reddy JN, RK Sarkar, SSC Patnaik, DP Singh, US Singh, AM Ismail, DJ Mackill (2010). Improvement of rice germplasm for rainfed lowlands of eastern India. SABRAO 13th International Congress. Cairns, Australia. http://open.irri.org/sabrao/images/stories/conference/site/papers/apb09final00211.pdf

Sarker, M.R.A., Newaz, M.A. , de Ocampo, M.P. , Thomson, M.J and Ismail, A.M. (2014). Introgression of the saltol qtl into brri dhan29, a popular variety in bangladesh, using marker assisted backcrossing. Abstracts of 4th international rice congress, at Bangkok from 27th October to November, 2014.

Septiningsih EM, AM Pamplona, DL Sanchez, CN Neeraja, GV Vergara, S Heuer, AM Ismail, DJ Mackill (2009). Development of submergence tolerant rice cultivars: The *Sub1* Gene and beyond. Ann. Bot. 103:151-160.

Singh, N.K., Singh, Rk.K, Singh, Y., Yadav, N, Singh, A.K et al. (2013). Genomics assisted transfer of QTLs for tolerance to drought, flooding and salinity into Indian mega rice varieties 2013. Abstracts of 7th Interantional rice genetic Symposium, 5-8th November, 2013 at Philippines.

Singh, A., Dixit, S. and Kumar, A. (2014). Using marker-assisted backcrossing (mab) to develop drought and submergence tolerant versions of popular variety TDK1. Abstracts of 4th international rice congress, at Bangkok from 27th October to November, 2014.

Siangliw, M and Siangliw, J. (2014). Molecular breeding of rice variety sin-thwe-latt for submergence and salinity-prone areas in Myanmar. Abstracts of 4th international rice congress, at Bangkok from 27th October to November, 2014.

Thomson MJ, AM Ismail, SR McCouch, DJ Mackill. (2010). Marker assisted breeding. p.451-469. In: Abiotic stress adaptation in plants: Physiological, molecular and genomic foundation. Pareek A, SK Sopory, HJ Bohnert, Govindjee (eds.) Springer, Berlin.

Toshio Yamamoto Search for articles by this author Affiliations National Institute of Agrobiological Sciences, Tsukuba, Japan , Fukuoka, S., Wu , J., Ando, T., Ordonio, R.L., Hirano, **x** Ko Hirano Search for articles by this author Affiliations Bioscience and Biotechnology Center, Nagoya University, Nagoya, Japan K and Matsuoka, M. (2014). Isolation of a novel lodging resistance QTL gene involved in strigolactone signaling and its pyramiding with a QTL gene involved in another mechanism. Molecular plant. DOI: http://dx.doi.org/10.1016/j.molp.2014.10.009.

Venuprasad R, CO Dalid, M Del Valle, D Zhao, M Espiritu, MTS Cruz, M Amante, A Kumar, GN Atlin (2009). Identification and characterization of large-effect quantitative trait loci for grain yield under lowland drought stress in rice using bulk-segregant analysis. Theor. Appl. Genet. 120:177-190.

Venuprasad, R., Fofana, M., Kazuki, S. Dramé, K.N et al. (2014). *Oryza glaberrima*: a source of abiotic stress tolerance for rice breeding. Abstracts of 4th international rice congress, at Bangkok from 27th October to November, 2014.

❑❑❑

# Section 2

## Climate Change and Mitigation Strategies

# 7

# Climate Change and its Effect on Water Availability and Mitigation Strategies

V.K. Choudhary, Manish Kanwat, P. Suresh Kumar and R. Bhagawati

## Introduction

"Food insecurity is an enormous challenge at a global scale, with strong implications both for environmental management and for socio-economic development" (Rockstrom, 2003). This is exaggerated by persistent drought and uneven distribution of rainfall resulting in low crop yield. As the results, millions of lives are threatened with starvation caused by food shortages particularly in the arid and semi arid regions where majority of population still rely on rain-fed agriculture to secure food security (Ntsheme, 2005).

The majority of the affected population threatened with starvation resides in the rural areas with little or no income at all. It is projected that large number of population will be without food in coming years. This is exacerbated by slow economic growth and poor performance in the agricultural sector. It is likely that the increment of the food gap will occur in the near future intensifying the shortage of food in the region (Kundlande *et al,* 2004).

In this context climate change is playing crucial role in food and nutritional security with less availability of water. Climate change is arguably the most severe long term threat to development facing present and future generations across the globe. The past 50 years have witnessed unprecedented changes in the eco-system. Eco-system changes on global and regional scales have already affected natural resource base in diverse conditions and environments. It adversely affects not only the living conditions of people who depend upon these eco-systems for their livelihood but also influences the whole socioeconomic system at the macro level.

The natural resource base, including land and water, that support and sustain the livelihoods of masses is degrading at accelerated rates. Melting Himalayan glaciers pose a direct threat to the water and food security of 1.6 billion people in South Asia region as per recent estimates of Asian Development Bank (ADB). The situation is likely to worsen in the water scarce regions in terms of severe drought and floods. Such conditions are likely to disrupt the balance in the pattern of water supply and demand for water across agriculture, domestic and industry sectors. This will lead to reduction in the choice of crops and cropping system, posing threats to food security and increasing frequency of water induced disasters (Anon, 2009).

## Assessment of climate change in South Asian countries

The climate change can be easily visualized by analyzing table 1, that how temperature and precipitation are changing and even shifting of the vents over the period of time. Though the average annual rainfall is almost with no change, but the numbers of rainy days shows drastic changes. The numbers of rainy days are getting few in numbers which is clear indication that in few day high volume of water are poured. This is main cause for flood and even for prolonged dry spell.

**Table 1.** Summary of key observed Past and Present Climate Trends and Variability (IPCC, 2007)

| Country | Change in Temperature | Change in Precipitation |
|---|---|---|
| India | 0.68°C increase per century, increasing annual mean temperature | Increased extreme rains in north-west in recent decades, lower number of rainy days along east coast |
| Nepal | 0.09°C rise per year in Himalaya, and 0.04°C in Terai region, more in winter | No distinct long term trends in precipitation records for 1948-1994 |
| Pakistan | 0.6°C-1.0°C rise in mean temperature in coastal areas | 10-14% decrease in coastal belt and hyper arid plains, increase during summer and winter over last 40 years in Northern Pakistan |
| Bangladesh | Increasing trend of about 1.0°C in May and 0.5°C in November during 1985-98 | Decadal rain anomalies above long term averages since 1960s |

*contd...*

| | | |
|---|---|---|
| Sri Lanka | 0.016°C increase per year between 1961-90 over entire country, 2°C increase per year in central highlands | Increasing trend in February and decreasing trend in June |

## Impact of climate change in South Asian countries

The impact of climate change can be visualized from the table 2 with the data presented of the socioeconomic damages caused due to flood events in South Asian countries. Due to climate change flood situation are more common and are causing heavy losses in terms of deaths, injuries, loss of house etc.

**Table 2.** Socio-Economic Damages caused due to Flood Events in South Asian Countries (1960-2008)

| Damages | Bangladesh | India | Nepal | Pakistan | Sri Lanka |
|---|---|---|---|---|---|
| Deaths | 52033 | 55656 | 5637 | 8877 | 1050 |
| Affected | 304628928 | 763986965 | 2977703 | 37687043 | 7957127 |
| Homeless | 4219724 | 13210000 | 84925 | 4234415 | 2746601 |
| Injured | 102390 | 1561 | 1072 | 1981 | 1002 |
| Total Affected | 308951042 | 777198526 | 3063700 | 41923439 | 10704730 |
| Estimated Cost (US$ '000) | 12038400 | 29417188 | 977213 | 2865178 | 374364 |

Flood Disaster in South Asia (Anon, 2009)

## Climate change scenario

In parts of India, temperature increases and decreases in precipitation, along with increasing water use, have caused water shortages that have led to drying up of lakes and rivers. Similarly, water shortages have been attributed to issues such as rapid urbanization and industrialization, population growth and inefficient water use, which are all aggravated by changing climate and its adverse impacts on demand, supply and water quality.

## Water availability in India

The gross per capita water availability in India is projected to decline from about 1,820 $m^3/yr$ in 2001 to as little as 1,140 $m^3/yr$ in 2050, as a result of population growth. Another study indicates that India will reach a state of water stress before 2025, when the availability is projected to fall below 1,000 $m^3$ per capita. These changes are due to climatic and demographic factors. The relative contribution of these factors is not known. The projected decrease in winter precipitation over the Indian subcontinent would imply less storage and greater water stress during the lean monsoon period. Intense rain occurring over fewer days, which implies increased frequency of floods during the monsoon, may also result in reduced groundwater recharge potential. Expansion of areas under severe water stress will be one of the most pressing environmental problems in India in the foreseeable future, as the number of people living under severe water stress is *likely* to increase substantially in absolute terms.

## Potential impact of climate change

Impacts of climate change on agriculture which are: (a) Increase in $CO_2$ to 550 ppm, this will increase yields of rice, wheat, legumes and oilseeds by 10- 20%; (b) A 1°C increase in temperature may reduce yields of wheat, soybean, mustard, groundnut, and potato by 3-7%; (c) Productivity of most crops to decrease marginally by 2020 and 10- 40% by 2100. Increased droughts, floods and heat waves will increase production variability; (d) Length of growing period in rainfed areas is likely to reduce, especially in peninsular regions and southern India; (e) the availability of water will be come down; (f) the soil fertility of hilly region will decrease due to decrease in potential organic carbon; (g) considerable effect on microbes, pathogens, and insects; (h) Possibly some improvement in yields of chickpea, rabi maize, sorghum and millets; and (i) Less loss in potato, mustard and vegetables in north-western India due to reduced frost damage.

## Adoption strategies

Some of the potential strategies are started to follow to cope up with the sudden change, and by using traditional wisdom, location specific technology, participatory research and development, capacity building, proper governance, sharing of information at various levels the impact of climate change may be normalize.

**Table 3.** Guiding principles for adaptation to climate change

| | |
|---|---|
| 1. Sustainable Development | Adaptation must be addressed in a broader development context, recognizing climate change as an added challenge to reducing poverty, hunger, diseases and environmental degradation |
| 2. Resilience | Building resilience to ongoing and future climate change calls for adaptation to start now by addressing existing problems in land and water management |
| 3. Governance | Strengthening institutions for land and water management is crucial for effective adaptation and should build on the principles of participation of civil society, gender equality, subsidiarity and decentralization |
| 4. Information | Information and knowledge for local adaptation must be improved, and must be considered a public good to be shared at all levels |
| 5. Economics and Financing | The cost of inaction, and the economic and social benefits of adaptation actions, calls for increased and innovative investment and financing |

## Relationship between water supply and crop yield

Agriculture is important to our economy, culture, and environment but is subject to mounting pressure from uncontrolled urbanization, global market pressures, and threats to the reliability and availability of fresh water. Actions are needed to both ensure a sustainable agricultural sector and to reduce the amount of water required for it.

Water conservation and efficiency improvements can reduce water use and improve water quality while maintaining or increasing crop yield. Yet these improvements often entail significant investment which can be a barrier to implementation. Smart policies can reduce this barrier.

Crop water requirements vary throughout the crop life cycle and depend on weather and soil conditions. Irrigation scheduling provides a means to evaluate and apply an amount of water sufficient to meet crop requirements at the right time. While proper scheduling can either increase or decrease water use, it will likely increase yield and/or quality, resulting in an improvement in water-use efficiency. It is likely that further incremental of water supply above the required water supply units to achieve maximum crop yield might cause decline in crop yield and eventual decrease in water productivity this can be clearly visualized from figure 1 & 2 (Zhang, 2003).

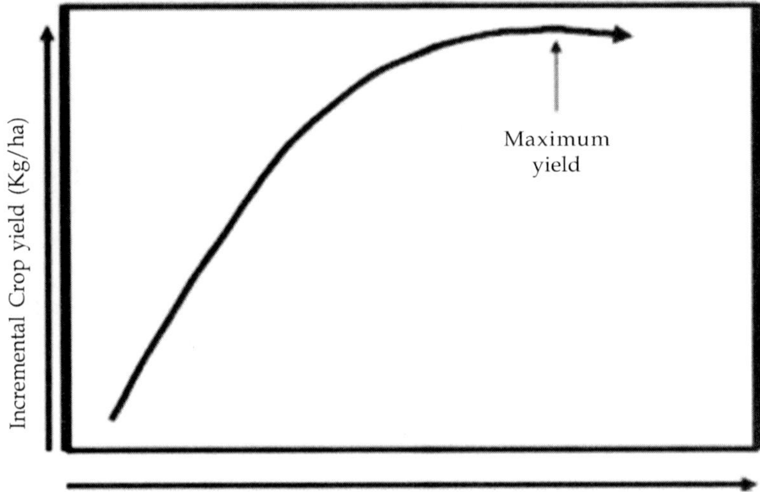


**Figure 1.** Crop yield and applied water relationship

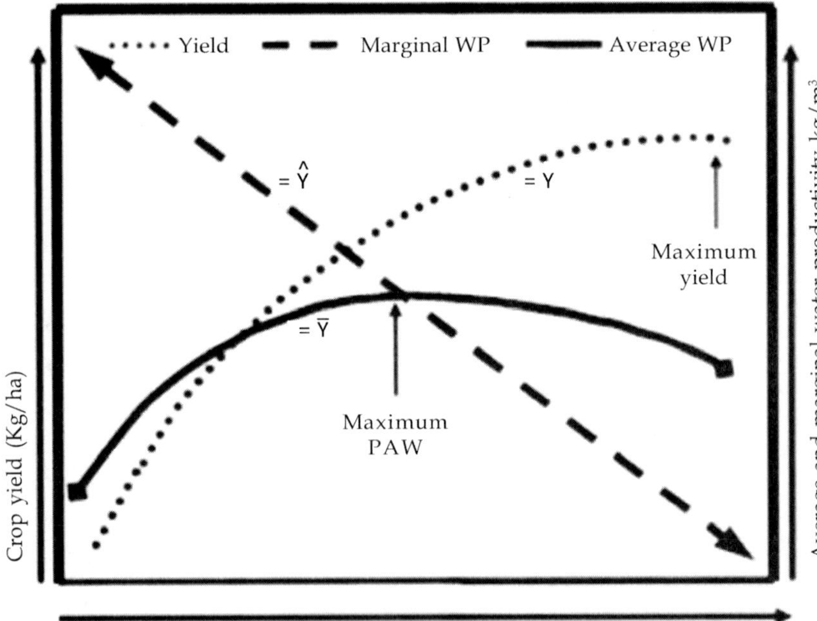


**Figure 2.** Relation of crop production, productivity of applied water (PAW) and marginal productivity to the crop water supply

## Rain water harvesting systems to increase water productivity

The challenge is to solve the problem of food insecurity by exploring the impact of rain water management on improving food security. On average, less than 30% of rainfall in rain-fed agriculture contributes to crop growth, while at least 70% is lost to the crop as evaporation, interception, drainage and surface runoff". While large proportion of rainfall is lost to run-off, evaporation, and drainage, the inter-seasonal prolonged dry spell causes crop failures. The rainwater management can provide the potential to increase crop yield (Figure 3). Rainwater harvesting can help to improve soil moisture especially during the period of the increased dry spell in the growing season (Rockstrom, 2003).

Micro-catchment systems (Within field/internal catchments systems)

## Rain water harvesting systems

*In-situ* water conservation (Tillage and cultural practices)

Runoff-based systems (Catchments and/or storage)

Direct application systems (Runoff diversion into cropland where soil profile provided moisture storage )

Storage systems (Distinct storage structures for supplemental irrigation and other uses)

Small catchment systems (Runoff generated from small external catchments and diverted to cropland)

Macro-catchment systems (flood diversion and spreading i.e.spate irrigation)

## Adaptation and mitigation strategies

Adaptation and mitigation of climate change impacts through community mobilization and capacity building of communities on planning and management of water resources through innovative technologies which can save energy and water.

## Low cost water saving technologies

Immediate application of appropriate eco-friendly low cost technologies and management practices for saving water at national, regional and community and household level is an imperative need. Various such technologies have been designed to cope-up with the water crises. For example;

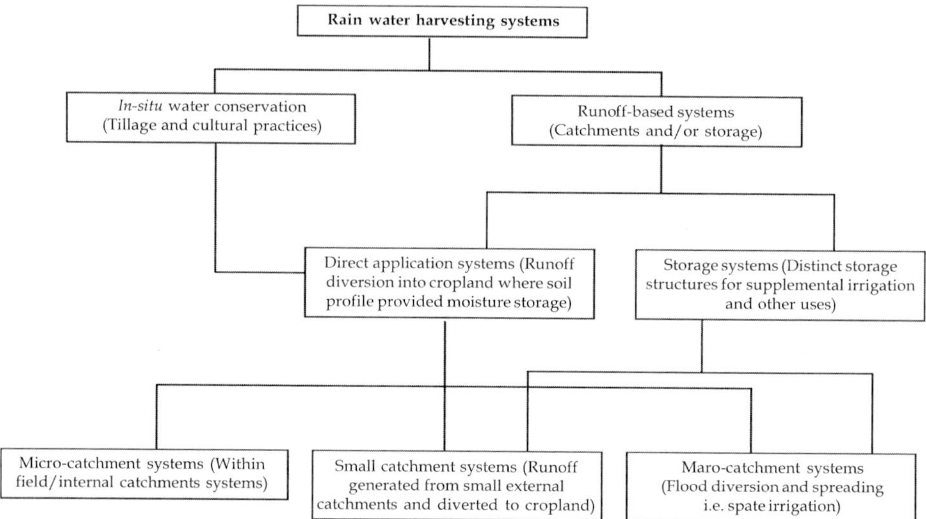


**Figure 3.** Adopted classification of rainwater harvesting technologies and systems (Ngigi, 2003).

A. *Land shaping technology* is an improved agro technology (Figure 4). The benefits of this technology are; (a) three dimensional (land, water, air) crops; (b) Option for integration of agriculture; (c) Aquaculture with duck rearing; (d) Introduction of double and triple crops; (e) Additional crop in pond & land embankment; (f) Harvested water utilized in 2nd and 3rd crops; (g) Conservation of ground water; and (h) Energy saving module.

B. In Gangetic plains of Eastern India, *moisture conservation technology* has been used to enhance and sustain farm production. The other technology is Micro Level Water Resource Development through *Tank-cum-Well Technology* designed by Water Technology Centre, Bhubaneshwar. The technology involves a system of tanks and dug wells in sequence. While tanks store run-off water which is recycled for irrigation, the open dug wells harvest water seeped in from tanks.

C. In Arunachal Pradesh, bunds which used to demarcate plots are raised and broadened and used for finger millet and vegetable cultivation. Similarly, in compartment bunds plot may be used for rice cum fish culture by following the traditional wisdom of *Apatani* plateau.

D. Community level check dams/water distributaries/ Nallas needs to be constructed to store flowing water during rainy season. The stored and harvested water may be used for, irrigation, domestic, small scale industries, navigation etc. with proper allocations.

E. Need to follow efficient and water saving irrigation techniques as per the topography, soil types and crops allocated. By this water utilization efficiency can be improved and saved water may be used for other important work.

F. Drip irrigation is one of the important tools to cope up from water scarcity during climate change. This technology has potential difference from conventional use of water for irrigation on various crops. For example; the percentage yield of Banana crop using conventional method would be 57.5 t/ha, while by using drip irrigation, it will be 87.5 t/ha resulting an increase of 52% yield and water saving to the tune of 45%. Adoption of drip irrigation will solve the problems of energy security, water security, food security and rural to urban migration in global change and population resource imbalance scenario. But drip irrigation has also some limitation that this can only be used for wide spacing row crops. The details can be visualized from Table 4.

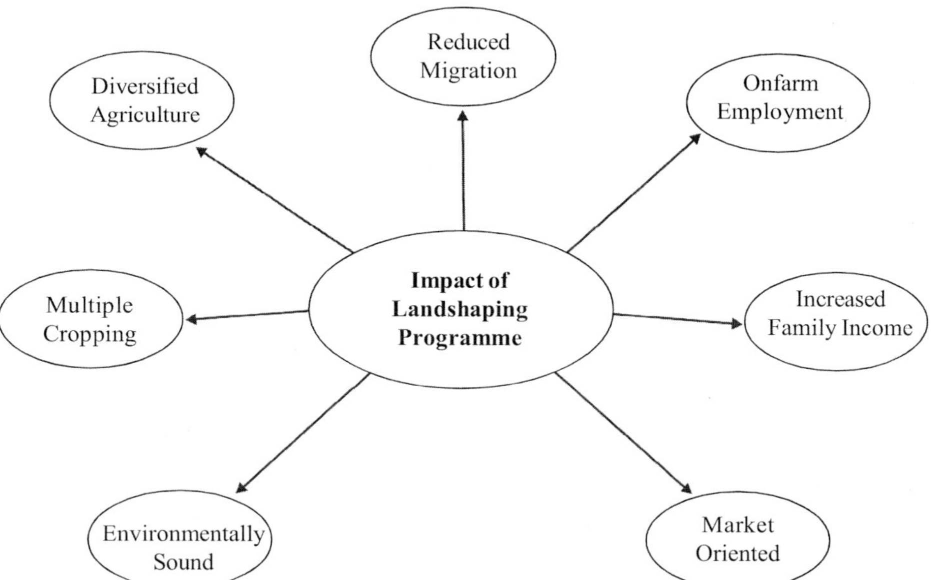


**Figure 4.** Impact of land-shaping programme

**Table 4.** Benefits of Drip Irrigation over Conventional Use of Water for Irrigation

| Crop | Conventional (Yield [t/ha]) | Drip (Yield [t/ha]) | % yield increase | Water savings (%) | Increase in water use efficiency (%) |
|---|---|---|---|---|---|
| Banana | 57.5 | 87.5 | 52 | 45 | 176 |
| Grapes | 26.4 | 32.5 | 23 | 48 | 136 |
| Sweet Lime | 100 | 150.0 | 50 | 61 | 289 |
| Pomegranate | 56.0 | 109.0 | 98 | 45 | 167 |
| Tomato | 32.0 | 48.0 | 50 | 31 | 119 |
| Watermelon | 24.0 | 45.0 | 88 | 36 | 196 |
| Chillies | 4.2 | 6.1 | 44 | 63 | 291 |
| Sugarcane | 128.0 | 170.0 | 33 | 56 | 204 |

It is well aware that already 80.0 per cent of basins with 60.0 per cent of the farm area are facing physical water scarcity. Projecting the food demand in 2050 at 400 million tonnes requires an additional irrigation of 60 mha, and this would create livelihood opportunities but will also demands huge investment initiatives. In the present context, the approach for climate change adaptation should be through promotion of National and local institutions, integrating integrated water resource management (IWRM), focusing on supply management-infrastructure, demands management-institutions, etc. However, the present institutions, management system, and infrastructure are inadequate to deal with the exigencies of climatic change. Thus, in addressing potential water shortages, as much attention should be given to managing demands as to increasing supply, by introducing more efficient technologies as well as simply promoting a culture of conservation (Anon, 2009).

To increase the water use efficiency by proper water management following things need to be followed, those are:

(i) Comprehensive water data base in public domain and assessment of the impact of climate change on water resource; (ii) Promotion of citizen and state actions for water conservation, augmentation and preservation; (iii) Focused attention on over-exploited areas; (iv) Increasing water use efficiency by 20%; and (v) Promotion of basin level integrated water resources management.

## Efficient water use efficiency

Water savings achieved through conservation and efficiency improvements are just as effective as new, centralized water storage and are often far less expensive. Many proven technologies and practices can improve water-use efficiency. Strengthen and expand efforts to promote the use of these technologies and practices.

- Revise and expand "Efficient Water Management Practices" for agricultural water agencies.
- Make agricultural "Efficient Water Management Practices" mandatory and enforceable by the State Water Resources Control Board.
- Develop institutional mechanisms to increase the reliability of agricultural water deliveries to users meeting high standards of water-use efficiency.

Expand water-efficiency information, evaluation programs, and on-site technical assistance provided through Agricultural Extension Services and other agricultural outreach efforts. Four scenarios for improving the water-use efficiency of the agricultural sector are evaluated:

*Modest Crop Shifting* - shifting a small percentage of lower-value, water-intensive crops to higher-value, water-efficient crops

*Smart Irrigation Scheduling* - using irrigation scheduling information that helps farmers more precisely irrigate to meet crop water needs and boost production

*Advanced Irrigation Management* - applying advanced management methods that save water, such as regulated deficit irrigation

*Efficient Irrigation Technology* - shifting a fraction of the crops irrigated using flood irrigation to sprinkler and drip systems

## Water conservation and efficiency scenario

Today, the challenge is to envision an agricultural sector that continues to supply food to the state and nation, to support rural livelihoods, and remains consistent with the goal of long-term sustainable water use for the state as a whole. There are many different ways for irrigators to use water productively. Farmers have long shown themselves to be flexible, dynamic, and innovative in response to water constraints. But rapid and unplanned changes in water availability can result in labor dislocations, debt, and production losses.

Water is only one of many constraints and incentives farmers must balance; constraints that may indirectly affect water use include fluctuating market conditions; agricultural policies; local soils and climates; and previous investment in irrigation technologies, farm equipment, and processing machinery. In general, farmers make economically rational decisions to maximize profits. Farmers also make choices independent of profit maximization: experience, family traditions, and community values all factor into their decisions.

There is an urgent need to give the full support in decision making at water user groups to set the decision about crop type, irrigation method, and management practices. Regular monitoring is required to correct the problems coming across and also facilitate them to improve their capability by capacity building programme. Support the water user to keep coordination between different catchment management group and efficient implementation of policies.

## Sprinkler Irrigation Systems

Sprinkler irrigation, introduced in the 1930s, delivers water to the field through a pressurized pipe system and distributes water via rotating sprinkler heads, spray nozzles, or a single gun-type sprinkler. The sprinklers can be either permanently mounted (solid set) or mounted on a moving platform that is connected to a water source (traveling). Although they have the poorest overall water-use efficiency among the sprinklers, traveling sprinklers are well-suited to irregularly sized or shaped fields and can be easily moved between fields (Evans *et al.*, 1998). Low-energy precision application (LEPA) and low elevation spray application (LESA) sprinklers are an adaptation of center pivot systems that use drop tubes that extend down from the pipeline to apply water on the ground or a few inches above the ground. LEPA and LESA systems can conserve both water and energy by applying the water at a low-pressure close to the ground, which reduces water loss from evaporation and wind, increases application uniformity, and decreases energy requirements.

Sprinklers provide a number of important advantages. If managed properly, they can improve water-use efficiency. Sprinklers often result in less ineffective runoff than a surface system, thereby reducing erosion, pollution of downstream water sources, and the economic cost of dealing with drainage. In addition, sprinklers tend to require less labor, thereby reducing labor costs and vulnerability to labor shortages (Burt *et al.* 2003).

## Drip/Micro-irrigation Systems

Drip irrigation refers to the slow application of low pressure water from plastic tubing placed near the plant's root zone. Drip systems commonly consist of buried PVC pipe mains and sub-mains attached to surface polyeth lateral lines a less expensive, but also less durable, option is drip tape. Water is applied through drip emitters placed above- or below-ground, referred to as surface and subsurface drip, respectively. Micro-irrigation systems are similar to drip systems with the exception that water is applied at a higher rate (5 to 50 gallons per hour) by a small plastic sprinkler attached to a stake (Evans *et al.* 1998).

Drip irrigation has been in use since ancient times when buried pots were filled with water that slowly seeped into the soil. Modern drip was facilitated by the advent of plastics during World War II and was first introduced in Israel. Although traditionally applied to specialty crops such as vegetables and grapes, drip irrigation systems are increasingly applied to row crops, and there are examples of use on field crops such as cotton, corn, alfalfa, and potatoes Drip irrigation allows for the precise application of water and fertilizer to meet crop needs and can increase crop yield and/or quality.

In recent past growers that with drip irrigation, "We consistently use less water, less fertilizer, and find tillage and ground preparation less costly. In addition, yields are higher and the quality of the product we grow is better. Drip irrigation pays, it doesn't cost!" (AWMC 2006a). Furthermore, "the potential for improved water and chemical management can benefit water quality, reduce potential runoff, and reduce potential leaching of nutrients and chemicals" (Evans *et al.*, 1998). Drip systems can be automated, thereby reducing labour costs. With drip systems, diseases are less likely to develop because water does not come into contact with crop leaves, stems, or fruit (Shock, 2006). Drip systems can be used on oddly shaped or hilly terrain.

## Comparison of irrigation technologies

In the present context it is very much required that as per the crop, we can use the various type of irrigation methods, to save water. The methods may vary with crop to crop but we need to use or recommend such methods which can be easily used with higher efficiency. Here some of the important methods along with their efficiencies are presented in table 5 so that growers and extension functionaries can promote the different methods and water can be saved for other uses.

**Table 5.** Irrigation System Efficiency

| Type of Irrigation System | Efficiency |
|---|---|
| **Flood** | |
| Basin | 85% |
| Border | 77.5% |
| Furrow | 67.5% |
| Wild Flooding | 60% |
| Gravity | 75% |
| **Average** | **73%** |
| **Sprinkler** | |
| Hand Move or Portable | 70% |
| Center Pivot and Linear Move | 82.5% |
| Solid Set or Permanent | 75% |
| Side Roll Sprinkler | 70% |
| LEPA (Low Energy Precision Application) | 90% |
| **Average** | **78%** |
| **Drip/Micro irrigation** | |
| Surface Drip | 87.5% |
| Buried Drip | 90% |
| Sub irrigation | 90% |
| Micro Sprinkler | 87.5% |
| **Average** | **89%** |

*Source:* Salas *et al.,* 2006

## Adoption

Adaptive capacity and adaptation related to water resources are considered very important. Historically, migration in the face of drought and floods has been identified as one of the adaptation options. Migration has also been found to present a source of income for those migrants, who are employed as seasonal labour. Other practices that contribute to adaptation include traditional and modern water-harvesting techniques, water conservation and storage, and planting of drought-resistant and early-maturing crops. The importance of building on traditional knowledge related to water harvesting and use has been highlighted as one of the most important adaptation requirements, indicating the need for its incorporation into climate change policies to ensure the development of effective adaptation strategies that are cost-effective, participatory and sustainable.

Generally, the frequency of occurrence of more intense rainfall events in many parts of country has increased, causing severe floods, landslides, and debris and mud flows, while the numbers of rainy days and total annual amount of precipitation have decreased. However, there are reports that the frequency of extreme rainfall in some areas has exhibited a decreasing tendency.

However, with climate change, an array of serious threats is apparent.

- Sea-level rise could impact on the river basin and on people living in the delta and other coastal areas.
- Temperature rises will be *likely* to reduce the productivity of major crops and increase their water requirements, thereby directly decreasing crop water-use efficiency.
- There will probably be a general increase in irrigation demand.
- There will also be a high degree of uncertainty about the flow of the various rivers.
- North East of India will be *likely* to experience an increase in water stress, with a projected decline in precipitation by 2050. This will increase water stress in all sectors.

In some parts of India, the conversion of cropland to forest (grassland), restoration and re-establishment of vegetation, improvement of the tree and herb varieties, and selection and cultivation of new drought-resistant varieties could be effective measures to prevent water scarcity due to climate change. Water saving schemes for irrigation could be used to avert the water scarcity in regions already under water stress. There are urgent needs to recycling and reuse of municipal wastewater and increasing efficiency of water use for irrigation and other purposes will be *likely* to help avert water scarcity.

Various parts of India to minimize the impacts of climate change on water resources, several of which address the existing inefficiency in the use of water:

- Modernization of existing irrigation schemes and demand management aimed at optimizing physical and economic efficiency in the use of water resources and recycled water in water-stressed areas;

- Public investment policies that improve access to available water resources, encourage integrated water management and respect for the environment, and promote better practices for the sensible use of water in agriculture;
- The use of water to meet non-potable water demands. After treatment, recycled water can also be used to create or enhance wetlands and riparian habitats.

## Impact and Vulnerability

- The imbalance of species loss and replacement leads to an initial loss in diversity.
- Species most affected by warming are restricted to the uppermost parts of mountains. For other species, the effect will mainly be upward migration.
- Species affected by cooling are those at lower altitudes.
- Changes will directly affect the indigenous biota. An even greater threat is that a warmer climate will increase the ease with which the islands can be invaded by alien species.

## Conclusion

There are urgent needs to implement some of the important practices in any area to have higher irrigation water use efficiencies:

(1) adopt a water management plan; (2) designate a water conservation coordinator; (3) support water management services such as mobile irrigation labs, irrigation scheduling, and water quality testing; (4) improve communication and cooperation among water suppliers and users; (5) identify institutional changes that will improve the potential for more flexible water deliveries and storage; and (6) improve pump efficiency to reduce operational costs.

The greatest challenge is to achieve food security in climate change scenario, through increased productive use of water used for agriculture, thereby leaving more water available for other users in the basin. The integrated approach to water management can improve rural livelihoods and simultaneously increase water productivity. The improved land and water management will lead to rainwater use efficiency, leaving more water available in the storage dams that can be used beneficially for crops grown in winter.

## References

Agricultural Water Management Council (AWMC) (2006). A Smaller Footprint: Managing Our Resources. Retrieved July 15, 2008 from http://www.agwatercouncil.org/Publications/menuid-86.html.

Annonymous (2009). Report of the Round Table Conference On Water, Livelihood and Adaptation to Climate Change in South Asia November 05–06, 2009 (ed: Dr Veena Khanduri) conference organized by India Water Partnership.

Burt, C., D. Howes, and G. Wilson (2003). California Agricultural Water Electrical Energy Requirements: Final Report. Prepared for the California Energy Commission by the Irrigation Technology Research Center, California Polytechnic State University, San Luis Obispo, California.

Evans, R.O., K.A. Harrison, J.E. Hook, C.V. Privette, W.I. Segars, W.B. Smith, D.L. Thomas, and A.W. Tyson (1998). Irrigation Conservation Practices Appropriate for the Southeastern United States. Georgia Department of Natural Resources Environmental Protection Division and Georgia Geological Survey. Project Report 32. Atlanta, Georgia.

Kundhlande, G., Groenewald, D.C., Baiphethi M.N., Viljoen, M.N., Botha J.J., van Rensburg, L.D., and Anderson, J.J. (2004). Socio-Economic Study on Water Conservation Techniques in Semi-Arid Areas, Water Research Commission Report, WRC Report No. 1267/1/04.

Ngigi, S.N. (2003). What is the Limit of Up-scaling Rainwater Harvesting in a River Basin? Physics and Chemistry of the Earth 28: 943–956.

Ntsheme, O.P. (2005). A Survey of The Current On-farm Agricultural Land And Water Management Practices In The Olifants Catchment-A Diagnostic Of Quaternary Catchment B72A, Unpublished MSc (WREM) report: University of Zimbabwe.

Rockstrom, J. (2003). Resilience Bulding and Water Demand Management for Drought Mitigation, Physics and Chemistry of the Earth, Volume 28, Pages 869-877

Salas, W., P. Green, S. Frolking, C. Li, and S. Boles (2006). Estimating Irrigation Water Use for California Agriculture: 1950s to Present. California Energy Commission, PIER Energy-Related Environmental Research. Sacramento, California.

Shock, C. (2006). Drip Irrigation: An Introduction. Oregon State University Extension Service. Retrieved July 31, 2008 fromhttp://extension.oregonstate.edu/umatilla/mf/sites/default/files/Drip_Irrigation_EM8782.pdf.

Zhang, H. (2003). Improving Water Productivity through Deficit Irrigation: Examples from Syria, the North China Plain and Oregon, USA, CSIRO Plant Industry, Wembley, Australia.

❑❑❑

# 8

# Potential of Conservation Agriculture in Drought Stress Management in Crop Plants

**R.L. Choudhary, Mahesh Kumar, Satish Kumar, Hardev Ram and Aradhana Kumari**

## Introduction

Global food production, already under the credit crunch, must double by 2050 to level off hunger. According to the UN World Food Programme (WFP), this will be possible only if food security is given top priority. The food security has become a complex issue influenced by climate and non-climate induced stressors which operate independently or in combinations are tackled appropriately. This climate induced stress is further augmented by climate change which affects agriculture and food production in complex ways. It affects food production directly through changes in agro-ecological conditions and indirectly by affecting growth and distribution of incomes, and thus demand for agricultural produce. Climate change has emerged as a major challenge in achieving goals of sustainable agriculture. Climate change refers to any change in climate over time, whether due to natural variability or/and as a result of human activity (IPCC, 2007). Agriculture impacts climate change with greenhouse gas emissions, and is at the same time impacted by effects of climate change as well. Amongst elements of climate change, the most important relate to increasing uncertainty in availability of water due to increasing frequency of drought and/or excess water events resulting in uneven water availability over time and space. Drought is a normal, recurring feature of the climate in most parts of the world. Drought is defined as the absence of rainfall or supplemental irrigation for a period of time sufficient to deplete soil moisture and injure plants. According to USDA drought occurs when annual precipitation is 75 percent of normal, or monthly precipitation is 60 per cent of normal. Drought stress results when water loss from the plant exceeds the ability of roots to absorb water and when the plant's water content is reduced enough to interfere with normal plant processes.

Periods of droughts can have significant environmental, agricultural, economic and social consequences. The effects of drought vary according to vulnerability of the agro-climatic zone subjected to the drought stress. Drought can also reduce water quality, because lower water flows reduce dilution of pollutants and increase contamination of remaining water sources. Even after a drought has ended, it may take months or even years for a plant to repair damaged root systems and regain its growth capacity it had prior to the drought. Drought is one of the most important constrains for crop production. The improvement of drought tolerance is very difficult because of the set of mechanisms involved.

## Types of Drought

1. **Meteorological drought** is brought about when there is a prolonged period with less than average precipitation. Meteorological drought usually precedes the other kinds of drought.
2. **Agricultural droughts** are droughts that affect crop production or the ecology of the range. This condition can also arise independently from any change in precipitation levels when soil conditions and erosion triggered by poorly planned agricultural endeavours cause a shortfall in water available to the crops. However, in a traditional drought, it is caused by an extended period of below average precipitation.
3. **Hydrological drought** is brought about when the water reserves available in sources such as aquifers, lakes and reservoirs fall below the statistical average. Hydrological drought tends to show up more slowly because it involves stored water that is used but not replenished. Like an agricultural drought, this can be triggered by more than just a loss of rainfall.

## Cause of Drought

Generally, rainfall is related to the amount and dew point (determined by air temperature) of water vapour carried by regional atmosphere, combined with the upward forcing of the air mass containing that water vapour. If these combined factors do not support precipitation volumes sufficient to reach the surface, the result is a drought. This can be triggered by high level of reflected sunlight (high albedo) and above average prevalence of high pressure systems, winds carrying continental, rather than oceanic air masses (i.e. reduced water content), and ridges of high pressure areas from behaviours which prevent or restrict the developing of thunderstorm activity or rainfall over one certain region.

Oceanic and atmospheric weather cycles such as the El Niño-Southern Oscillation (ENSO) make drought a regular recurring feature.

## Classification of drought resistance mechanism

Drought resistance is the ability of the plant to carry all essential physiological functions to the optimum even in condition of dehydration. Plants are able to cope up with drought either of desiccation postponement (the ability to maintain tissue hydration) and desiccation tolerance. Earlier literature quite often has used the term *drought avoidance* in place of the - *drought tolerance* but this term is an ambiguous in the sense that drought is a meteorological condition that can be tolerated by all plants that survive it but avoided by none. However, some plants that complete their life cycles during the wet season, before the onset of drought are known to display drought escape in true sense. Drought tolerance tends to be the greatest during early seedling stages and progressively decreases through later stages of development until the flowering and early seed-filling stages. Other categories of the plants which display ability of defer the symptoms of drought and so called '*desiccation postponers*'. These dissication *postponers* are usually water savers which use water conservatively, preserving some in the soil for use late in their life cycle.

## Effect of drought on crops

Favourable southwest summer monsoon in India is critical in securing water for irrigating Indian crops. In some parts of the country, the failure of the monsoons result in water shortages, resulting in below-average crop yields. This is particularly true of major drought-prone regions such as southern and eastern Maharashtra, northern Karnataka, Andhra Pradesh, Orissa, Gujarat, and Rajasthan. Three main mechanisms reduce crop yield by soil water deficit: (i) reduced canopy absorption of photosynthetically active radiation, (ii) decreased radiation-use efficiency and (iii) reduced harvest index (Earl and Davis, 2003).

The impacts of drought on crop growth and yield depend on the severity and duration of stress and the stage of plant development at which the stress occurs. The most sensitive stages of development to drought stress are generally during panicle development and during flowering in cereals and the period just before flowering and during flowering in legumes. The various effects of drought stress on plants are as follows:

## Crop growth and yield

Water is a major component of plant cells, and is the medium in which growth processes occur. The first and foremost effect of drought is impaired germination and poor stand establishment (Harris *et al.,* 2002). Drought stress has been reported to severely reduce germination and seedling stand (Kaya *et al.,* 2006). Drought stress during the vegetative stage greatly reduced the plant growth and development. Growth is accomplished through cell division, cell enlargement and differentiation, and involves genetic, physiological, ecological and morphological events and their complex interactions. These all events are affected by water deficit. Cell growth is one of the most drought-sensitive physiological processes due to the reduction in turgor pressure (Taiz and Zeiger, 2006). Under severe water deficiency, cell elongation of higher plants can be inhibited by interruption of water flow from the xylem to the surrounding elongating cells (Nonami, 1998). Impaired mitosis, cell elongation and expansion result in reduced plant height, leaf area and crop growth under drought.

Many yield-determining physiological processes in plants respond to water stress. Yield integrates many of these physiological processes in a complex way. Water stress applied at pre-anthesis reduced time of anthesis, while at post anthesis shortened the grain-filling period. Drought stress reduced grain yield by reducing the number of tillers, spikes and grains per plant and individual grain weight. Post-anthesis drought stress was detrimental to grain yield regardless of the stress severity (Samarah, 2005). Drought at flowering commonly results in barrenness. A major cause of this, though not the only one, was a reduction in assimilate flux to the developing ear below some threshold level necessary to sustain optimal grain growth (Yadav *et al.,* 2004). As conclusion we can say that prevailing drought reduces plant growth and development, leading to hampered flower production and grain filling, and thus smaller and fewer grains. A reduction in grain filling occurs due to a reduction in the assimilate partitioning and activities of sucrose and starch synthesis enzymes.

## Photosynthesis

Biological processes, such as photosynthesis, are greatly reduced without adequate water which arises by a decrease in leaf expansion, impaired photosynthetic machinery, premature leaf senescence and associated reduction in food production (Wahid and Rasul, 2005). Reduced photosynthesis means reduced plant growth, including root growth. Drought can influence photosynthesis either through pathway regulation by stomatal closure and decreasing flow of $CO_2$ into mesophyll tissue (Chaves *et al.*, 2003; Flexas *et al.*, 2004) or by directly impairing metabolic activities (Farquhar *et al.*, 1989). In general, during the initial onset of drought stress, decreased conductance through stomata is the primary cause of decline in photosynthesis (Cornic, 2000). At later stages with increasing severity, drought stress causes tissue dehydration, leading to metabolic impairment. Recent studies suggest that both diffusive limitation through stomatal closure and nonstomatal limitation (such as oxidative damage to chloroplast) are responsible for decline in photosynthesis under drought stress (Zhou *et al.*, 2007).

Lower photosynthesis rate is the typical effect of water stress in plants which is principally attributed to the stomatal limitation and the metabolic impairment. Apart from restricted diffusion of $CO_2$ in the leaves, other biochemical changes leads to reduced rate of photosynthesis as a manifestation of decrease in photosynthetic pigment levels and reduced activity of photosynthetic enzymes in the chloroplast effecting both light and dark reactions. Prolonged water stress leads to decrease in chlorophyll content of leaves due to inhibition of *denovo* chlorophyll synthesis. This inhibition of chlorophyll synthesis is brought about as a consequence of inhibition at all the four consecutive stages of chlorophyll synthesis. These four stages are: (1) formation of 5-aminolevulinic acid (ALA) from glutamic acid , (II) ALA condensation into porphobilinogen and primary tetrapyrrol (III) Light dependent conversion of protochlorophyllide into chlorophyllide and (IV) synthesis of chlorophyll a and b along with their inclusions into developing pigment protein complexes of photosynthetic apparatus (Lisar *et al*, 2012). Contrary to this, water stress is always accompanied with increase in xanthophyll pigments as many of the intermediates of xanthophyll cycle have their role in inhibition of ROS production which increases during abiotic stress. Water stress also leads to decrease in ATP and NADPH intermediates later required in dark reaction of photosynthesis. This reduction in ATP and NADPH is a result of direct negative effect of water stress on cyclic and noncyclic electron transport chain during light reaction leading to its disruption and ultimately effecting photophosphorylation as a whole. Decrease in ATP synthase during water stress is also a well established fact.

Water stress also leads to rapid decrease in the amount of RuBisCO, a key enzyme in reductive $CO_2$ assimilation pathway and also leaves negative impact on activity of this enzyme. Though there is strong evidence that photosynthetic assimilation of $CO_2$ by stressed leaves is not limited primarily by $CO_2$ diffusion alone but by inhibition of ribulose biphosphate (initial substrate for carboxylation in calvin cycle) synthesis as a result of less ATP production due to impairments of electron transport chains during light reaction. In addition to RuBisCO, other enzymes of reductive $CO_2$ assimilation are also negatively affected during water stress leading to decreased activity. These enzymes are NADP-dependent glyceraldehyde phosphate dehydrogenase, PEP carboxylase, Phosphoribulose kinase, NAD-Dependent malate dehydrogenase, fructose-1,6 bisphosphatase.

## Membrane stability

Cell membranes are major targets of environmental stresses. Lipids are important membrane components, and changes in their composition may help to maintain membrane integrity and preserve cell compartmentalization under water stress conditions. Strong water deficit leads to a disturbance of the association between membrane lipids and proteins as well as to a decrease in the enzyme activity and transport capacity of the bilayer. The most striking changes in lipid composition are an increase in fatty acid unsaturation and variations in the balance between the two galactolipid classes monogalactosyl-diacylglycerol (MGDG) and digalactosyl-diacylglycerol (DGDG). An increase in fatty acids having less than 16 carbons and lipid peroxidation is well known effect of water stress in plants. Lipolytic activities are always higher in sensitive cultivars than resistant one.

## Mineral nutrition

The management of plant nutrients is very helpful to develop plant tolerance to drought. Better plant nutrition can effectively alleviate the adverse effects of drought by a number of mechanisms. Drought results in increased generation of the reactive oxygen species (ROS) due to energy accumulation in stressed plants which increases the photo-oxidative effect and damage the chloroplast membrane. Application of macro-nutrients like N, K and Ca reduce the toxicity of ROS by increasing the concentration of antioxidants like superoxide dismutase (SOD); catalase (CAT) and peroxidase (POD) in the plant cells. These antioxidants scavenge the ROS and reduce the photo-oxidation and maintain the integrity of chloroplast membrane and increase the photosynthetic rate in the crop plants. Similarly, the application of some micro- nutrients like Zn and Si also increase antioxidants concentration and improves drought tolerance in plants. In other mechanism, nutrients like P, K, Mg and Zn improve the root growth which in turn increases the intake of water which helps in stomatal regulation and enhances the drought tolerance. Application of nutrients like Potassium and Calcium help to maintain high tissue water potential under drought condition and improve drought tolerance by osmotic adjustment. The micronutrients like Cu and B alleviate the adverse effects of drought indirectly by activating the physiological, biochemical and metabolic processes in the plants. Increasing evidence suggests that mineral-nutrient status of plants plays a critical role in increasing plant resistance to environmental stress factors.

*Nitrogen-* Nitrogen absorption and utilization by plants under water stress is very critical for plant growth and productivity. Nitrogen affects carbon partitioning and improves accumulation of soluble sugars and especially starch which in turn improve leaf growth. Exogenous application of nitrogen increases antioxidative defence mechanisms which results in reduced photo-oxidation of chloroplast pigments, and reduced leaf senescence. Water stress leaves a direct negative impact on activity of enzymes Nitrate reductase (NR), the first enzyme in the pathway of nitrogen assimilation. Drought induced nitrogen deficiency largely contributes to growth inhibition under water deficit (Heckathorn *et al.,* 1997) mainly affecting the leaf size through decreasing the cell number and cell size (MacAdam *et al.,* 1989). Inorganic fertilizer application has been reported to mitigate the adverse effects of water stress on crop growth and development.

*Calcium-* is a major controller of plant metabolism and development. Calcium is considered to play a role in mediating stress response during injury, recovery from injury, and acclimation to stress (Palta, 2000). Calcium plays an essential role in structural and functional integrity of plant membranes and other structures. Calcium, being a part of calcium pectate thus plant cell wall, has a very prominent role in the maintenance of cell structure. Its activates the plasma membrane enzyme ATPase which pumps back the nutrients lost during cell membrane damage due to Ca deficiency and recover the plant from injury (Fig. 1). Calcium is component of calmodulin which controls the plant metabolic activities and enhances the plant growth under drought.

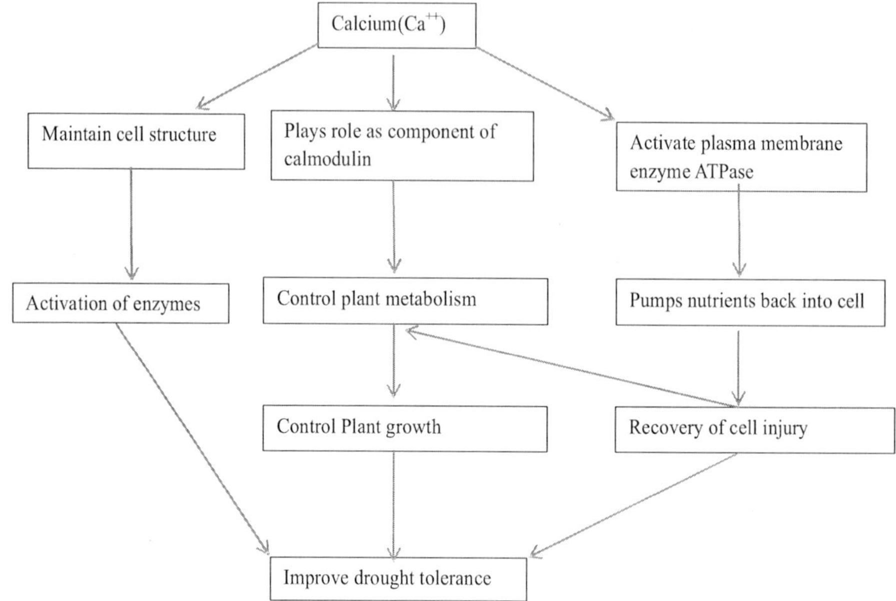


**Figure 1.** Possible mechanism through Ca nutrition can help plant to minimise the detrimental effects of drought. (*Source*: Waraich *et al.*, 2011).

*Potassium*- Potassium (K), which is conventionally known to play important role in regulation of stomatal activity, has multidimensional role in survival of plants under environmental stress conditions. Potassium is essential for many physiological processes, such as photosynthesis, translocation of photosynthates into sink organs, maintenance of turgor, activation of enzymes, and reducing excess uptake of ions such as Na and Fe in saline and flooded soils (Marschner, 1995; Mengel and Kirkby, 2001). Plants suffering from environmental stresses like drought have a larger internal requirement for K. Environmental stress factors cause oxidative damage to cells by inducing formation of ROS, especially during photosynthesis. As drought stress is associated with stomatal closure resulting in reduced photosynthetic rate and thereby with decreased $CO_2$ fixation which leads to enhanced need for K by plants suffering from environmental stresses. Further, Formation of ROS is intensified because of inhibited $CO_2$ reduction as a result of drought stress and high light intensity. When plants are grown under low supply of K, drought-stress induced ROS production can be additionally enhanced, at least due to K-deficiency-induced disturbances in stomatal opening, water relations, and photosynthesis. In addition to this, under drought conditions chloroplasts lose high amounts of K to further depress photosynthesis (Sen Gupta and Berkowitz, 1987) and induce further ROS formation. This strongly support the idea that increases in severity of drought stress result in corresponding increases in K demand to maintain photosynthesis and protect chloroplasts from oxidative damage.

## Assimilate partitioning

Assimilate translocation to reproductive sinks is vital for seed development. Seed set and filling can be limited by availability or utilization, i.e., assimilate source or sink limitation, respectively (Asch *et al.,* 2005). Drought stress frequently enhances allocation of dry matter to the roots, which can enhance water uptake (Leport *et al.,* 2006). The export rate of sucrose from source to sink organs depends upon the current photosynthetic rate and the concentration of sucrose in the leaves (Komor, 2000). Drought stress decreases the photosynthetic rate, and disrupts the carbohydrate metabolism and level of sucrose in leaves that spills over to a decreased export rate. Apart from source limitation, the capacity of the reproductive sinks to utilize the incoming assimilates is also affected under drought stress and may also play a role in regulating reproductive abortion (Zinselmeier *et al.,* 1999).

## Respiration

Drought tolerance is a cost-intensive phenomenon, as a considerable quantity of energy is spent to cope with it. The fraction of carbohydrate that is lost through respiration determines the overall metabolic efficiency of the plant (Davidson *et al.,* 2000). In wheat, depending on the growth stage, cultivar and nutritional status, more than 50% of the daily accumulated photosynthates were transported to the root, and around 60% of this fraction was respired (Lambers *et al.,* 1996). The alternative pathway branches from the cytochrome pathway and donates electrons to oxygen directly by alternative oxidase. When plants are exposed to drought stress, they produce reactive oxygen species, which damage membrane components (Blokhina *et al.,* 2003). In this regard, alternative oxidase activity could be useful in maintaining normal levels of metabolites and reduce reactive oxygen species production during stress. Water deficit in the rhizosphere leads to an increased rate of root respiration, leading to an imbalance in the utilization of carbon resources, reduced production of adenosine triphosphate and enhanced generation of reactive oxygen species.

## Water absorption through roots

An extensive root system is advantageous to support plant growth during the early crop growth stage and extract water from shallow soil layers that is otherwise easily lost by evaporation. In the root system of a tree or shrub, tiny, delicate root hairs extending from tender feeder roots at the extremities of the root system are responsible for the bulk of water uptake. Confined to the upper 15 inches or so of the soil profile, they are the first part of the root system affected by dry soil conditions. With the loss of root hairs, the water absorbing capacity of the plant is severely reduced. A limited root system will accelerate the rate at which drought stress develops. A root system may be limited by the presence of competing root systems, or by site conditions such as compacted soils. Newly installed plants and poorly established plants may be especially susceptible to drought stress because of the limited root system or the large mass of stems and leaves in comparison to roots. Generally, when water availability is limited, the root: shoot ratio of plants increases because roots are less sensitive than shoots to growth inhibition by low water potentials. Under drought stress conditions roots induce a signal cascade to the shoots via xylem causing physiological changes eventually determining the level of adaptation to the stress. Abscisic acid (ABA), cytokinins, ethylene, malate and other unidentified factors have been implicated in the root–shoot signaling.

## Water relations

Relative water content (RWC), leaf water potential, stomatal resistance, rate of transpiration, leaf temperature and canopy temperature are important characteristics that influence plant water relations. Exposure of plants to drought stress substantially decreased the leaf water potential, relative water content and transpiration rate. Crop plants have developed many mechanisms to survive water deficit, including escape, tolerance, and avoidance of tissue and cell dehydration (Turner, 1986). Plants tolerate drought by maintaining sufficient cell turgor to allow metabolism to continue under increasing water deficits. Tolerance to stress involves at least two mechanisms, osmotic adjustment and changes in the elastic properties of tissues. RWC is considered a measure of plant water status, reflecting the metabolic activity in tissues and used as a most meaningful index for dehydration tolerance. A decrease in the RWC in response to drought stress has been noted in wide variety of plants. Although components of plant water relations are affected by reduced availability of water, stomatal opening and closing is more strongly affected. Moreover, change in leaf temperature may be an important factor in controlling leaf water status under drought stress. Drought-tolerant species maintain water-use efficiency by reducing the water loss. However, in the events where plant growth was hindered to a greater extent, water-use efficiency was also reduced significantly.

### Nutrient availability, uptake, and metabolism

Drought stress due to decreased water availability strongly influences nutrient absorption and uptake by plants. Although nutrient and water absorption processes are independent processes, the need for water for absorption and transport makes them highly dependent on each other. Most nutrients are absorbed by plant roots as ions and water is the medium of transport. Under fully irrigated conditions when soil water potential is high, the absorption and transport of water and nutrient are higher. Drought stress decreases nutrient transport by diffusion and mass flow to the root surfaces and nutrient absorption by roots, which is influenced by water potential. Under water stress, roots are unable to take up nutrients from the soil because of lack of activity of fine roots, water movement, and ionic diffusion of nutrients.

### Other effects

Other physiological effects of drought on plants are the reduction in vegetative growth, in particular shoot growth. Reduced cyclin dependent kinase activity results in slower cell division as well as inhibition of growth under water deficit condition. Leaf growth is generally more sensitive than the root growth. Reduced leaf expansion is beneficial to plants under water deficit condition, as less leaf area is exposed resulting in reduced transpiration. In accordance, many mature plants subjected to drought respond by accelerating senescence and abscission of the older leaves. This process is also known as leaf area adjustment. Regarding root, the relative root growth may undergo enhancement, which facilitates the capacity of the root system to extract more water from deeper soil layers. Besides the direct effects of drought, a plant under stress becomes more susceptible to insect and disease problems that can attack a weakened plant. The time required for drought injury to occur depends on the water-holding capacity of the soil, environmental conditions, stage of plant growth, and plant species. Plants growing in sandy soils with low water-holding capacity are more susceptible to drought stress than plants growing in clay soils.

## Conservation Agriculture (CA) and climate change

In an era where climate change is central in development policies and practice, conservation agriculture (CA) appears to potentially contribute in addressing the challenge of adapting agricultural practices to climate change (FAO, 2011). Over the past four decades or so, internationally, rapid strides have been made to evolve and spread resource conservation technologies (RCTs) like zero and reduced tillage systems, better management of crop residues and planting systems which enhance water and nutrient conservation. CA which has its roots in universal principles of providing permanent soil cover (through crop residues, cover crops and agroforestry), minimum soil disturbance and crop rotations is now considered the principal road to sustainable agriculture. Thus it is a way to achieve goals of higher productivity while protecting natural resources and environment. The widespread adoption of no-tillage under a great range of different ecological and socioeconomic conditions on more than a 124 million ha world-wide shows, that the system can be made to work and function under extremely varied conditions. The faster adoption of this sustainable production system should be encouraged in order to reverse the process of soil degradation into a process of rehabilitating or building up its health, fertility and productive capacity. CA has come up as a new paradigm to achieve goal of sustained agricultural production. It is a major step toward transition to sustainable agriculture. CA practices hold the promise of providing both a strategy for mitigating climate change and also working as an adaptive mechanism to cope with climate change. CA practices have a great potential to sequester significant quantities of atmospheric $CO_2$ in the form of soil organic matter, as well as significant reduction in GHGs emissions through improved use efficiency of the production system while building and maintaining good soil structure and health compared to intensive tillage systems that does exactly the opposite (FAO, 2008; Kassam *et al.*, 2009; Friedrich *et al.*, 2009).

## Potential of Conservation Agriculture in Drought Stress Management

A significant cause of low production and crop failure in rainfed agriculture is lack of water in the soil. This is caused by a combination of low and erratic rainfall, and poor utilisation of the water that is available. Soil moisture management is, therefore, a key factor when trying to enhance agricultural production. Increasing the amount of water stored in the soil can result in improved yields, reduced risk of yield losses due to drought, recharge of groundwater, securing the water level in wells and the continuity of river and stream flows. CA is one way of improving soil moisture management. Permanent soil cover, either plant residues or growing crops, protects the soil surface from the negative effect of raindrop impacts. It reduces crust formation and susceptibility to erosion, and enhances porosity on the soil surface. It also reduces direct water loss through evaporation from the upper layers of the soil and establishes better conditions. Reducing or stopping tillage means that the soil is not disturbed and that the moisture loss and soil compaction that follows tillage is avoided. This increases the infiltration and percolation of water through the soil, leading to better root development and crop growth. Decomposition of organic matter and subsequent loss to the atmosphere is also reduced. One of the most important impacts of minimizing soil disturbance is that it improves the living conditions of beneficial organisms and so enhances their activity. Crop roots and soil organisms are responsible for the creation of a network of interconnected pores. Soil compaction should be restricted to defined areas, helps in maximizing soil porosity and water infiltration. The rotation of different crops, with their different root systems, optimizes the network of root channels in the soil, leading to increased water penetration, increased water holding capacity and more water being available for crop use, to deeper soil depths. The overall impact of CA is a productive soil system, able to carry crops through dry conditions because of the enhanced soil water store, the deep rooting of the crops, the biological activity and the high content of organic matter.

Farmers generally believe that rice prefers continuous pond water conditions during the growing season for maximum yields. Such a practice results in very low irrigation efficiency. Due to over-exploitation of groundwater in rice-wheat belt, there is immense pressure on the agricultural sector to reduce its water consumption. Researchers have been evaluating if cereals such as rice and wheat can be grown with micro-irrigation systems. Very recently direct seeded rice crop was grown with drip and sprinkler system in Haryana and Punjab. Compared to flood puddled transplanted rice, drip and sprinklers saved nearly 60% and 48% irrigation water, respectively. With micro irrigation yield gains averaged to 19%. Similarly, micro-irrigation systems saved 42% irrigation water in wheat and improved crop yields by 28%. The results presented in clearly brings out that there is huge potential for growing cereals with micro-irrigation systems. The direct seeded rice eliminates puddling reduced GHGs emissions, saves on water (~30%) and labour, and also improves soil health and system productivity.

Limited experimental results and farmers experience indicate that considerable saving in water (up to 20-30 %) and nutrients are achieved with zero-till planting and particularly in laser levelled and bed planted crop (Singh, 2011). In maize-wheat cropping system economic yield of maize increased by 19 and 29% with furrow-irrigated raised-bed (FIRB) planting (5.66 t $ha^{-1}$) compared to flat no-till and conventional-till planting systems, respectively. The productivity of wheat was higher by 7 and 9%, respectively, under flat no-till (5.56 t $ha^{-1}$) compared to FIRB and conventional till flat planting. The water productivity (kg grain $m^{-3}$ water) of maize and wheat was higher in FIRB planting (2.79 and 1.98) followed by flat no-till (1.74 and 1.89) and the lowest (1.36 and 1.38) in conventional till system. The net profit was more in flat no-till (15640 and 21290 Rs $ha^{-1}$) followed by FIRB (15270 and 20280 Rs $ha^{-1}$) and least in conventional-till (6620 and 15500 Rs $ha^{-1}$), respectively, for maize and wheat (Gangwar, 2012).

Planting of crops in furrow irrigated raised beds led to an increase in yield by 15-24% in legumes, 11-18% in vegetables, and 15% in maize with 13-25% saving in irrigation water. The double no-till practice in RW helped in similar productivity level with less water and cost of production that helped in improving water productivity and profitability. Furrow irrigated raised bed planting (FIRB) technology was evaluated through on-station as well as on-farm trials for improving water productivity and crop diversification. A yield increase of 15-24% in legumes, 11-18% in vegetables and 15% in maize was recorded with 13-25% saving in irrigation water in legumes, cereals, vegetables. In aerobic rice-based systems the yield of succeeding *rabi* crops viz. wheat (5.72 t/ha), chickpea (2.05 t/ha) and mustard (1.93 t/ha) were greater than after transplanted rice with yield of 5.02, 1.51 and 1.5 t/ha in wheat, chickpea and mustard, respectively. The system's productivity of aerobic rice-wheat, aerobic rice-chickpea and aerobic rice-mustard were 16.84, 16.78 and 14.02 t/ha, respectively, whereas under transplanted rice based systems the corresponding productivity were only 15.27, 14.03 and 12.25 t/ha (Gangwar, 2012).

## Other options for drought stress management

Strategies for drought protection, mitigation or relief include:

1. Dams - many dams and their associated reservoirs supply additional water in times of drought.
2. Watershed- It is very helpful in providing irrigation during drought spell.
3. Cloud seeding - a form of intentional weather modification to induce rainfall.
4. Desalination - of sea water for irrigation or consumption.
5. Drought monitoring - Continuous observation of rainfall levels and comparisons with current usage levels can help prevent man-made drought. Careful monitoring of moisture levels can also help predict increased risk for wildfires, using such metrics as the Keetch-Byram Drought Index or Palmer Drought Index.

6. Land use - Carefully planned crop rotation can help to minimize erosion and allow farmers to plant less water-dependent crops in drier years.
7. Mulching – It helps in retention of moisture in situ, which may utilise my succeeding crops.
8. Outdoor water-use restriction - Regulating the use of sprinklers, hoses or buckets on outdoor plants, filling pools, and other water-intensive home maintenance tasks.
9. Rainwater harvesting - Collection and storage of rainwater from roofs or other suitable catchments.
10. Recycled water - Former wastewater (sewage) that has been treated and purified for reuse.
11. Transvasement - Building canals or redirecting rivers as massive attempts at irrigation in drought-prone areas.
12. Paira or utera cropping – This method is also helpful for use of residual water for succeeding crops.
13. Use of PGRs – Use of different plant growth regulators (promoter and retardant), depends on different growth stages of plants may helpful in mitigating drought stress.
14. Use of Osmo protectants - Use of different osmoprotectants such as glycine betain, proline, trehalose etc may help plants to retain water inside the cell and help plants to cope up under drought conditions.
15. Anti-transpirants - Anti-transpirants are compounds applied to the leaves of plants to reduce transpiration. They preserve and protect plants from drying out too quickly.

## Conclusion

Adopting CA system (includes planting on raised beds) offers opportunities for crop diversification. Cropping sequences/rotations and agroforestry systems when adopted in appropriate spatial and temporal patterns can further enhance natural ecological processes which contribute to system resilience and reduced vulnerability to yield reducing disease/pest problems. Limited studies indicate that a variety of crops like mustard, chickpea, pigeon pea, sugarcane, etc., could be well adapted to the new systems with advantage. RCTs could very well be supplemented by rainwater conservation, increased bund height, irrigation scheduling, land leveling etc for in-situ moisture/water conservation and conjunctive use of rain, surface and groundwater to enhance water productivity.

Moreover, agroforestry systems can be useful in maintaining production during drier years, a common phenomenon in arid regions. In drought situations, deep root systems of trees are able to explore a large volume of water and nutrients, which help to maintain depleting soil moisture conditions to some extent which act as an insurance against drought.

## References

Asch, F., Dingkuhnb, M., Sow, A. and Audebert, A. (2005). Drought-induced changes in rooting patterns and assimilate partitioning between root and shoot in upland rice. Field Crops Research 93: 223–236.

Blokhina, O., Virolainen, E. and Fagerstedt, K.V. (2003). Antioxidants, oxidative damage and oxygen deprivation stress: a review. Annals of Botany 91: 179–194.

Chaves, M.M., Maroco, J.P. and Pereira, J.S. (2003). Understanding plant responses to drought—From genes to the whole plant. Functional Plant Biology 30: 239–264.

Cornic, G., Bukhov, N.G., Wiese, C., Bligny, R. and Heber, U. (2000). Flexible coupling between light-dependent electron and vectorial proton transport in illuminated leaves of C-3 plants. Role of photosystem-I dependent proton pumping. Planta 210: 468–477.

Davidson, E.A., Verchot, L.V., Cattanio, J.H., Ackerman, I.L. and Carvalho, H.M. (2000). Effects of soil water content on soil respiration in forests and cattle pastures of eastern Amazonian. Biogeochemistry 48: 53–69.

Earl, H. and Davis, R.F. (2003). Effect of drought stress on leaf and whole canopy radiation use efficiency and yield of maize. Agronomy Journal 95: 688–696.

FAO, (2008). An International Technical Workshop on Investing in Sustainable Crop Intensification: The Case for Improving Soil Health. In: Proceedings of FAO, Rome. 22-24 July 2008.

FAO, (2011). Climatic Risk Analysis in Conservation Agriculture in Varied Biophysical and Socio-economic Settings of Southern Africa. Rome: Food and Agriculture Organisation of the United Nations.

Farquhar, G.D., K.T. Hubick, A.G. Codon, and R.A. Richards. (1989). Carbon isotope fractionation and plant water use efficiency. p. 220–240. In P.W. Rundell et al. (ed.) Stable isotopes in ecological research. Springer-Verlag, New York.

Flexas, J., Bota, J., Loreto, F. (2004). Diffusive and metabolic limitations to photosynthesis under drought and salinity in $C_3$ plants. Plant Biology 6: 269–279.

Friedrich, T., Kassam, A.H. and Shaxson, F. (2009). Conservation Agriculture. In: Agriculture for Developing Countries. Science and Technology Options Assessment (STOA) project. European Technology Assessment Group, Karlsruhe, Germany.

Gangwar, B. (2012). An over view of PDFSR and Winter School, in: Singh, K.K., Singh, V.K. and Mishra, R.P. (Eds.), System Based Conservation Agriculture for Sustained Productivity and Soil Health, Project Directorate for Farming Systems Research, Modipuram, Meerut, U.P. pp. 408.

Harris, D., Tripathi, R.S. and Joshi, A. (2002). On-farm seed priming to improve crop establishment and yield in dry direct-seeded rice. In: Pandey S., Mortimer M., Wade L., Tuong T.P., Lopes K., Hardy B. (Eds.), Direct seeding: Research Strategies and Opportunities, International Research Institute, Manila, Philippines, pp. 231–240.

Heckathorn, S.A., De-Lucia, E.H. and Zielinki, R.E. (1997). The contribution of drought-related decreases in foliar nitrogen concentration to decreases in photosynthetic capacity during and after drought in prairie grasses. Physiologia Plantarum 122: 62–67.

IPCC, (2007). Climate Change 2007: Impacts, Adaptation and Vulnerability. Summary for Policymakers, IPCC AR4 WGII, Cambridge University Press, Cambridge, UK.

Kassam, A.H., Friedrich, T., Shaxson, F. and Pretty, J. (2009). The spread of Conservation Agriculture: Justification, sustainability and uptake. International Journal of Agricultural Sustainability 7(4): 292-320.

Kaya, M.D., Okçub, G., Ataka, M., Çýkýlýc, Y. and Kolsarýcýa, Ö. (2006). Seed treatments to overcome salt and drought stress during germination in sunûower (*Helianthus annuus* L.). European Journal of Agronomy 24: 291–295.

Komor, E. (2000). Source physiology and assimilate transport: the interaction of sucrose metabolism, starch storage and phloem export in source leaves and the effects on sugar status in phloem, Australian Journal of Plant Physiology 27: 497–505.

Lambers, H., Atkin, O.K. and Scheureater, I. 1996. Respiratory patterns in roots in relation to their function. In: Waisel Y. (Ed.), Plant Roots, The Hidden Half. Marcel Dekker, New York.

Leport, L., Turner, N.C., French, R.J., Barr, M.D., Duda, R. and Davies, S.L. (2006). Physiological responses of chickpea genotypes to terminal drought in a Mediterranean-type environment. European Journal of Agronomy 11: 279–291.

Lisar, S.Y.S., Rouhollah Motafakkerazad, Mosharraf, M. Hossain and Ismail, M.M.Rahman (2012). Water Stress in Plants: Causes, Effects and Responses, Water Stress, Ismail Md. Mofizur Rahman (Ed.), ISBN: 978-953-307-963-9.

MacAdam, J.W., Volenec, J.J. and Nelson, C.J. (1989). Effects of nitrogen on mesophyll cell division and epidermal cell elongation in tall fescue leaf blades. Plant Physiology 89: 549–556.

Marschner, H. (1995). Mineral nutrition of higher plants. 2nd edition, Academic Press, London.

Mengel, K. and Kirkby, E.A. (2001). Principles of plant nutrition. 5th edition, Springer Netherlands.

Nonami, H. (1998). Plant water relations and control of cell elongation at low water potentials. Journal of Plant Research 111: 373–382.

Palta, J.P. (2000). Stress Interactions at the Cellular and Membrane Levels. *Hort. Sci.* 25(11): 1377.

Roy, M.M., Tiwari, J.C. and Ram, M. (2011). Agroforestry for climate change adaptations and livelihood improvement in Indian hot arid regions. International Journal of Agriculture and Crop Sciences 3(2): 43-54.

Samarah, N.H. (2005). Effects of drought stress on growth and yield of barley, Agronomy for Sustainable Development 25: 145–149.

Sen Gupta, A. and Berkowitz, G.A. (1987). Osmotic adjustment, symplast volume, and non-stomatally mediated water stress inhibition of photosynthesis in wheat. Plant Physiology 85: 1040–1047.

Singh, K.K., (2011). Long-term effect of RCTs and crop residue management on productivity, profitability and soil health in rice-wheat cropping system. *In*: PDFSR Annual Report 2010-11.

Taiz, L. and Zeiger, E. (2006). Plant Physiology, 4th edition, Sinauer Associates Inc. Publishers, Massachusetts.

Turner, N.C. (1986). Crop water deficits: a decade of progress. Advances in Agronomy 39: 1–51.

Wahid, A. and Rasul, E. (2005). Photosynthesis in leaf, stem, ûower and fruit. In: Pessarakli M. (Ed.), Handbook of Photosynthesis, 2nd edition, CRC Press, Florida, pp. 479–497.

Waraich, E.A, Rashid, A., Saifullah, Ashraf, M.Y. and Ehsanullah (2011). Role of mineral nutrition in alleviation of drought stress in plants, Australian journal of crop science 6: 764-777.

Yadav, R.S., Hash, C.T., Bidinger, F.R., Devos, K.M. and Howarth, C.J. (2004). Genomic regions associated with grain yield and aspects of post ûowering drought tolerance in pearl millet across environments and tester background. Euphytica 136: 265–277.

Zhou, Y., Lam, H.M. and Zhang J. (2007). Inhibition of photosynthesis and energy dissipation induced by water and high light stresses in rice. Journal of Experimental Botany 58: 1207–1217.

Zinselmeier, C., Jeong, B.R. and Boyer, J.S. (1999). Starch and the control of kernel number in Maize at low water potentials. Plant Physiology 121: 25–35.

□□□

# 9

# Climate Change, Water Scarcity and Use of Waste Water for Sustainable Agriculture and Food Security

**V.D. Meena and M.L. Dotaniya**

## Introduction

Ground water (GW) resources in most areas of world are shrinking at an alarming rate and may not meet the ever increasing demands from agriculture and industry in future. Large decreases in GW levels (up to 200 m in some places) have been observed due to over exploitation (Toze, 2006). It accounts for more than 65% irrigation water and 85% of drinking water supplies lies in a critical state. Many areas of the world, already experiencing stress on water availability, which placed severe strains on existing resources with resulting environmental impacts. Half of the population of the developing world suffers from at least one disease caused by insufficient water supply and sanitation. In Australia, over-abstraction of fresh water for intensive irrigation activities has caused 33% of the land area to be at risk of salination. The Saudi Arabia is another example with demonstrated impacts on natural water resources due to increasing demands on GW by the agricultural sector (Bushnak, 2002). Estimates reveled that agriculture sector consumes maximum of the GW and ~80% of actual water resources are utilized in agriculture for irrigation purpose only. At majority of the locations in India, water level has declined at the rate of 1 meter per year (CGWB, 2011). Unsustainable GW depletion is a very serious issue and it is a very difficult problem to address. Reuse of the wastewater (WW) for agricultural irrigation purposes would reduces the amount of water that needs to be extracted from water resource. It is not only the potential solution to reduce the freshwater demand for zero water discharge, avoiding the pollution load in the receiving sources but also has a potential to supply essential nutrients to support plant growth (Weber *et al.* 1996). Estimates on the basis of 70% of the sewage available from large cities, shows that these effluents have the potential for irrigating (7.5 cm), about 21000 hectares of land on daily basis or alternately about 7.8 million hectares on annual basis (Minhas and Samra, 2004).

In India, encountering the problems of water scarcity and high cost of fertilizers, domestic WW is the better option to successfully use for irrigation. Many farmers in the areas near to urban/peri-urban localities are even compelled to use WW to irrigate their crops, due to absence of better alternatives (Ghimire, 1994). WW utilization for crop production, is an economic alternate that could substitute nutrient needs (Chaw and Reves, 2001; Rusan *et al.*, 2007) and water requirement of crop plants. Findings indicate that, the use of waste water could increase water resources potential for irrigation may prove to be beneficial for agricultural production.

## Status of wastewater generation, treatment capacity and current use

Estimates revealed an annual production of ~30 million tons of WW in the World (Cheraghi *et al.*, 2009). The per capita water supply in India is about 188.73 liter per day; whereas production of WW is 137.82 liter per person per day demonstrate the seriousness of the problem. With urban water use, only 15–25% of water diverted or withdrawn is consumed, the rest being returned as wastewater to the urban hydrologic system (Qadir *et al.*, 2010) that means almost 75-85% of the water supplied for domestic use, comes out as a wastewater. In India, an estimated 38,254 million litres per day (MLD) sewage is generated in major cities of India, but the sewage treatment capacity is only of 11,786 MLD (Kaur *et al.*, 2012). According to CPCB report, estimated sewage water generation from Class-I cities and Class-II towns together is 38,254 MLD, out of which only 11,554 MLD (32%) is being treated with a capacity gap of 26,467 MLD (68% of total generation), which needs urgent attention of all concerned as it has large number of open and covered channels which carry a mixture of WW generated by domestic, municipal and industrial activities. The projected WW generation estimates of 1,22,000 MLD for the country by 2050 (Bhardwaj, 2005). Cities around Ganga basin are generating 2637.7 MLD of sewage but are in a position to treat 1174.4 MLD *i.e.*, 44.2% only. The remaining sewage goes off in Ganga River without any treatment, which pollutes the Ganga. Thus, there is a large gap between generation and treatment of WW in India. Even the treatment capacity existing is also not effectively utilized due to operation and maintenance problem. Operation and maintenance of existing plants and sewage pumping stations is not satisfactory, as nearly

39% plants are not conforming to the general standards prescribed under the Environmental (Protection) Rules for discharge into streams as per the CPCB's survey report. In India, there are 234 sewage water treatment plants (STPs). Most of these were developed under various river action plans and are located in (just 5% of) cities/towns along the banks of major rivers (CPCB, 2005a). In a number of cities, the existing treatment capacity remains underutilized while a lot of sewage is discharged without treatment in the same city. Due to this, pollutants entering to the groundwater, rivers, and other water bodies. Apart from domestic sewage, about 13,468 MLD of WW is generated by industries (mostly large scale) of which only 60% is treated (Kaur *et al.,* 2012).

In India, 498 class I cities (cities having population more than one lac) generates 35558.12 MLD, which is 93% of total WW produced. All these cities having capacity to treat only 11,553.68 MLD, which is 32% of the sewage generated. There are 410 class II towns (population between 50,000 and 1,00,000) which generate 2,696.20 MLD sewage water and having capacity to treat just 233.7 MLD sewage *i.e.* less than 10% only. Nearly 52% class-I cities (260 out of 498) are located in Andhra Pradesh, Maharashtra, Tamil Nadu, Uttar Pradesh and West Bengal. The class-II towns are mostly located in Andhra Pradesh, Maharashtra, Tamilnadu, Uttar Pradesh and Gujarat (total 225, nearly 50%).

There are 35 metropolitan cities (with more than 10 lac population) generates 15,644 MLD of sewage. Out of 11,553.68 MLD sewage treatment capacities, 8,040 MLD is treated in these cities only *i.e.* 69%. This indicates that other than metropolitan cities, the capacity of remaining 463 Class-I cities is only 31%. Among the metropolitan cities, Delhi has the maximum treatment capacity that is 2,330 MLD (30% of the total treatment capacity of metropolitan cities) and next, Mumbai has the capacity of 2,130 MLD, which is 26% of total capacity. Delhi and Mumbai therefore in combination have 55% of treatment capacity of the metropolitan cities.

Worldwide more than 800 million farmers are engaged in urban agriculture. Of this group, about 200 million practice market oriented farming on open spaces, often using poor-quality irrigation water in the absence of good-quality water (Qadir *et al.,* 2010). Reliable estimates of projected WW use in agriculture are needed for better planning and managing risks, but limited information makes estimating future use difficult (Qadir *et al.,* 2007a). At present there are no clear estimates of the extent of irrigation with urban WW. Some people say it is an insignificant source of water for agriculture because the amounts of water diverted to cities and later disposed as WW are small in relation to the amount of water needed for agriculture in most developing countries. Jimenez and Asano (2004) and IWMI (2006) suggest that at least 3.5Mha are irrigated globally with untreated, partly treated, diluted, or treated wastewater. Others claim that worldwide, more than 20Mha are irrigated with urban WW, and that WW has an important impact on agricultural productivity and livelihoods (Van der Hoek, 2002). In India, estimates revealed that ~73,000 ha were irrigated with WW during early nineties and presently the area under WW irrigation is on the rise (Strauss and Blumenthal, 1990). It has been estimated that in India sewage waters can annually irrigate about 1.0 Mha (Sengupta, 2008) to 1.5 Mha of land area and have a potential to contribute about one million tons of nutrients and 130 million man-days of employment (Minhas and Samra, 2004). The estimates of WW irrigated area for direct use were about 6 ha per MLD and for indirect use 39 ha per MLD (IWMI, 2013). The area under indirect use accounts for mixing with non-wastewater sources of irrigation. Using these volume-area relationships, the data for class-I cities and class-II towns indicate that the potential irrigable land can be estimated to be around 1.1Mha (Table 1). In Hyderabad alone the approximately 40,500 ha area irrigated with WW through Musi river (Van der Hoek, 2004). It has been reported that irrigation with sewage or sewage mixed with industrial effluents results in saving of 25-50% of N and P fertilizer and leads to 15-27% higher crop productivity, over the normal water (Anon, 2004). In peri-urban areas, farmers usually adopt year round, intensive vegetable production systems (300-400% cropping intensity) or other perishable commodity like fodder and earn up to 4 times more from a unit land area compared to freshwater (Minhas and Samra, 2004). Besides crop farming, WW is used also for aquaculture in Africa, and in Central, South, and Southeast Asia (Bangladesh, Cambodia, China, India, Indonesia, and Vietnam). In many areas, treated WW is used for fodder production, groundwater recharge or other environmental purposes, such as enhancing water supply for wetlands, wildlife refuges, riparian habitats, and urban lakes and ponds. A recent global survey found that vegetables (32% frequency of responses) are besides cereals (27%) the most common crops produced with diluted or raw WW (Raschid-Sally and Jayakody, 2007).

**Table 1.** Estimates on potential irrigable land with wastewater in Class-I cities and Class-II towns (YUVA, 2006)

| | Volume of WW (MLD) | Ratio of direct versus indirect use | Potential irrigable land (ha) |
|---|---|---|---|
| Treatment capacity | 11,787 | 6 | 70,722 |
| Untreated | 26,467 | 39 | 1,032,213 |

## Composition of sewage water

The composition of sewage water is quite variable depending upon the contributing source, mode of collection and treatment provided. Although a large proportions of these sewage water is organic in nature and contains essential plant nutrients but sometimes toxic metals are also present in appreciable amounts. The sewage water generated in India contains more than 90% water. The solid portion contains 40-50% organics, 30-40% inert materials, 10-15% bio-resistant organics and 5-8% miscellaneous substances on oven dry weight basis (Antil and Narwal, 2008). The composition of sewage water is not constant and changes within the year due to several factors. The extent of the contaminants will be least during the rainy seasons. The variation in the composition of sewage water within season has also been reported by Singh and Kansal (1985b) for different cities of Punjab. During the same year, the pH of the sewage water of one location of Hisar varied from 7.3 to 8.7 and EC from 0.52 to 1.55 dS $m^{-1}$. The concentration of metals in sewage water is season specific. For example, Pb content varied from 8.1 to 30 mg $L^{-1}$ during monsoon and 11.0 to 41.3 mg $L^{-1}$ during summer. Similarly, Cd ranged from 4.0 to 22.0 mg $L^{-1}$ in summer season while it was not detected during monsoon and Cr content in raw sewage during summer was 7.2 mg $L^{-1}$ while in monsoon it was 1.6 mg $L^{-1}$ (Antil and Narwal, 2008).

## Nutritional value of sewage water

Sewage water has been considered as low price fertilizer because of its high NPK content (Chaw and Reves, 2001). Although a wide variation in nutrient concentrations, WW contained 48.3, 7.6, 72.4 and 34.6 mg $L^{-1}$ of N, P, K and S respectively; besides their micronutrient manurial value of 0.34 mg Zn, 10.8 mg Fe, 0.2 mg Cu and 0.36 mg Mn $L^{-1}$ (Chhabra, 1989). Thus, five irrigations of 7.5 cm each with SW could add about 181, 29, 270 and 130 kg $ha^{-1}$ of N, P, K and S, respectively which is adequate to meet the nutrient requirement of the crops. Later on Minhas and Samra (2004) estimated an average content of N, P and K in SW and ~70% utilization in agriculture sector shows that WW can annually contribute 380, 60, 520 and 1.4 thousand tons of N, P, K and Zn, respectively, in addition to other micro-nutrients. The total value of these nutrients would be around 1.78 million US$ annually. Yadav *et al.* (2003) also reported a nutrient potential of 8,100, 1,200 and 11,000 tons of N, P and K through sewage discharges of 485 MLD that can be used for irrigation (supplemented) to an area of 16,000-32,000 ha/annum in peri-urban areas in Haryana. Phung *et al.* (2009) proved that WW was rich in N, P, K and analysis of soil samples at harvest time showed that total N, P and K in rice irrigated with WW were significantly higher than plots without WW application. The concentration of Fe (4.81-7.26 mg $L^{-1}$), Mn (0.45-1.17 mg $L^{-1}$), Zn (0.63-2.00 mg $L^{-1}$), Cu (0.024-1.18 mg $L^{-1}$) and B (2.11-5.75 mg $L^{-1}$) was higher in sewage effluents indicating that they are good source of plant micronutrients which may help in mitigating emerging problems of their deficiency that are otherwise overcome by application of costly chemical fertilizers (Kharche *et al.* 2011). Wastewater irrigation in the Tula Valley in Mexico provides 2400kg organic matter, 195kg N, and 81kg P $ha^{-1}$ $yr^{-1}$, contributing to significant increases in crop yields (Jimenez, 2005). In an assessment done by CPCB, for coastal class-I cities and class-II towns, the annual value of the N, P and K loads from a total of about 5,000 MLD of WW was estimated at INR 1,091 million.

## Impact of sewage water irrigation

In India, land around the cities receiving sewage waters containing both the domestic and industrial wastes. These wastes are suitable for crop production provided the content of major plant nutrients is high and that of toxic elements is low. Its long-term application would affect the physical, chemical and biological properties of soils (Antil *et al.,* 2007).

## Impact on physical properties of soil

Wastewater irrigation has shown improvement in the physico-chemical properties of the soil as compared to that of resulted from the groundwater application. Kharche *et al.* (2011) find out that bulk density and hydraulic conductivity of soils irrigated with WW was higher as compared to soils irrigated with ground water (Table 2). In contradictory, Mathan (1994) reported significantly lower bulk density and enhanced hydraulic conductivity with sewage irrigation for 15 years. This might be due to improvement in total porosity and aggregate stability in the sewage-irrigated soils due to addition of organic matter; which plays an important role in improving soil physical environment. The soil moisture retention of sewage-irrigated soils was slightly higher due to addition of organic matter through sewage. Rattan *et al.* (2001) observed enhanced available water content in the soils due to increased water holding capacity of soil.

**Table 2.** Mean values of physical properties of sewage and well-irrigated soils (Kharche et al. 2011)

| Soil depth (cm) | Bulk density ($Mg\ m^{-3}$) | | Hydraulic conductivity ($cm\ h^{-1}$) | | Water retention (%) | | | |
|---|---|---|---|---|---|---|---|---|
| | | | | | 33 kPa | | 1500 kPa | |
| | WW | GW | WW | GW | WW | GW | WW | GW |
| 0-30 | 1.28 | 1.40 | 1.22 | 1.12 | 43.2 | 38.0 | 15.4 | 13.4 |
| 30-60 | 1.30 | 1.38 | 1.19 | 0.94 | 45.5 | 42.6 | 18.5 | 15.7 |

## Impact on chemical properties of soil

### Soil pH

There are different views associated with effects of long-term WW application on soil pH. There are studies (Rusan *et al.*, 2007; Osaigbovo *et al.* 2006), which confirms increase in soil pH with WW irrigation of soil. A long-term study on the use of sewage WW for 50-60 years, 0.5 unit increase in soil pH has been reported than soils irrigated with GW (Gupta and Mitra, 2002; Saravanamoorthy and Kumari, 2007). Such effect may be attributed to the high content of basic cations *viz.* $Na^+$, $Ca^{2+}$ and $Mg^{2+}$ in the WW. On contradictory, the soil pH decreased by 0.38 units as a result of sewage water irrigation (Rana *et al.*, 2010). The decrease of about one unit in pH with sewage water irrigation was also observed in soils of Haryana (Narwal *et al.*, 1993). Ramesh (2003) reported that soils irrigated with sewage water recorded higher pH and EC as compared to soils irrigated with normal water.

## Electrical conductivity (EC)

Significantly maximum EC value of soil irrigated with municipal WW application consecutively for 10-years has been reported (Rusan *et al.*, 2007). A study in the arid climate on the use of municipal effluents in soils for 16-months recapitulated increased EC value by 0.5 dS $m^{-1}$, and the ex-changeable sodium percentage (ESP) by 6.8 (Hayes *et al.*, 1990) . In a potato cultivated soils of Jalandhar (India), an increase in EC from 0.28 to 0.64 dS $m^{-1}$ has been reported with the use of municipal WW contaminated by leather effluents (Brar *et al.*, 2002). Ambika *et al.* (2010) suggested that WW application increases the soil salinity, organic carbon, N, K, Ca, Mg cations to a lot. Soil is a bio-filter that can reduce a large part of domestic WW pollutants, but this filtering increased EC, SAR, Na, Ca and Mg of soil (Darvishi *et al.*, 2010). Bhise *et al.* (2007) reported that wastewater irrigation had adverse effect on some soil properties like EC, ESP and soluble cations (Na/K and Na/Mg) ratios.

## Soil organic carbon (SOC)

Due to sewage irrigation increase in the soil organic carbon content is a positive sign of improved soil health to certain extent. Long-term use of sewage water to crops results in significant increase in SOC than soils irrigated with GW (Brar *et al.,* 2002; Rusan *et al.*, 2007). In sub-tropical Indian soils, an increase of 38-79% in SOC content with 20-years of sewage WW irrigation as compared to GW irrigated soils has been reported (Rattan *et al.*, 2005). Similarly, after 36-years of domestic sewage effluent irrigation, significant improvement in the soil organic matter (OM) from 1.24-1.78%, especially down to a distance of one Km along the disposal channel, has been reported (Yadav *et al.,* 2002). Sewage WW irrigation for 50-60 years in soils of Kolkatta has exhibited an increase in SOC from 0.19 to 0.37% (Gupta and Mitra, 2002) and from 1.24 to 1.73% after 25 years in soils of Kurukshetra (Yadav *et al.*, 2002). Large additions of OM from sewage WW and anaerobic conditions developed due to heavy loading of OM had reduced OC decomposition and have resulted in build-up of soil OC. Therefore, long-term WW irrigation can be a good means of carbon sequestration in soil and can thus be referred as a soil quality sustaining practice. Sewage irrigated soil has 0.882% organic carbon compared to 0.707% of tube well irrigated soil at Rohtak (Rana *et al.*, 2010). Untreated sewage water contained a significant amount of C (116-257 mg/$L^{-1}$), and its application for 5 years resulted significant increase in soil organic C (SOC) in the soil profile up to 30 cm (Saha *et al.*, 2010). The organic carbon of sewage irrigated soil has been increased from 5.0 to 5.90 g $kg^{-1}$ and higher than well water irrigated soil (Singh *et al.*, 2013).

## Impact on soil fertility parameters

Improvement in soil fertility with WW irrigation over a period of time has been well documented (Table 3-6). After 10-years of municipal WW application, significant increase in soil available N, P and K content in surface (0-20 cm) soil has been reported (Rusan *et al.*, 2007). In 36-years long-term experiment on the use of domestic sewage WW, the build-up of 2908 kg TN $ha^{-1}$, 58 kg Av-P $ha^{-1}$, 2115 kg TP $ha^{-1}$, 305 kg Av-K $ha^{-1}$ and 4712 kg TK $ha^{-1}$ in surface soil has also been reported (Rusan *et al.*, 2007).

**Table 3.** Effect of waste water irrigation on soil fertility parameters

| Location | Time period of study | Waste water type | Parameter | Content WW | Content GW | Reference |
|---|---|---|---|---|---|---|
| Pasupathy Palayam, Karur (TN) | 50 days | Textile waste water | pH | 7.92 | 7.85 | Saravanamoorthy and Ranjitha-Kumari (2007) |
| | | | N (kg $ha^{-1}$) | 118 | 117 | |
| | | | P (kg $ha^{-1}$) | 13.45 | 11.10 | |
| | | | K (kg $ha^{-1}$) | 56.0 | 55.0 | |
| | | | OM (%) | 1.29 | 0.29 | |
| Kolkatta | 50-60 years | Sewage Effluents | pH | 7.7 | 7.2 | Gupta and Mitra (2002) |
| | | | OC (%) | 0.37 | 0.19 | |
| | | | TN (%) | 0.10 | 0.06 | |
| | | | TP (%) | 0.10 | 0.05 | |
| | | | TK (%) | 0.13 | 0.07 | |
| Kurukshetra | 25 years | Sewage Effluents | pH | 8.1 | 8.3 | Yadav *et al.* (2002) |
| | | | OC (%) | 1.73 | 1.24 | |
| | | | TN (%) | 0.15 | 0.08 | |
| | | | TP (%) | 0.88 | 0.56 | |
| | | | TK (%) | 0.24 | 0.18 | |
| Bhopal | 5 years | Sewage Effluents | pH | 7.74 | 7.76 | Saha *et al.* (2010) |
| | | | OC (%) | 0.6 | 0.5 | |
| | | | TN (%) | 0.71 | 0.66 | |
| | | | TP (%) | 0.19 | 0.14 | |
| | | | TK (%) | 1.87 | 1.94 | |
| Rohtak | 35 years | Sewage Effluents | pH | 7.84 | 8.15 | Rana *et al.* (2010) |
| | | | OC (%) | 0.88 | 0.70 | |
| | | | P (kg $ha^{-1}$) | 13.69 | 11.27 | |
| | | | K (ppm) | 20.14 | 8.13 | |

**Table 4.** Impact of Sewage wastewater and groundwater (GW) irrigation on chemical properties of soil after harvest (Ladwani *et al.*, 2012)

| Crops | Irrigation sources | pH | EC (dS $m^{-1}$) | Organic matter (g $kg^{-1}$) | Avail. Nitrogen (Kg $ha^{-1}$) | Avail. Phosphorus (Kg $ha^{-1}$) | Avail. Potassium (Kg $ha^{-1}$) |
|---|---|---|---|---|---|---|---|
| Wheat | SW | 7.62 | 0.50 | 5.92 | 296.26 | 26.92 | 343.25 |
| | GW | 7.45 | 0.45 | 5.53 | 266.51 | 18.13 | 323.26 |
| Gram | SW | 7.66 | 0.55 | 5.49 | 238.92 | 22.87 | 317.82 |
| | GW | 7.57 | 0.47 | 5.38 | 248.78 | 17.76 | 345.20 |
| Palak | SW | 7.62 | 0.53 | 5.46 | 238.65 | 23.87 | 332.92 |
| | GW | 7.49 | 0.44 | 5.24 | 205.92 | 20.13 | 310.15 |

**Table 5.** Effect of sewage and groundwater (GW) irrigation on chemical properties of soil after harvest (Singh *et al.*, 2013)

| Crops | Irrigation sources | pH | EC (dS $m^{-1}$) | Organic matter (g $kg^{-1}$) | Avail. Nitrogen (kg $ha^{-1}$) | Avail. Phosphorus (kg $ha^{-1}$) | Avail. Potassium (kg $ha^{-1}$) |
|---|---|---|---|---|---|---|---|
| Wheat | SW | 7.62 | 0.48 | 5.90 | 283.26 | 23.92 | 343.25 |
| | GW | 7.59 | 0.45 | 5.51 | 265.51 | 19.13 | 322.26 |
| Gram | SW | 7.60 | 0.50 | 5.79 | 228.92 | 21.87 | 327.82 |
| | GW | 7.57 | 0.47 | 5.38 | 248.78 | 17.76 | 345.20 |
| Palak | SW | 7.62 | 0.51 | 5.76 | 222.65 | 21.87 | 325.92 |
| | GW | 7.59 | 0.48 | 5.34 | 205.92 | 19.13 | 315.15 |
| Methi | SW | 7.63 | 0.51 | 5.59 | 268.64 | 19.13 | 324.79 |
| | GW | 7.60 | 0.48 | a5.17 | 288.50 | 23.23 | 342.13 |
| Berseem | SW | 7.52 | 0.55 | 5.83 | 250.87 | 19.13 | 322.27 |
| | GW | 7.50 | 0.52 | 5.42 | 216.38 | 13.66 | 299.41 |

**Table 6.** Nutrient additions to soil, when irrigating with treated wastewater (Lazarova and Bahri, 2005)

| **Nutrient** | **Concentration (mg $L^{-1}$)** | **Fertilizer** | |
|---|---|---|---|
| | | **Irrigation at 3000 $m^3$ $ha^{-1}$** | **Irrigation at 5000 $m^3$ $ha^{-1}$** |
| Nitrogen | 16–62 | 48–186 | 80–310 |
| Phosphorus | 4–24 | 12–72 | 20–120 |
| Potassium | 2–69 | 6–207 | 10–345 |
| Calcium | 18–208 | 54–624 | 90–1040 |
| Magnesium | 9–110 | 27–330 | 45–550 |
| Sodium | 27–182 | 81–546 | 135–910 |

(*Note*: Data describing nutrient concentrations in treated wastewater and the volume of irrigation water applied)

## Impact on yield and quality of crops

Wastewater contains essential plant nutrients and thus its reuse in agriculture serves as an important source of nutrients and irrigation water for crops. Several studies reported that application of sewage water to agricultural land improve the physical, chemical and biological characteristics of soils and provide essential plant nutrients such as N, P, S and micronutrients, which increased the crop yield (Katterman and Day, 1989; Panicker, 1995). The results of many studies on the use of WW for long period of time have recapitulated significant increased in crop yields. The use of WW for irrigation to maize, sunflower, groundnut and soybean registered 19.3, 29.9, 5.9 and 4.8% higher grain yield, respectively and from 6.9 to 13.9% increase in different sugarcane varieties (Udayasoorian *et al.*, 1999). A long term experiment have shown that sewage WW irrigation had highest grain yield of wheat, rice and cotton by 23, 46 and 50% than GW irrigation (Sharma and Kansal, 1984). The highest yield of cauliflower (28.53 t ha$^{-1}$) and red cabbage (46.86 t ha$^{-1}$) were obtained with the untreated wastewater (Kiziloglu *et al.*, 2008). Short duration (5 years) of municipal sewage water irrigation did not have any harmful effect on soil quality as well as crop productivity except for slight build-up in coliform population; rather, it proved beneficial in improving soil fertility, productivity, and produce quality and significantly increased grain and straw yield of the wheat crop by 31% and 43%, respectively (Saha *et al.*, 2010). Application of domestic wastewater increased the yield of wheat, gram, palak, methi, berseem compared to irrigation with ground water and it also increases total N, P, K and organic carbon content of soil (Ladwani *et al.*, 2012, Singh *et al.*, 2013) (Table 7).

**Table 7.** Effect of sewage and groundwater irrigation on crop yield and quality after harvest

| Crop | Irrigation sources | Test weight (1000 seeds) | Grain yield (q ha$^{-1}$) | Straw yield (q ha$^{-1}$) | References |
|---|---|---|---|---|---|
| Wheat | SW | 49.1 | 38.98 | 51.51 | Saha *et al.* (2010) |
| | GW | 45.1 | 29.77 | 35.99 | |
| | SW | 3.70 | 17.37 | 22.81 | Ladwani *et al.* (2012) |
| | GW | 3.53 | 17.14 | 22.33 | |
| Gram | SW | 16.70 | 13.18 | 25.03 | |
| | GW | 16.23 | 13.88 | 25.42 | |
| Palak | SW | 1.40 | 9.36 | 8.57 | |
| | GW | 1.13 | 9.02 | 8.06 | P.K. Singh *et al.* (2013) |
| Methi | SW | 1.41 | 9.11 | 10.28 | |
| | GW | 1.31 | 9.68 | 10.51 | |
| Berseem | SW | 0.29 | 2.68 | 8.49 | |
| | GW | 0.25 | 2.15 | 8.06 | |
| Foxtail millet | WW | 1.96 | 61.37 | 74.43 | Asgharipour and Azizmoghaddam, (2012) |
| | 50% diluted | 2.01 | 67.96 | 82.73 | |
| | Raw Sewage | 2.03 | 75.11 | 91.09 | |

Application of sewage water resulted remarkable increase in the mean plant height (3.4%), number of tiller plant$^{-1}$ (31.8%), length of ear (18.8%) in wheat. As a result, straw as well as grain yields also were increased significantly by 43.1 and 34.3%, respectively due to application of sewage water (ISSS Annual Report, 2006-07). Mahida *et al.* (1981) also reported increased yield of different vegetable crops with the use of sewage water (Table 8).

**Table 8.** Effect of sewage irrigation on yield (t ha$^{-1}$) of vegetable crops (Mahida, 1981)

| Crop | Source of irrigation | |
|---|---|---|
| | Canal water | Sewage water |
| Beet root | 8.75 | 16.27 |
| Carrot | 9.71 | 11.75 |
| Radish | 7.26 | 8.33 |
| Potato | 6.12 | 9.33 |
| Ginger | 6.04 | 9.80 |
| Knolkhol | 9.70 | 16.57 |
| Cabbage | 9.27 | 12.13 |
| Cauliflower | 6.96 | 9.09 |
| French beans | 6.63 | 8.06 |
| Tomato | 10.01 | 13.38 |
| Tobacco | 1.12 | 1.25 |
| Groundnut | 2.88 | 2.90 |

Yield attribute parameters of raw sewage effluent treatments were generally higher than those found with the 50% diluted sewage effluent treatments, whereas the well water treatments exhibited the lowest values. The 1000 grain weight was slightly higher whereas, the grain and biological yields for the plants irrigated with raw sewage were 10.52 and 10.46% greater, respectively; than the yields of the plants irrigated with 50% diluted sewage. The grain and biological yields of the plants irrigated with 50% diluted sewage were also 10.73 and 11.15% higher, respectively, than the well water irrigation (Asgharipour and Azizmoghaddam, 2012) (Table 7). Similarly, the effects of different concentrations of sewage water (0, 25, 50 and 100%) on growth performance, leaf area, biomass and nutrient accumulation of *Hardwickia binata* growing under nursery conditions was found maximum at 50% concentration over others whereas, the sewage water concentration at 75 and 100% retarded the growth of *H. binata* seedlings (Paliwal et al.

1998). Textile waste water application increased germination, chlorophyll a, b and total content, growth parameters, yield and yield contributing parameters (Saravanamoorthy and Ranjitha Kumari, 2007).

## Groundwater recharge

The leaching and drainage of wastewater, applied for crop irrigation, to groundwater aquifer may serve as a source of groundwater recharge. In some regions, 50-70 percent of irrigation water may percolate to groundwater aquifer (Rashed *et al.*, 1995). The influence of percolated wastewater on groundwater quality and its recharge is thus likely to be substantial. Despite poor quality, groundwater recharge through wastewater application can be a vital environmental and economic service in regions where freshwater supplies are limited and groundwater removal rates exceed replenishment rates.

## Ecological impacts

When drainage water from WW irrigation schemes drains particularly into small confined lakes and water bodies and surface water, and if phosphates in the orthophosphate form are present, the remains of nutrients may cause eutrophication. This causes imbalances in plant microbiological communities of water bodies (Smith *et al.*, 1999). This may in turn affect other higher forms of aquatic life and influence the presence of water birds and reduce biodiversity. Sofar as these water bodies serve local communities for their needs, the ecological impacts can be translated into economic impacts, which can be quantified.

## Social impacts and poverty

General concerns such as nuisance, poor environmental quality, poor hygiene, odor, noise, higher probability of accidents, etc. Social concerns such as food safety, health and welfare, impaired quality of life, loss of property values, and sustainability of land use. Natural resource concerns such as pollution of vital water resources, loss of fish, wildlife, exotic species, etc. These concerns may be addressed with appropriate educational and public awareness programs.

Wastewater is a critical resource for livelihoods in peri-urban areas. More than 10% of the world's population consumes foods produced by irrigation with wastewater (WHO, 2006). The benefits of wastewater irrigation are both direct and indirect. The direct benefits accrue due to more water for irrigation and inexpensive nutrients in wastewater, which together may result in higher cropping intensity, more cropped area, higher yield and production, additional employment, and enhanced food security for the local population (Hanjra and Gichuki, 2008). For instance, peri-urban areas that use wastewater for crop production also provide employment for women and other landless laborers. It also enables crop specialization and year-round production. Landless farmers who lease agricultural land for horticultural nurseries can afford a better standard of living for their families and contribute towards improving the quality of urban environment by supplying nursery plants for landscaping. Wastewater irrigation may have secondary or indirect benefits due to spin off at regional or national level. The indirect impacts of wastewater irrigation on employment, income levels and its distribution and social effects such as human capital and equity may be as equally important as the direct and immediate benefits.

### Future research needs

i) To develop eco-friendly technology for the use of sewage and industrial effluents to improve crop productivity and soil quality and to protect quality of farm produce and environment from degradation.

ii) Research should be focused to study the long term effects of wastewater on human, animal health and soil biodiversity.

iii) The studies on the socioeconomic impacts of wastewater reuse for crop production are rare. Future research is needed to explore the impacts of reuse on poverty and social equity.

iv) To work out the environmental cost due to the use of wastewater irrigation.

## Conclusion

Based on the literatures thus reviewed, it can be concluded that sewage WW could be a valuable, nutritive, economic and alternative irrigation source to fresh water and as a source of fertilizers, since it has high contents of both organic matter and nutrients (N, P and K). Continuous use of sewage irrigation recorded improvement in water retention, hydraulic conductivity, organic C and buildup of available N, P, K and micronutrient status and soil microbial count. These findings thus, recapitulates that WW have great potential as manure when used for irrigation to crops. Periodic monitoring of chemical composition of sewage water, soil and crop produce is however, suggested for its safe and long term use and to prevent the entry of toxic metals in food chain.

For avoiding the possibility of contamination entering into humans through the food chain, further studies are needed to determine the extent of heavy metals in edible plants and crops, which will be helpful to evaluate the health and safety of crop quality in sewage irrigation areas.

## References

Ambika SR, Ambica PK, Govindaiah (2010). Crop growth and soil properties affected by sewage water irrigation: A review. Agril Reviews 31(3): 203-209.

Anonymous (2004). Annual Progress Report (2000-03) NATP-MM Project on "Use of urban and industrial effluents in agriculture". CSSRI, Karnal-Haryana, India.

Antil RS (2012). Impact of sewage and industrial effluents on soil-plant health, industrial waste. DOI: 10.5772/37403. Available at http://www.intechopen.com/books/industrial-waste/impact-of-sewer-water-and-industrial-wastewaters-on-soil-plant-health.

Antil RS, Dinesh, Dahiya SS (2007). Utilization of sewer water and its significance in INM. Proceedings of ICAR sponsored Winter School on Integrated Nutrient Management, Department of Soil Science and Directorate of Human Resource Management, CCS Haryana Agricultural University, Hisar, India, Dec. 4-24 pp 79-83.

Antil RS, Narwal RP (2008). Influence of sewer water and industrial effluents on soil and plant health. In: Puranik VD, Garg VK, Kaushik A, Kaushik CP, Sahu SK, Hegde AG, Ramachandarn TV, Saradhi IV, Prathibha P, (ed) Groundwater resources: Conservation and management, Department of Environmental Science and Engineering, GJU Science and Technology, Hisar, India, pp 37-46.

Asgharipour MR, Azizmoghaddam HR (2012). Effects of raw and diluted municipal sewage effluent with micronutrient foliar sprays on the growth and nutrient concentration of foxtail millet in southeast Iran. Saudi J Biol Sci 19: 441-449.

Bhardwaj RM (2005). Status of wastewater generation and treatment in India. IWG-Env, International Work Sessionon Water Statistics, Vienna, June 20-22, 2005 pp 9. Available at http://unstats.un.org/unsd/environment/envpdf/ pap_wasess3b6india.pdf

Bhise PM, Challa O, Venungopalan MV (2007). Effect of waste water irrigation on soil properties under different land use systems. J Indian Soc Soil Sci 55(3): 254-258.

Brar MS, Khurana MPS, and Kansal BD (2002). Effect of irrigation by untreated sewage effluents on the micro and potentially toxic elements in soils and plants. In Proc 17th World Congress of Soil Science held at Bangkok, Thailand from August 14-21, Vol.-IV, Symposium no.-24 198: 1-6.

Bushnak AA (2002). Future strategy for water resources management in Saudi Arabia. In: Proceeding of A Future Vision for the Saudi Economy Symposium, Riyadh, 12–23 October, p 37.

Central pollution control board (CPCB) (2009). Sewage treatment in India. CPCB, New Delhi.

CGWB (2011). Ground Water Year Book-India 2010-11. Central Ground Water Board, Ministry of Water Resources. Government of India.

Chaw R, Reves AS (2001). Effect of waste water on *Menthe piperita and Spinaceal Oleraceae*. J Environ Biol 51: 131-145.

Cheraghi M, Lorestani B, Yousefi N (2009). Effect of waste water on heavy metal accumulation in Hamedan Province Vegetables. Int J Bot 5: 190-193.

Chhabra R (1989). Sewage water utilization through forestry. Na-tional Printers, New Delhi, India. pp 1-9.

CPCB (1978). The water (prevention and control of pollution) cess rules, 1978 ministry of works and housing. Central Pollution Control Board, New Delhi, India.

CPCB (1988). Assessment of Vehicular Pollution in Metropolitan Cities Part - I, Abridge Report, Central Pollution Control Board, New Delhi, India.

CPCB (1995). Status of Water supply and wastewater generation, collection, treatment and disposal in metro cities (1994-95), Central Pollution Control Board, New Delhi, India.

CPCB (2005a). Parivesh Sewage Pollution – News Letter. Central Pollution Control Board, Ministry of Environment and Forests, Govt.of India, NewDelhi.

CPCB (2007). Annual Report 2006-2007-Central Pollution Control Board, Ministry of Environment and Forests, New Delhi, India.

Darvishi HH, Manshouri M, Farahani HA (2010). The effect of irrigation by domestic waste water on soil properties. J Soil Sci Environ Manage 1(2): 030–033.

Ghimire SK (1994). Evaluation of industrial effluents toxicity in seed germination and seedling growth of some vegetables, M.Sc. Dissertation, Central Department of Botany (1994), Tribhuvan University, Kirtipur, Kathmandu, Nepal.

Government of Jordan (2003). Technical Regulations for Reclaimed Domestic Waste Water JS893/2002, Jordan Institution for Standards and Meteorology, Amman, Jordan.

Gupta SK, Mitra A (2002). *In:* Advances in Land Resource Management for 21st Century, Soil Conservation Society of India, New Delhi, India, pp 446-460.

Hanjra M.A, Gichuki F (2008). Investments in agricultural water management for poverty reduction in Africa: case studies of Limpopo, Nile, and Volta river basins. Natural Resources Forum 32 (3): 185-202.

Hayes AR, Mancino CF, Pepper LL (1990). Irriga-tion of turf grass with secondary sewage effluent: I. Soil and leachate water quality. Agron J 82: 939-943.

IISS Annual Report (2006-07). Recycling and rational use of different waste in agriculture and remediation of contaminated soils. Annual Report of Indian Institute of Soil science (ISSS), Bhopal, India, pp 72-83.

IWMI (2006). Recycling realities: managing health risks to make wastewater an asset. Water Policy Briefing 17. IWMI, Colombo, Sri Lanka.

IWMI (2013). Wastewater and agricultural reuse challenges in India. IWMI research report 147.

Jimenez B (2005). Treatment technology and standards for agricultural wastewater reuse: a case study in Mexico. Irrig Drain 54 (Suppl. 1): S22-S33.

Jimenez B, Asano T (2004). Acknowledge all approaches: the global outlook on reuse. Water 21: 32–37.

Katterman FRH, Day AD (1989). Plant growth factors in sewage sludge. Biocycle 3: 64-65.

Kaur R, Wani SP, Singh AK, Lal K (2012). Wastewater production, treatment and use in India, presented at the second regional workshopof the project 'Safe use of wastewater in agriculture', 16–18 May2012, New Delhi.

Kharche VK, Desai VN, Pharande AL (2011). Effect of sewage irrigation on soil properties, essential nutrients and pollutant element status of soils and plants in a vegetable growing area around Ahmednagar city in Maharashtra. J Indian Soc Soil Sci 59: 177-184.

Kiziloglu FM, Turan M, Sahin U, Kuslu Y, Dursun A (2008). Effects of untreated and treated wastewater irrigation on some chemical properties of cauliflower (*Brassica olerecea* L. var. *botrytis*) and red cabbage (*Brassica olerecea* L. var. *rubra*) grown on calcareous soil in Turkey. Agricultural Water Manage 95: 716-724.

Ladwani Kiran D, Ladwani Krishna D, Vivek S and Manik DS Ramteke (2012). Impact of domestic wastewater irrigation on soil properties and crop yield. Int J Scientific Res. Publications 10 (2): 1-7.

Lazarova V, Bahri A (2005). Water reuse for irrigation: agriculture, landscapes, and turf grass. CRC Press, Boca Raton, USA.

Mahida UN (1981). Influence of sewage irrigation on vegetable crops. In: Water pollution and disposal of waste water on land, Tata Mcgrew Hill Pub., New Delhi.

Mathan KK (1994). Studies on the influence of long-term municipal sewage effluent irrigation on soil properties. Bio Technol 48: 275-276.

Minhas PS, Samra JS (2004). Wastewater use in peri-urban agriculture impacts and opportunities. *Technical Bulletin* 02/2004. Central Soil Salinity Research Institute (CSSRI), Karnal, Haryana, India.

Narwal RP, Gupta AP, Anoop-Singh, Karwasra SS (1993). Composition of some city waste waters and their effect on soil characteristics. Ann Biol 9:239-45.

Osaigbovo A, Ulamen O, Ehi-Robert (2006). Influence of pharmaceutical effluent on some soil chemical properties and early growth of maize (*Zea mays L)*. African J Biotech 5: 612-617.

Paliwal K, Karunaichamy KSTK, Ananthavalli M (1998). Effect of sewage water irrigation on growth performance, biomass and nutrient accumulation in *Hardwickia binata* under nursery conditions. Bioresour Technol 66: 105-111.

Panicker PVRC (1995). Recycling of human waste in agriculture. In: Tandon HLS (eds) Recycling of waste in agriculture. Fert Dev, Consultation Organization, New Delhi India, pp 68-90.

Phung Cao van, Nguyen BP, Tran KH, Bell RW (2009). Technical report irrigating rice crops with waste water to reduce environmental pollution from catfish production in the Mekong Delta, Australia.

Qadir M, Wichelns D, Raschid-Sally L, McCornick PG, Drechsel P, Bahri A, Minhas PS (2010). The challenges of wastewater irrigation in developing countries. Agricultural Water Manage 97: 561–568.

Qadir M, Wichelns D, Raschid-Sally L, Minhas PS, Drechsel P, Bahri A, McCornick P (2007a). Agricultural use of marginal-quality water-opportunities and challenges. In: Molden D (eds) Water for Food, Water for Life: A Comprehensive Assessment of Water Management in Agriculture. Earthscan, London,UK.

Ramesh M (2003). Soil and water resource characteristics in relation to land disposal of sewage effluents and suitability of sewage water for irrigation. M.Sc. *(Ag.)* Thesis Acharya N.G. Ranga Agricultural University, Hyderabad.

Rana L, Dhankhar R, Chhikara S (2010). Soil characteristics affected by long term application of sewage wastewater. Int J Environ Res 4(3): 513-518.

Raschid-Sally L, Jayakody P (2007). Understanding the drivers of wastewater agriculture in developing countries-results from a global assessment. In: Comprehensive Assessment Research Report Series, IWMI, Colombo, Sri Lanka.

Rashed M, Awad SR, Salam MA, Smidt E (1995). Monitoring of groundwater in Gabal el Asfar wastewater irrigated area (Greater Cairo). Water Sci Technol 32(11): 163-169.

Rattan RK, Datta SP, Chhonkar PK, Suribabu K, Singh AK (2005). Long-term impact of irrigation with se-wage effluents on heavy metal content in soils, crops and groundwater-A case study. Agric Ecosys Environ 109: 310-322.

Rattan RK, Datta SP, Singh AK, Chonkar PK, Suribau K (2001). Effect of long-term application of sewage effluents on available nutrient and available water status in soils under Keshopur effluent irrigation scheme in Delhi. J water Manage 9: 21-26.

Rusan MJ, Hinnawi M, Rousan L (2007). Long term effect of waste water irrigation of forage crops on soil and plant quality parameters. Desalinization 215: 143-152.

Saha JK, Panwar N, Srivastava Ajay, Biswas AK, Kundu S, Subba Rao A (2010). Chemical, biochemical, and biological impact of untreated domestic sewage water use on Vertisol and its consequences on wheat (*Triticum aestivum*) productivity. Environ Monit Assess 161: 403-412.

Saravanamoorthy MD, Ranjitha-Kumari BD (2007). Effect of textile water on morphology and yield on two varieties of peanut (*Arachis hypogea* L.). *J* Agric Tech 3: 335-343.

Sengupta AK (2008). WHO guidelines for the safe use of wastewater, excreta and greywater. In: National Workshop on Sustainable Sanitation, New Delhi, May 19-20.

Sharma VK, Kansal BD (1984). Effect of N, farm yard manure, town refuse and sewage water on the yield and N content of Maize. J Ecol 11: 77-81.

Singh J, Kansal BD (1985b). Effect of long-term application of municipal waste water on some chemical properties of soils. J Res Punjab Agric Univ 22: 235-42.

Singh KP, Mohon D, Sinha S, Dalwani R (2004). Impact assessment of treated/ untreated wastewater toxicants discharge by sewage treatment plants on health, agricultural, and environmental quality in wastewater disposal area. Chemos 55: 227-255.

Singh PK, Deshbhratar PB and Ramteke DS (2013). Effects of sewage wastewater irrigation on soil properties, crop yield and environment. Agricultural Water Manage 103: 100-104.

Smith VH, Tilman GD, Nekola JC (1999). Eutrophication: impacts of excess nutrient inputs on freshwater, marine, and terrestrial ecosystems. Environ Pollution 100:179–196.

Strauss M, Blumenthal U (1990). Human waste use in agriculture and aquaculture: Utilization practices and health perspectives. IRCWD Report 09/90. International Reference Center for Waste Disposal (IRCWD). Duebendorf, Germany.

Toze S (2006). Reuse of effluent water-benefits and risks. Agricultural Water Management 80: 147–159.

Udayasoorian C, Devagi P, Ramaswami PP (1999). Case study on the utilization of paper and pulp mill effluent irrigation for field crop. In: Proceedings of workshop on 'Bioremediation of polluted habitats', pp 71-73.

Van der Hoek W (2004). A framework for a global assessment of the extent of wastewater irrigation: The need for a common wastewater typology, In: Scott CA, Farunqui NI, Raschid-Sally L (ed) Wastewater use in irrigated agriculture, confronting the livelihood and environmental realities, CAB International, International Water Management Institute, International Development Research Centre, Trowbridge, pp 11-24.

Weber B, Avnimelech Y, Juanico M (1996). Salt enrichment of municipal sewage new prevention approaches in Israel. Environ Manage 20 (4): 487-495.

WHO (2006) WHO Guidelines for the safe use of wastewater, excreta and greywater-Policy and regulatory aspects, Vol. 1. World Health Organization, Geneva.

Wim van der Hoek (2002). A framework for a global assessment of the extent of wastewater irrigation: The need for a common wastewater typology. International Water Management Institute (IWMI), Bierstalpad, Netherlands. Wastewater use in irrigated agriculture. http://hdrnet.org/364/1).

Yadav RK, Goyal B, Sharma RK, Dubey SK, Minhas PS (2002). Post-irrigation impact of domestic sewage ef-fluent on composition of soils, crops and ground water-A case study. Environ Int 28: 481-486.

Yadav, R.K., Chaturvedi, R.K., Dubey, S.K., Joshi, P.K., Minhas, P.S. (2003). Potential and hazards associated with sewage irrigation in Haryana. Indian J. Agric. Sci. 73(5), 249-255.

YUVA (2006). National assessment of wastewater generation and utilization: A case of India. Youth for Unity and Voluntary Action, Mumbai.

❑❑❑

# 10

# Impact of Climate Change on Crop and Livestock Sectors and Strategies for Mitigation and Adaptation

**H.M. Meena, R.K. Singh, Dipak Kumar Gupta and A.K. Mishra**

## Introduction

Climate change and climate variability have become a reality. Recent climate changes have widespread impacts on human and natural systems including agriculture. According to Inter-Governmental Panel on Climate Change (IPCC, 2007), climate change refers to a statistically significant variation in either the mean state of the climate or in its variability, persisting for an extended period (typically decades or longer). For example, it could show up as a change in climate normal (expected average values for temperature and precipitation) for a given place and time of year, from one decade to the next. Climate change may be due to natural internal processes or external forcing, or to persistent anthropogenic changes in the composition of the atmosphere or in land use. In last 130 years (1880-2012), the global averaged combined land and ocean surface temperature has increased by 0.85 [0.65 to 1.06] °C (IPCC, 2014). This increase in temperature is disturbing global climate system as a result we are observing climate change on entire continent. IPCC has projected the temperature increase between 1.1 °C and 6.4 °C by the end of the 21$^{st}$ Century (IPCC, 2007).

In Indian context, mean temperatures show significantly increasing trends over all the states except Chhattisgarh, Haryana, Jammu and Kashmir, Meghalaya, Orissa, Punjab, Uttarakhand and West Bengal. Punjab (-0.01 $^{0}$C/year) has shown significant decreasing trends in mean temperature, while no trends were observed in Chhattisgarh, Haryana, Meghalaya, Orissa, Uttar Pradesh and West Bengal during 1951-2010 (Fig.1a). Spatially coherent increasing and decreasing trends in annual rainfall are found in many states of India, though not statistically significant. However, annual rainfall is significantly increasing in West Bengal (+3.63 mm/year) and significantly decreasing in Uttar Pradesh (-4.42 mm/year) and Andaman and Nicobar (-7.77 mm/year) during 1951-2010 (Fig.1b) (Rathore *et al.,* 2013).

The main cause of observed global warming and as a result climate change is increasing atmospheric concentration of Greenhouse Gases (GHG) such as carbon dioxide ($CO_2$), Methane ($CH_4$), Nitrous oxide ($N_2O$) and Chlorofluorocarbon (CFCs) etc. These GHG are radioactively active gases that allow incoming short wave radiation from sun to pass and reach earth surface while they partially absorb out going long wave thermal radiation (infrared) emitted from earth surface and again reemit it in all direction thus making earth surface warmer. GHG concentration has increased since the pre-industrial era, driven largely by economic and population growth, and is now higher than ever and is at levels that are unprecedented in at least 800,000 years. The concentrations of $CO_2$, $CH_4$ and $N_2O$ have increased by 40%, 150% and 20%, respectively since 1750 (IPCC, 2014). The global increases in $CO_2$ concentration are primarily due to fossil fuel combustion (Coal, Petroleum and Natural gas) in industry, transportation, electricity generation and other uses and land use change, while those of $CH_4$ and $N_2O$ are primarily due to agriculture. The enteric fermentation in livestock (cattle, buffalos, sheep, goat and pigs etc.) and puddle transplanted rice are major sources of $CH_4$ while application of nitrogenous fertilizer in soil led to emission of $N_2O$ after irrigation or flash rainfall.

Due to global warming, the frequency of weather related extreme events like floods, droughts, frost, cold and heat waves has considerably increased in the recent past. Continuation of such trends associated with rise in temperature is expected to melt ice, glaciers, re-distribute water flow in rivers, raise see levels, sub-merge coastal habitats, islands, generate tsunamis and dislocate human and livestock settlements. It has been reported that over the period 1901–2010, global mean sea level has been risen by 0.19 (0.17 to 0.21) m (IPCC, 2014). Predicted spatial redistribution of precipitation, droughts, floods and water balance will change land use, pests, diseases and other ecological parameters. Attaining food security is the major challenge to India in the 21$^{st}$ century in climate change scenario.

**Table 1.** Daily potential evapotranspiration (mm) of arid Rajasthan during monsoon season

| District | Normal | At elevated air temperatures by | | | | Percentage increase in PET by 21st Century |
|---|---|---|---|---|---|---|
| | | 2020 | 2050 | 2080 | 2100 | |
| Banner | 5.3-8.9 | 5.5-9.2 | 5.6-95 | 5.8-9.8 | 5.9-10.2 | 11-15 |
| Bikaner | 6.1-10.3 | 6.4-10.7 | 6.6-11.1 | 6.8-11.3 | 7.0-11.5 | 12-15 |
| Churu | 5 3-8.1 | 5.5-8.3 | 5.6-86 | 5.8-8.9 | 6.0-9.2 | 13-14 |
| Ganganagar | 5.3-7.5 | 5.5-7.7 | 5.7-8.0 | 5.9-8.2 | 6.1-8.4 | 12-15 |
| Hanumangarh | 5.4-7.6 | 5.6-7.8 | 5.7-8.1 | 59-8.3 | 6.1-8.5 | 12-13 |
| Jaisalmer | 4.7-11.4 | 5.0-11.9 | 5.3-12.5 | 5.5-13.0 | 5.8-13.5 | 18-23 |
| Jalore | 4.8-8.5 | 5.0-8.8 | 5.1-9.1 | 52-9.4 | 5.4-9.7 | 13-14 |
| Jhunjhunu | 4.9-8.0 | 5.1-8.3 | 5.2-8.5 | 5.3-8.8 | 5.4-9.0 | 10-13 |
| Jodhpur | 5.0-10.8 | 5.3-11.0 | 5.5-11.2 | 5.7-11.6 | 5.8-12.1 | 12-16 |
| Nagaur | 3.8-7.8 | 3.9-8.0 | 4.1-8.2 | 4.2-8.5 | 4.4-8.7 | 12-16 |
| Pali | 4.6-10.1 | 4.7-10.5 | 4.8-10.9 | 5.1-11.3 | 5.2-11.7 | 13-16 |
| Sikar | 4.6-6.1 | 4.7-6.2 | 4.8-6.4 | 4.9-6.6 | 5.0-6.7 | 9-10 |

(*Source:* Poonia and Rao, 2013)

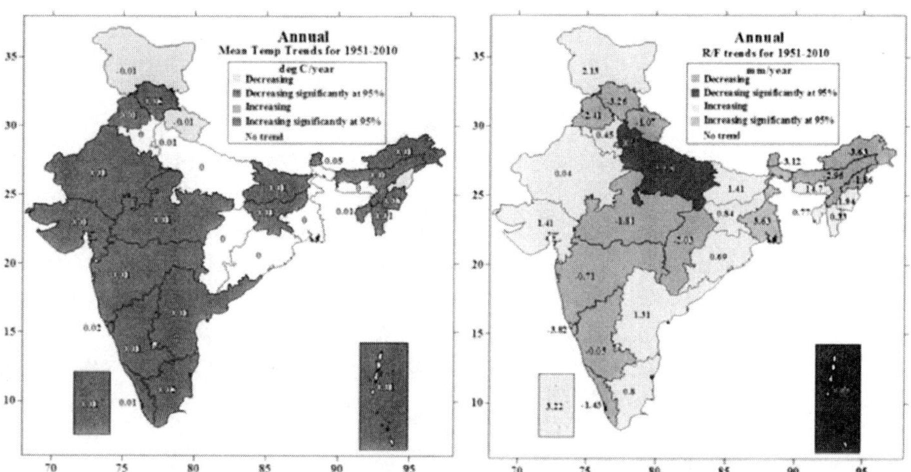


**Figure 1.** State level annual mean temperature (a) and rainfall trend (b) during 1951-2010. (*Source:* Rathore *et al.*, 2013).

The livestock sector alone contributes nearly 25.6% of value of output at current prices of total value of output in Agriculture, Fishing & Forestry sector. The overall contribution of Livestock Sector in total GDP is nearly 4.11% at current prices during 2012-13 (Anon., 2012). Little is known about the interactions of climate and increasing climate variability with other drivers of change in livestock systems and in broader development trends. In many places in the tropics and subtropics, livestock systems are changing rapidly, and the spatial heterogeneity of household response to change may be very large (Thornton *et al.*, 2009). Houghton *et al.* (2001)

concluded that direct effects from air temperature, humidity, wind speed and other climate factors influence animal performance: growth, milk production, wool production and reproduction.

## Impact of climate change on crop and livestock sector

The climate change has and will have direct or indirect impact on all form of life including crops and livestock. Temperature increase, $CO_2$ fertilization, abnormal weather events and frequent natural calamities will directly affect while occurrence of new diseases and pests, change in crop pollinators dynamics, change in soil fertility and change in forage quality will have indirect effect. In recent decades, changes in climate have caused impacts on natural and human systems on all continents and across the oceans (IPCC, 2014).

## Impact on agricultural crops

With the threat of climate change looming large on crop productivity, the most vulnerable regions of the world are the tropics. In semi-arid regions higher temperatures and rainfall variability could also significantly affect agricultural production with large implications on food security (IPCC, 2007). Assessment of many studies covering a wide range of regions and crops shows that negative impacts of climate change on crop yields have been more common than positive impacts (IPCC, 2014). Climate change has projected to increase climatic stresses on crops; and a global analysis has projected that an increase in temperatures is likely to affect crop production by 10–40% in India by 2080-2100 (IPCC, 2007). Droughts, floods, tropical cyclones, heavy precipitation events, hot extremes and heat waves are known to negatively impact agricultural production, and farmers' livelihood (Naresh Kumar *et al.*, 2011). A study conducted by Poonia and Rao (2013) during major cropping season of monsoon period, the impact of projected temperatures shows that the PET may increase by 0.1 to 0.4 mm day$^{-1}$ by 2020, 0.2 to 0.8 mm day$^{-1}$ by 2050, 0.4 to 1.2 mm day$^{-1}$ by 2080 and 0.5 to 1.6 by 2100 (Table 1). Thus, by the end of 21$^{st}$ century, the PET requirements during monsoon period increases by 9 to 20% from the current levels of PET.

Recent research indicates that monsoon rainfall became less frequent but more intense in India during the latter half of the Twentieth Century, thus increasing the risk of drought and flood damage to the country's wet-season (kharif) rice crop. Auffhammer *et al.* (2012) have reported that drought and extreme rainfall negatively affected rice yield in predominantly rainfed areas during 1966–2002, with drought having a much greater impact than extreme rainfall. Production of pearl millet, the major cereal crop of the arid Rajasthan grown during kharif, was reduced by 10-30% during mild drought, 35-60% during moderate drought and 75-90% during severe droughts.

The production of wheat, a crop sensitive to weather, may be influenced by climate change. The regional vulnerability of wheat production to climate change in India was assessed by quantifying the impacts and adaptation gains in a simulation analysis using the InfoCrop-WHEAT model. The study projects that climate change will reduce the wheat yield in India in the range of 6 to 23% by 2050 and 15 to 25% by 2080 (Naresh Kumar *et al.,* 2014). A study conducted revealed a progressive reduction in irrigated rice yields due to climate change towards the end of the century, if no adaptation is followed; while for rainfed rice, negative impacts likely to reduce with time due to projected increase in rainfall in many areas. However, spatio-temporal variations exist for the magnitude of impacts. On an aggregated scale, the mean of all emission scenarios indicate that climate change is likely to reduce irrigated rice yields by ~4 % in 2020 (2010–2039), ~7 % in 2050 (2040–2069), and by ~10 % in 2080 (2070–2099). On the other hand, rainfed rice yields in India are likely to be reduced by ~6 % in the 2020 scenario, but in the 2050 and 2080 scenarios they are projected to decrease only marginally (<2.5 %) (Naresh Kumar *et al.,* 2013). These changes raise concerns about food security. Numerous studies have demonstrated that the *kharif* harvest is lower when total June–September rainfall is lower (Webster *et al.,* 1998; Selvaraju, 2003; Krishna Kumar *et al.,* 2004). A drought during the summer of 2009 was one of the most severe in decades, with rice harvest declining by 14% (Commission for Agricultural Costs and Prices, 2010). Flooding associated with heavy rain events can also damage crops (Goswami *et al.,* 2006).

In an another study conducted at Hisar, Haryana showed the highest decreasing trend during annual and monsoon rainfall at the rate of 28.9 and 20.5 mm/year, respectively, which also observed the highest coefficient of variation of annual and monsoon rainfall among the study sites. Sunshine hours decreased over the years during the rice season at Faizabad, Ludhiana and Samastipur while decreasing trend was noticed for Ludiana and Hisar. The significant negative trend of the potential yield of rice at the rate of 0.029t/ha/year at Ludhiana is a serious concern with the decreasing trend of sunshine hours and increase in trend of minimum temperature scenario (Subash and Ram Mohan, 2012). Possible impact of climate change on wheat production in India has been worked out by the climate scientists for the period during 2000 to 2070 shown in Fig. 2 (Singh, 2009). Droughts in 1965-66 and 1966-67, 1979-80, 1987-88, 2002-03 and recent drought in 2009-10 (Fig. 3) affected the country food grains production to a large extent (Gopakumar, 2011).

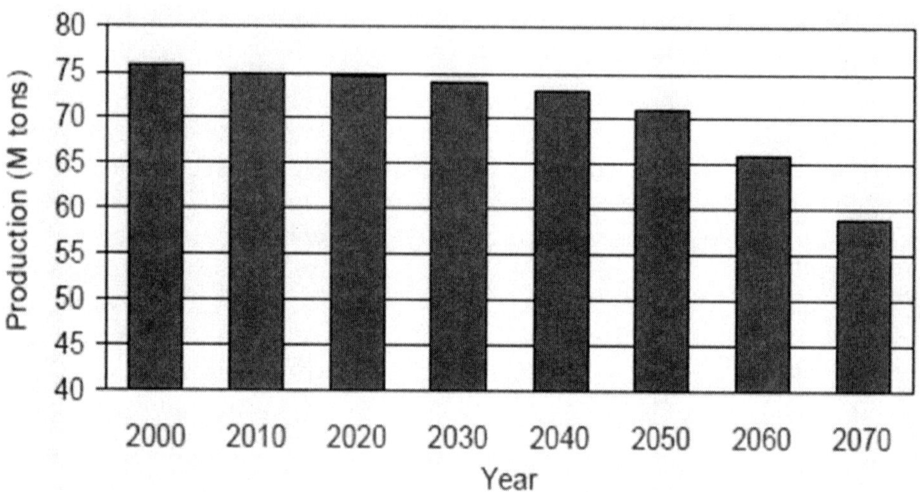


**Figure 2.** Possible Impact of Climate Change on Wheat Production in India (*Sources:* Singh, 2009)

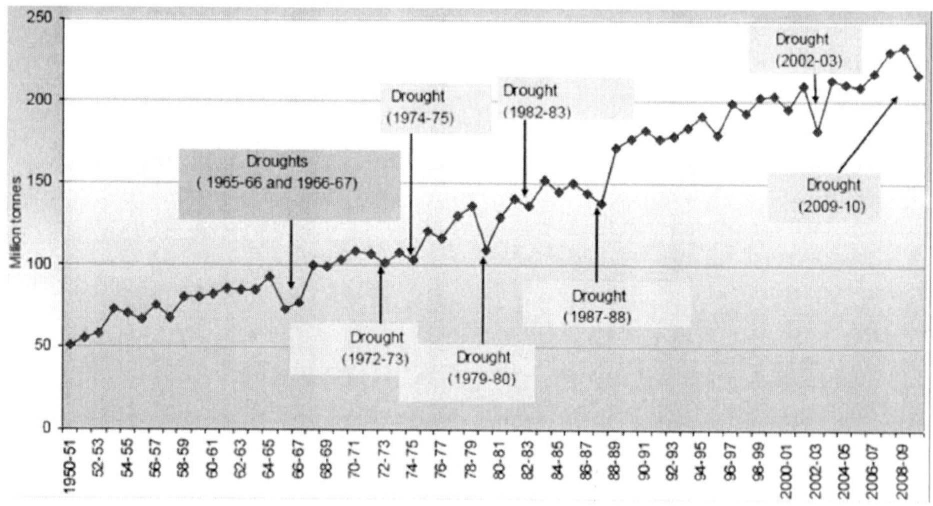


**Figure 3.** Impact of droughts on Indian foodgrains production from 1950- 51 to 2009-10 (*Source:* Gopakumar, 2011)

## Impact on Livestock

The total livestock population consisting of Cattle, Buffalo, Sheep, Goat, Pig, Horses & Ponies, Mules, Donkeys, Camels, Mithun and Yak in the country is 512.05 million numbers in 2012 (Livestock Census, 2012). The total livestock population has decreased by about 3.33% over the previous census (Figure 4).

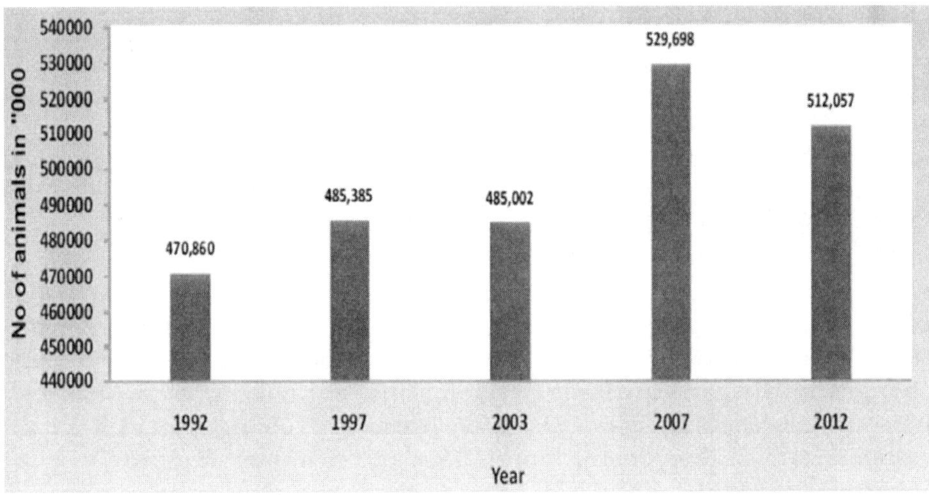


**Figure 4.** Trends of total livestock population in India during 1992-2012 (*Source:* Anonymous, 2012)

In India, animals are reared under extensive or semi- intensive system of management with a constant expo-sure to natural climatic conditions. In such conditions, heat stress is caused by a combination of environmental factors (temperature, relative humidity, solar radiation, air movement and pre-cipitation). Stress has been defined, as the inability of an animal to cope with its environment, a phenomenon which is often reflected in a failure to achieve genetic potential.

Climate change poses formidable challenge to the development of livestock sector in India. It can directly affect livestock by heat stress, increased spread existing vector born diseases and macro-parasite of accompanied by emergence of new diseases and indirectly by scarcity of water, feed and competition of natural resources with demand of rising human population (Thorntone *et al.*, 2008). The anticipated rise in temperature between 2.3 and 4.8°C over the entire country together with increased precipitation resulting from climate change is likely to aggravate the heat stress in dairy animals, adversely affecting their productive and reproductive performance, and hence reducing the total area where high yielding dairy cattle can be economically reared. The predicted negative impact of climate change on Indian agriculture would also adversely affect livestock production by aggravating the feed and fodder shortages (Sirohi and Michaelowa, 2007). It has been reported that fodder scarcity which is generally shorter by 20-30% of the demand during normal years, touches 80-100% during severe droughts (Rao *et al.*, 2012). The increase in $CO_2$ concentration in atmosphere as consequence rate of photosynthesis will increase C:N ration and thus affect digestibility and quality of fodder.

Animal responses vary according to the duration and the intensity of the thermal challenge. In cold arid region, short-term changes in physiological, behavioural and immunological functions are required to survive acute stressful events such as summer heat waves. The severity of these short thermal challenges depends on the magnitude (intensity x duration) of heat wave events and the possibility of recovering during the cool night time period. In contrast, under hot arid conditions, livestock are heat challenged most of the time. The long-term thermoregulatory responses underlying heat acclimation (or acclimatization) increase the physiological strains, which in most cases are accompanied by reduced performance. These responses include a reduction of metabolic rate, changes in the cardiovascular system, efficient alteration in heat loss (vasomotor response: vasodilatation response), changes in behaviour response and in the general morphology of the animal (Figure 5).

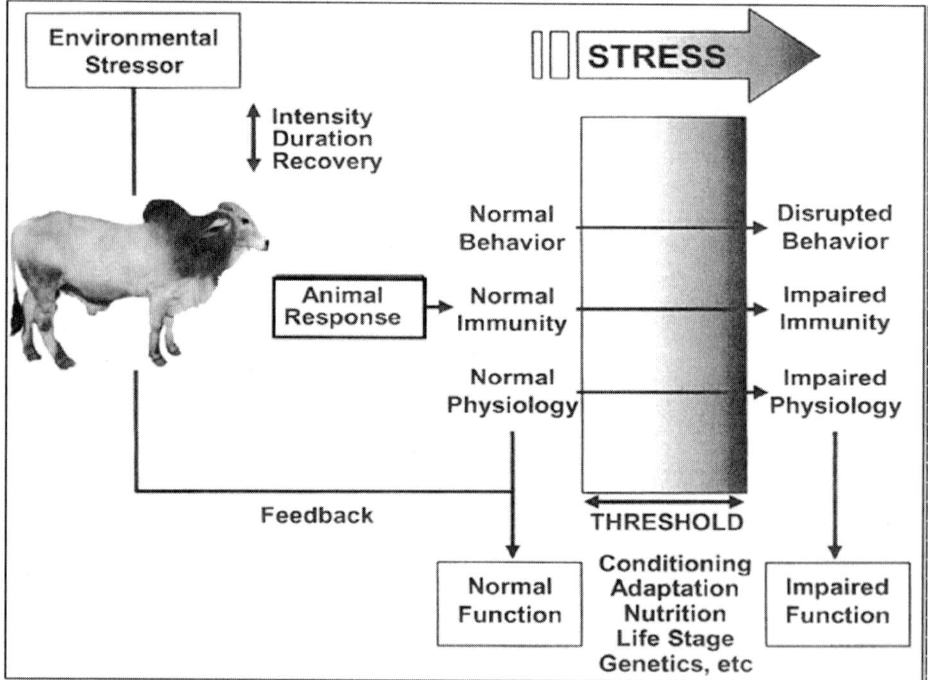


**Figure 5.** Response of animals' environmental stressors on performance and health (Hahn, 1999)

There is a two-way relationship between livestock production and environmental health. On the one hand, livestock contribute to climate change and other environmental problems, and at the same time livestock health and productivity can be adversely affected by these same environmental upsets (Sherman, 2010).

## Adaptation and mitigation strategies

Adaptation and mitigation are two complementary strategies for responding to climate change. Adaptation is the process of adjustment to actual or expected climate and its effects, in order to either lessen or avoid harm or exploit beneficial opportunities. Mitigation is the process of reducing emissions or enhancing sinks of greenhouse gases, so as to limit future climate change. Both adaptation and mitigation can reduce and manage the risks of climate change impacts.

## Mitigation strategies

The emission of $N_2O$ from crop can be reduced by enhancing the nitrogen use efficiency. Of total nitrogen applied in soil hardly 20-30% nitrogen is utilized by plants other get waste in the form of nitrate leaching, nitrous oxide emission and run-off loss. To avoid the wastage and reduce nitrous oxide emission the crop demand and nitrogen application should be matched. The fertilizer application should be soil test based, in right amount on right time and place. The use of slow release urea and nitrification inhibitor has also reported to reduce nitrous oxide emission. Methane is emitted from flooded puddle rice field due to anaerobic decomposition of organic matter by specialised bacteria known as methanogens. To reduce methane emission strategies should be to avoid anaerobic condition in root zone of the rice plants so the emission can be reduced by efficient management of water. The intermittent wetting and drying, mid season drainage, system of rice intensification (SRI) and direct seeded rice (DSR) has been reported to reduce methane emission. Methane is also formed in rumen of ruminants by the process of fermentation by microbes during the feed digestion. This process is not only responsible for methane emission but also leads to significant loss of feed energy. To reduce emission strategies should be focused on reducing number of livestock by increasing feed/fodder conversion efficiency into milk and meat and by developing breed of high milk and meat yield. It has been suggested that emission can also be reduced by changing feed quality and feed mixture that leads to less C: N ratio in feed and increased digestibility. However the mitigation strategies should be planed in such way that lead to potential reduction in GHG emission while also ensuring food security.

## Adaptation strategies for agricultural crops

According to IPCC strategies should be focussed on enhancing drought and pest resistance; enhancing yields, providing financial safety net for farmers to ensure continuation of farming enterprises and suppressing opportunistic agricultural pests and invasive species. The following strategies may be adapted;

*Developing new plant genotypes for drought, heat and cold tolerance adapted to climatic variability and ranges*: A wide range of crop improvement options for enhancing adaptation to climate variability are available. These include the choice of the type of cultivar, direct selection in multiple environments, including farmer-participatory testing, indirect selection for individual adaptation traits using conventional or genomic selection methods, and selection for responsiveness or compatibility to improved crop and soil management techniques. To maximize impact, several options need to be combined.

*Promotion of conservation agriculture:* Conservation agriculture practices can also contribute to making agricultural systems more resilient to climate change. In many cases, conservation agriculture has been proven to reduce farming systems GHG s emissions and enhance their role as carbon sinks. A case study on Zambia by Samuel Bells showed that combining conservation agriculture with tree cultivation increases crop yields by 240–400 per cent enhances soil carbon sequestration and improves resilience to climate shocks. In addition, the biomass that is produced can be used for feeding livestock.

*Development of contingency plans:* It seeks to identify effective institutional, participatory and collaborative processes involved in designing agricultural adaptation strategies at the national and sub-national levels in the country. Its methodology involves review of agricultural adaptation policy documents, research initiatives, stakeholder engagement processes, and inter-sectoral collaborations in adaptation planning.

*Developing precision and accurate forewarning mechanism*: Early-Warning Systems include the provision of timely and effective information, through identified institutions, that allows individuals exposed to hazard to take action in order to avoid or reduce risk and prepare for effective response. There should be a mechanism to develop medium and long range forecasting systems (15-20 days in advance) so that farmers have reasonable time to respond to risks.

*Carbon sequestration in different land use systems:* Carbon sequestration has been suggested as a means to help mitigate the increase in atmospheric carbon dioxide concentration. Silvipastoral systems can better sequester carbon in soil and biomass and help to improve soil conditions. In a study conducted at RRS, Bhuj of Central Arid Zone Research Institute showed that the silvipastoral system sequestered 36.3% to 60.0% more total soil organic carbon stock compared to the tree system and 27.1–70.8% more in comparison to the pasture system. The soil organic carbon and net carbon sequestered were greater in the silvipastoral system. Thus, silvipastoral system involving trees and grasses can help in better sequestration of atmospheric system compared with systems containing only trees or pasture (Mangalassery *et al.,* 2014).

*Water and nutrient management:* Managing water and nutrient and increasing its utilization efficiency not only increases agricultural productivity and food security, but also improves resilience to climate change and reduction of greenhouse gas emissions. In a study conducted in semi arid sub-African region it was found that stone bunds, zaï and half-moon techniques combined with the application of organic and/or mineral fertilizers were sustainable land management practices that have increased agricultural productivity, vegetative cover and carbon sequestration (Zougmoré *et al.* 2014).

*Modification of crop micro climate*: Modification of crop micro climate by means such as use of agroforestry system, wind breaks, various low cost structures etc play a vital role in modifying the micro-climate for the field crops. Construction of small water harvesting structures may also be helpful in this regard. As per Fanish and Priya (2013) agroforestry has many potential, such as enhance the overall (biomass) productivity, soil fertility improvement, soil conservation, nutrient cycling, micro-climate improvement, carbon sequestration, bio drainage, bio-energy and biofuel etc. Moreover, the important elements of agroforestry systems that can play a significant role in the adaptation to climate change include changes in the microclimate, protection through provision of permanent cover, opportunities for diversification of the agricultural systems, improving efficiency of use of soil, water and climatic resources, contribution to soil fertility improvement, reducing carbon emissions and increasing sequestration, and promoting gender equity (Rao *et al.,* 2007).

*Use of genetic engineering:* With the help of genetic engineering conversion of C-3 crops to the more carbon responsive C-4 crops can be done to achieve greater photosynthetic efficiency leading to enhanced carbon sequestration. Crops can be developed with better water and nitrogen use efficiency which may result in reduced emissions of greenhouse gases or greater tolerance to drought or submergence.

*Bio energy:* Bioenergy is energy derived from biomass, which can be deployed as solid, liquid and gaseous fuels for a wide range of uses, including transport, heating, electricity production, and cooking. Bioenergy systems can cause both positive and negative effects and their deployment needs to balance a range of environmental, social and economic objectives that are not always fully compatible. The integration of bioenergy systems into agriculture and forest landscapes can improve land and water use efficiency and help address concerns about environmental impacts. We conclude that the high variability in pathways, uncertainties in technological development and ambiguity in political decision render forecasts on deployment levels and climate effects very difficult. However, uncertainty about projections should not preclude pursuing beneficial bio energy options (Creutzig *et al.,* 2014).

*Organic farming*: as an adaptation strategy to climate change and variability, is a concrete and sustainable option and has additional potential as a mitigation strategy. The careful management of nutrients and carbon sequestration in soils are significant contributors in adaptation and mitigation to climate change and variability in several climate zones and under a wide range of specific local conditions. Organic farming as a systematic approach for sustained biological diversity and climate change adaptation through production management, minimizing energy randomisation of non-renewable resources; and carbon sequestration is a viable alternative. The purpose of potential organic farming is therefore to attempt a gradual reversal of the effects of climate change for building resilience and overall sustainability.

## Adaptation strategies for livestock

The livestock employ physiological mechanisms to counter the climate stress. The physiological mechanisms are also complemented by the behavioural process, for example buffaloes use wallowing during summer to reduce thermal loads and maintain thermal equilibrium. Goats are more adapted to harsh and drier conditions and have greater resistance to dehydration than temperate animals (Devendra 2007, Misra *et al.,* 2012, Misra and Kumawat, 2014). An integrated approach may be adopted for reducing climate stress of livestock. The experts have suggested many strategies for adaptation of livestock under changing climate these can be summarised below :

*Breeding*: There are clear genetic differences in resistance to climate stress, with tropically-adapted breeds experiencing lower body temperatures during heat stress than non-adapted breeds. The breeding should be promoted to develop breeds of cattle and other livestock that can tolerate heat stress and can survive, grow and reproduce in conditions of poor nutrition, parasites and disease. In tropical areas, the process of natural selection has favoured the emergence of breeds with a high ability to cope with thermal stress. Indigenous breeds, which have co-evolved in drylands over millennia and have adapted to the prevalent climatic and disease environments, will be essential. Many local breeds are already adapted to harsh living conditions like Tharparker breeds of Rajasthan. However, their adaptation includes not only heat tolerance but also their ability to survive, to grow and to reproduce in the presence of poor seasonal nutrition, high parasite and disease pressure. There is need to identify and utilize these hardy breeds in breeding programme.

*Adjustment in production practices*: The integrated farming system approach should be developed by diversification, intensification and/or integration of pasture management, livestock and crop production together. It will reduce risk and enhance resilience to climate change by modifying microclimate, enhancing soil productivity and availability of feed and fodder. The added value of integrating crops and livestock has been understood and practised by the dryland farmers of western Rajasthan for thousands of years and yet these systems can hold a key for climate resilient agriculture in the future (Misra *et al.* 2012). The approaches like mixed livestock farming systems, such as stall-fed systems and pasture grazing should be promoted. The integration that reflects a synergetic relationship among the components (the whole is greater than the sum of the parts) of crops, livestock and trees and that this synergetic relationship results in enhanced social, economic/productive and environmental sustainability of the systems and improves the livelihoods of those farmers who manage them.

*Livestock management systems*: To reduce the direct impact the focus must be given on (i) designing high quality, low-cost natural shade and improving water availability to reduce heat stress from increased temperature, (ii) changes in livestock/herd composition (selection of large animals rather than small); (iii) reduction of livestock numbers - a lower number of more productive animals leads to more efficient production and lower GHG emissions from livestock production (iv) developing infrastructure to harvest and store rainwater, such as tanks connected to the roofs of houses and small surface and underground dams.

*Restoration of Community Grazing Resources:* Community grazing resources (CPR) in most parts of the country has either drastically been reduced or has almost disappeared (Misra *et al.,* 2014). Restoration is most important not only for providing regulated grazing but also for protecting valuable resource. One option is restoration of such lands to village panchyat for their improvement and proper management through silvi pastoral development (Misra *et al.,* 2014). In order to maintain the vigour of the grass and legumes, and also to meet the needs of the livestock, CPR may be subjected to one of the grazing systems namely (i)

Continuous grazing, (ii) Deferred grazing, (iii) Rotational grazing and (iv) Deferred rotational grazing for a given site at a specific period as per requirement. For fixing stocking rate in the grazing land, information on grazing height/ animal weight relationships would be of much use to make adjustment to the traditional carrying capacity during the growing season, or year round, as the case may be.

*Alternative feed resources:* Inclusion of alternate feed resources in the animals' diet could be a useful strategy to minimize the nutritional stress during lean months. It has been demonstrated that feeding of Cactus to livestock reduced water requirement and increased nutrients digestibility without affecting health of the calves in arid or semi-arid regions (Mathur and Misra, 2014). Similarly, feeding of *Blepharis indica, Anabaena azollae* may also hold promise to provide feed as well as water to animals during summer feed scarcity, particularly in semi-arid and arid regions of the country (Sahoo *et al.*, 2013). Inclusion of *Prosopis juliflora* pods in concentrate mixture or feed blocks improved the production, reproduction of animals during summer without any adverse effect on health (Misra *et al.*, 2015).

*Capacity building for livestock keepers*: There is a need to improve the capacity of livestock producers and herders to understand and deal with climate change increasing their awareness of global changes. In addition, training in agro-ecological technologies and practices for the production and conservation of fodder improves the supply of animal feed and reduces malnutrition and mortality in herds.

## Conclusion

It is now well proven fact that climate change is happening and due to this, the frequencies of extreme weather events have increased and are influencing agricultural crops and livestock. Several studies done so far have suggested that impact of climate change is having a heavy toll on crop and livestock production. Agricultural crops and livestock are not only being negatively affected or will be affected in future due to climate change but also are significant sources of GHG. Agriculture and livestock being essential for human existence and livelihood for millions of farmer priority must be focused to make it climate resilient and environmental friendly. To mitigate/negate its impact strategies have to be formulated and executed at the earliest. In order to sustain our agricultural/livestock production with present day challenges we have to have more research on understanding the potential impact and developing more effective mitigation and adaptation strategies.

## References

Auffhammer, M., Ramanathan, V. and Vincent, J.R. (2012). Climate change, the monsoon, and rice yield in India .Climatic Change 111:411–424.DOI 10.1007/s10584-011-0208-4.

Creutzig F, Ravindranath NH, Berndes G, Bolwig S, Bright R, Cherubini F, Chum H, Corbera E, Delucchi M, Faaij A, Fargione J, Haberl H, Heath G, Lucon O, Plevin R, Popp A, Robledo-Abad C, Rose S, Smith P, Stromman A, Sangwon S and Masera O. (2014). Bioenergy and climate change mitigation: an assessment. Global Change Biology. Bioenergy. Available from: 10.1111/gcbb.12205.

Devendra, C. (2007). Perspectives on animal production systems in Asia. Livestock Science 106: 1–18.

Fanish SA and Priya RS (2013). Review on Benefits of Agro Forestry System. International Journal of Education and Research, 1(1): 1-12.

Gopakumar CS (2011). Impacts of Climate variability on Agriculture in Kerala. Thesis, Ph.D, Cochin University of Science and Technology, Kerala.

Goswami BN, Venugopal V, Sengupta D, Madhusoodanan MS, Prince KX (2006). Increasing trend of extreme rain events over India in a warming environment. Science, 314:1442–1445.

Hahn, GL (1999). Dynamic response of cattle to thermal heat loads. Journal of Animal Science 77:10-20.

Houghton JT, Ding Y, Griggs DJ, Noguer M, van der Linden PJ, Dai X, Maskell K, Johnson CA (2001). Climate change: The scientific basis. Contribution of working group I to the third assessment report of the Intergovernmental Panel on Climate Change. New York: Cambridge University Press.

IPCC (2007). IPCC fourth assessment report: climate change 2007(AR4). IPCC, Geneva.

IPCC (2014). Fifth assessment synthesis report of Intergovernmental Panel on Climate Change.

Krishna Kumar K, Rupa Kumar K, Ashrit RG, Deshpande NR, Hansen JW (2004). Climate impacts on Indian agriculture. Int J Climatol 24:1375–1393.

Livestock Census (2012). 19th Livestock census, All India report, Ministry of Agriculture, India. http://www.dahd.nic.in/dahd/WriteReadData/Livestock.pdf

Mangalassery S, Devi Dayal, Meena SL and Bhagirath Ram (2014). Carbon sequestration in agroforestry and pasture systems in arid northwestern India. Current Science, 107: 1290-1298.

Mathur BK, and Misra AK (2014). Feeding of thornless cactus (*Opentificus indica*) to livestock in arid regions. National workshop on Cactus Pear, CAZRI, Jodhpur, 21 March 2014.

Misra AK, Mathur BK, Kumawat RN, Patidar M, and Roy MM (2015). Strategies for enhancing livelihood of small holders through livestock production in drylands of India. Lead paper submitted for presentation in the National Seminar on Livestock Production Practices for Small Farms of Marginalized Groups and Communities in India. College of Veterinary Sciences & animal Husbandry, CAU, Aizawl, Mizoram. 28-30 January, 2015.

Misra AK and Kumawat RN (2014). Farmers coping strategies to climate change and contingency planning in drylands of India. ICAR Short Course on Climate change mitigation and adaptation under arid and semi-arid region. CAZRI, Jodhpur. December 8-17, 2014.

Misra AK, Raghuvansi MS, Tiwari J C and Roy MM (2014). Strategies for enhancing feed and fodder resources for improving livestock production in cold arid regions. In: e- Compendium of 3rd Interface meeting on improvement of yak husbandry and upliftment of socio-economic status of yak rearers in the country 84-102. Leh, Jammu & Kashmir: September 22-24, 2014.

Misra AK, Sirohi AS, and Mathur BK (2012). Strategies for Managing Livestock under Environmental Stress in Drylands of India. Annals of Arid Zone 51(3&4): 219-243.

Naresh Kumar S Aggarwal PK Saxena R Rani S Jain S and Chauhan N (2013). An ssessment of regional vulnerability of rice to climate change in India. Climatic Change. 118:683–699.DOI 10.1007/s10584-013-0698-3.

Naresh Kumar S, Aggarwal PK, Rani S, Jain S, Saxena R and Chauhan N (2011). Impact of climate change on crop productivity in Western Ghats, coastal and northeastern regions of India. Current Science, Vol. 101, NO. 3, 10.

Naresh Kumar S, Aggarwal PK, Swaroopa Rani DN, Saxena R, Chauhan N and Jain S (2014). Vulnerability of wheat production to climate change in India. Clim Res 59:173-187.

Poonia S and Rao AS (2013).Climate Change and Its Impact on Thar Desert Ecosystem . Journal of Agricultural Physics. 13(1): 71-79.

Rao AS, Poonia S and Choudhary S (2012). Climate change impact and drought scenario in arid Rajasthan. Ann.Pl.Soil Res. 14(1): 25-28.

Rao KPC, Verchot LV and Laarman J (2007). Adaptation to Climate Change through Sustainable Management and Development of Agroforestry Systems. An Open Access Journal published by ICRISAT (ejournal.icrisat.org), 4(1): 1- 30.

Rathore LS, Attri SD and Jaswal AK (2013). State Level Climate Change Trends in India under the Meteorological Monograph Series, ESSO/IMD/EMRC/ 02/2013.

Sahoo A, Kumar D and Naqvi, SMK. (eds) (2013). Climate resilient small ruminant production. NICRA, CSWRI, Avikanagar, India. P 1-106.

Selvaraju R (2003) Impact of El Niño-southern oscillation on Indian foodgrain production. Int J Climatol 23:187–206.

Sherman, DM (2010). A Global Veterinary medical Perspective on the concept of one health: Focus on livestock. ILAR Journal. 51(3): 281-287.

Singh G (2009). Climate change and Indian Agriculture: Issues and copping up strategies. 60th International Executive Council Meeting & 5th Asian Regional Conference, 6-11 December 2009, New Delhi, India.

Sirohi S and Michaelowa A (2007). Sufferer and cause: Indian livestock and climate change. Climatic Change 85:285–298, DOI 10.1007/s10584-007-9241-8.

Subash N and Ram Mohan HS (2012). Evaluation of the impact of climatic trends and variability in rice–wheat system productivity using Cropping System Model DSSAT over the Indo-Gangetic Plains of India. Agricultural and Forest Meteorology 164: 71-81.

Thornton P, van de Steeg J, Notenbaert M H, Herrero M (2009). The impacts of climate change on livestock and livestock systems in developing countries: A review of what we know and what we need to know. Agricultural Systems 101: 113–127.

Wani SA, Chand S, Najar GR and Teli MA (2013). Organic Farming: As a Climate Change Adaptation and Mitigation Strategy. Current Agriculture Research Journal 1(1): 45-50.

Webster PJ, Magana VO, Palmer TN, Shukla J, Tomas RA, Yanai M, Yasuna T (1998). Monsoons: processes, predictability, and the prospects for prediction. J Geophys Res 103:14451–14510.

Zougmoré R, Jalloh A and Tioro A (2014) .Climate-smart soil water and nutrient management options in semiarid West Africa: a review of evidence and analysis of stone bunds and zaï techniques. Agriculture & Food Security. http://www.agricultureandfoodsecurity.com/content/3/1/16.

❑❑❑

# 11

# System of Rice Intensification (SRI) - A Climate Smart Agriculture Approach

**Arnab Banerjee and Manoj Kumar Jhariya**

## Introduction

The vast majority of climate change impacts and the overall impact of climate change on rice production are likely to be negative. Overwhelming scientific research and evidence have shown that the climate is ongoing. According to the International Food Policy Research Institute (IFPRI) report Climate Change: Impact on Agriculture and Costs of Adaptation forecasts there will be an increase in the net price of rice between 32 and 37% as a result of climate change by 2050. They also show that yield losses in rice could be between 10 and 15%. While there is still ongoing scientific exploration into climate change, IRRI recognizes two universal trends predicted by all climate change models viz., temperatures will increase, resulting in more heat stress and rising sea levels and there will be more frequent and severe climate extremes.

## Climatic stress on Rice production

a) **Sea Level Rise:** Experts have predicted that, as a consequence of melting polar ice caps and glaciers due to rising temperatures, sea-water levels may rise on average by about 1 m by the end of the 21$^{st}$ century. Rice is grown in vast low-laying deltas and coastal areas in Asia; sea-level rise would therefore, severly affect rice production due to the influence of changes in the climate. As for example more than half of Vietnam's rice produce, for instance, is grown in the Mekong River delta—all of which would be affected due to sea-level rise. The entire hydrology of the delta will be affected; sediment discharge and shoreline gradients will change causing significant reduction in rice production in these low lying coastal areas.

b) **Flood:** Rice is unique in that it can thrive in wet conditions where other crops fail. Uncontrolled flooding is a problem, however, because rice cannot survive if submerged under water for long periods of time. Factors such as flooding caused by sea-level rises in coastal areas, predicted increased intensity of tropical storms with climate change will likely hinder rice production. About 20 million hectares of the world's rice-growing area is at risk of occasionally being flooded to submergence level, particularly in major rice-producing countries such as India and Bangladesh presently. Major flooding events are likely to increase in frequency with the onslaught of climate change and rice-growing areas, currently not exposed to flooding, will experience floods in near future.

c) **Salinity:** Salinity is also associated with higher sea levels as this will bring saline water further inland and expose more rice-growing areas to salty conditions. Rice is only moderately tolerant of salt and yields can be reduced when salinity is present. As with sea-level rises, the effects of salinity can permeate throughout whole deltas and fundamentally change hydrological systems, soil quality and therefore, hamper rice productivity.

d) **Rise in $CO_2$ and temperature level:** Increases in both carbon dioxide levels and temperature will also affect rice production. Resereaches has revealed that higher carbon dioxide levels typically increase biomass production, but not necessarily yield. Higher temperatures can decrease rice yields as they can make rice flowers sterile, meaning no grain is produced. Higher respiration losses linked to higher temperatures also make rice less productive. The different factors for elevated temperature, carbon dioxide levels, changes in humidity, and the interactions of these factors make future rice yields under these conditions challenging. IRRI research indicates that a rise in night time temperature by $1^0$C may reduce rice yields by about 10%.

e) **Water Scarcity:** Rice requires ample water to grow. Rainless days for a week in upland rice-growing areas and for about two weeks in shallow lowland rice-growing areas can significantly reduce rice yields. Average yield reduction in rainfed, drought-prone areas has ranged from 17 to 40% in severe drought years, leading to production losses and food scarcity. With the onset of climate change, the intensity and frequency of droughts are predicted to increase in rainfed rice-growing areas and droughts could extend further into water-short irrigated areas. Water scarcity affects more than 23 million hectares of rainfed rice production areas in South and Southeast Asia. In Africa, recurring drought affects nearly 80% of the potential 20 million hectares of rainfed lowland rice. Drought also affects rice production in Australia, China, USA, and other countries.

Water scarcity will have a significant impact on agriculture. Shortage of water for agricultural production has already become a major problem in some countries. In world's rice bowls – particularly China and India – the scarcity of water is acute, with competing demands on fresh water sources (human and industrial needs) triggering conflicts. It is estimated that close to 1.3 billion people worldwide are still without safe water supply and the situation continues to deteriorate. Fresh water species, reckoned as an indicator of ecosystem health, have declined by 40% since 1970's. Various studies and estimates lead us to expect demand for rice to increase by about 38% by 2040.

For increasing food production through intensive agriculture the farmers of our country using enormous input of inorganic fertilizers, particularly nitrogenous fertilizer (Subba Rao, 1979). In India the fertilizer cost is high and will become more expensive with increasing energy cost, diminishing raw material, expected gradual phasing out of subsidies on fertilizer and increasing demand. The most horrifying aspect of modern chemical input based intensive agriculture is the rapid fall of chemical fertilizer response. The rapid destruction of soil structure and microbial nutrient recycling system brought about by indiscriminate use of chemical fertilizer has resulted in decline of our soil productivity. More over application of excessive and unbalanced chemical fertilizer polluting our habitat in many ways and necessitates heavy pesticide application for crop protection, which in turn causes hazardous pollution of environment and build up toxic pesticide residues in crops (Kabi, 2006).

f) **Outbreaks of Weeds, Pests and Diseases:** Weed infestation and rice-weed competition are predicted to increase and will represent a major challenge for sustainable rice production. Also, extreme weather events have recently led to dramatic rodent population outbreaks in Asia due to unseasonal and asynchronous cropping. Surveys in hundreds of farmers fields over the last 10 years reflected rice diseases and pests are strongly influenced by climate change. Water shortages, irregular rainfall patterns, and related water stresses increase the intensity of some diseases, including brown spot and blast. On the other hand, new environmental conditions and shifts in production practices that farmers may adopt to cope with climate change could lead to reductions of diseases such as sheath blight or insects such as whorl maggots or cutworms.

US National Research Council (US NRC, 2011) assessed the literature on climate change and its impact on crop yields. Further they stressed on uncertainties in their projections of the changes in crop yields. Actual changes in yields may be above or below these central estimates. They also provided an estimated the "likely" range of changes in yields. "Likely" means a greater than 67% chance of being correct, based on expert judgement. The likely ranges are summarized in the image descriptions by following two graphs (Figure 1 and 2).

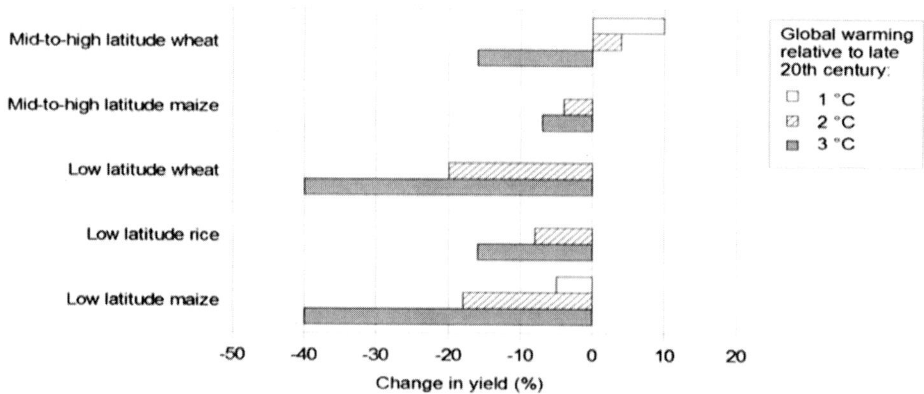


**Figure 1.** Projected changes in crop yields at different latitudes with global warming (US NRC, 2011).

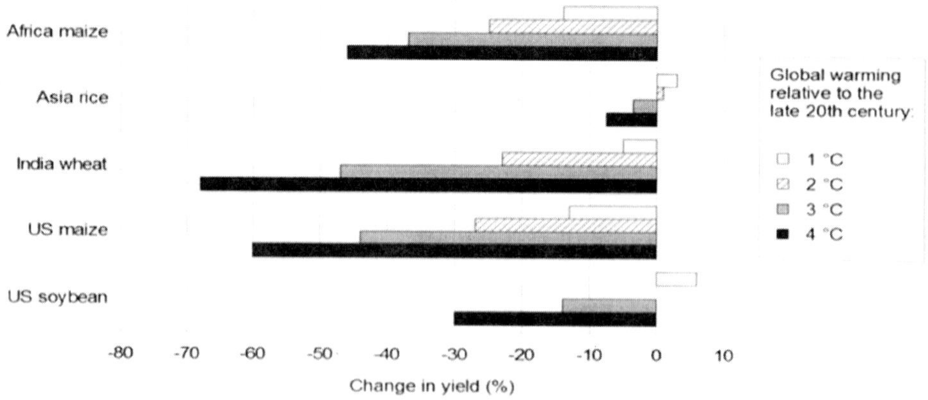


**Figure 2.** Projected changes in yields of selected crops with global warming (US NRC, 2011).

## SRI: A Civil-Society Innovation

The *System of Rice Intensification,* known as SRI - *le Système de Riziculture Intensive* in French and *la Sistema Intensivo de Cultivo Arrocero* (SICA) in Spanish - is a climate-smart, agroecological methodology for increasing the productivity of rice and more recently other crops by changing the management of plants, soil, water and nutrients.

The System of Rice Intensification (SRI) is an unusual innovation in several ways in that its methods can raise, concurrently, the productivity of the land, labor, water and capital invested in irrigated rice production. This positive-sum dynamic violates precept that, there are no free lunches, which assumes that there must always be some tradeoff. Of course, there are costs involved with SRI adoption, particularly increased labor from farmers during their initial learning phase; and there are some conditions where the methods will be inappropriate or impractical, e.g., where there is little water control and flooding creates anaerobic soil conditions.

But with skill and confidence as well as innovation, SRI can become labor-saving over time, saving water (by 25-50%) and seed (by 80-90%), reducing costs (by 10-20%), and raising paddy output at least 25-50%, and often 50-100% and sometimes even more. This sounds too good to be true, of course; but the productivity of SRI methods has been validated in 28 countries, from China to Cuba, Peru to Philippines, Gambia to Zambia, and even in Iraq, Iran and Afghanistan. SRI works by changing the management of the plants, soil, water and nutrients utilized in paddy rice production. Specifically, it involves transplanting single young seedlings with wider spacing, carefully and quickly into fields that are not kept continuously flooded, and whose soil has more organic matter and is actively aerated. These practices improve the growth and functioning of rice plants-root systems and enhance the numbers and diversity of the soil biota that contribute to plant health and productivity (Stoop *et al.*, 2002; Uphoff, 2003; Randriamiharisoa *et al.*, 2006; Mishra *et al.*, 2006).

The cumulative effect of these methods is raise not only the yield of paddy (kg of unmilled rice harvested per hectare) without relying on improved varieties or agrochemical inputs, but also to increase the outturn of milled rice, *i.e.*, kg of consumable rice per bushel of paddy, by 10-15%. This bonus on top of higher paddy yields is due to having fewer unfilled grains (less chaff) and fewer broken grains (less shattering).

In addition, farmers report - and researchers have verified - that SRI crops are more resistant to most pests and diseases, and better able to tolerate adverse climatic influences such as drought, storms, hot spells or cold snaps. The length of the crop cycle (time to maturity) is also reduced, with higher yields. Resistance to biotic and abiotic stresses will become more important in the coming decades as farmers around the world have to cope with the effects of climate change and the growing frequency of extreme events. The resistance of SRI rice plants to lodging caused by wind and/or rain, given their larger root systems and stronger stalks, can be quite dramatic. In general, one can say that use of SRI methods reduces the agronomic and economic risks that farmers face (Uphoff, 2007).

SRI differs from most agricultural technologies promoted in recent decades in that it is a civil-society innovation, originating not from research stations or laboratories, but from the dedicated work of a Jesuit priest, subsequently amplified and adapted through the efforts of farmers, NGOs and other non-state actors. Father Henri de Laulanié spent 34 years of his life, working with small scale farmers in Madagascar to devise better ways to raise paddy yields with the aim of reducing the pervasive poverty and hunger in that country (Laulanié, 2003). He sought low-cost methods that did not rely on expensive and environmentally-unfriendly external inputs and was able to succeed in his objective just by modifying the way that rice plants, soil, water and nutrients are managed (Laulanié, 1993).

India needs to increase rice production by 2.5 million tons a year in order to meet its requirement in 2050. This translates to almost 92% increase on its current production. Clearly, India can not produce so much rice using 3000-5000 liters of water per kg of rice production as it uses today. India can not expand its irrigated area to a level that can match goals in any land extension strategy, as there is paucity of additional land and water that can be mobilized. In the most intensively cropped areas under rice, where ground water is often used for irrigation, water table have been falling at the alarming rate of one meter per year or more.

Rice production in India increases 4.5 times during the last 57 years from 30.9 million tones in 1950 to 139.4 million tones in 2006. The rice productivity was also increased three times from 1.0 MT/ha in 1950 to 3.1 MT/ha in 2001-02. But rice productivity is now improving at a much slower rate compared to that recorded during earlier decades. Now all India mean annual growth rate is literally stagnant to 0.54%. In India the 'Green Revolution' has boosted the productivity of both rice and wheat but the revolution has given away technology that thirsts for enormous amount of water as well as chemical fertilizer. A Significant increase in production of rice is constrained not only by paucity of land for cultivation but also by shortage of water and need for application of chemical fertilizer has increased manifold.

Scarcity of water is more common in areas where the conventional water intensive method of irrigated rice cultivation through inundation is followed. About 70-80% of global fresh water withdrawals are for agricultural sector particularly for irrigation and rice accounts for about 85% of this, mainly due to inundated rice cultivation. In Asia, about 84% of water withdrawal is for agriculture, used mostly in flooded rice irrigation. At the global level around 63% of all rice is irrigated, although regional ratios vary. Most of the rice cultivation in major rice producing countries is in irrigated areas. In India and Indonesia, the proportion is approximately 50% and 47%, respectively. Irrigated ecosystem contributes to 75% of India's rice production.

So, now the countries like India has to adopt methods like System of Rice Intensification (SRI) through which rice can be cultivated with assured increase in productivity with 40-70% less water as well as chemical inputs in comparison with conventional inundated rice cultivation method. 'SRI' based on eight basic principles which are :

i) Preparing high quality land

ii) Developing nutrient rich and unflooded nurseries

iii) Using young seedlings for early transplantation

iv) Transplanting the seedlings singly

v) Ensuring wider spacing between seedlings

vi) Preferring compost or farm yard manure to synthetic fertilizers

vii) Managing water carefully so that the plant's root zone remain moisten but are not continuously saturated

viii) Weeding frequently.

SRI improves yield with less water, less seed and less chemical inputs than most conventional methods of inundated rice cultivation. The SRI is a methodology aimed at increasing the yield of rice produced in farming. It is a low water, labor-intensive, organic method that uses younger seedlings singly spaced and typically hand weeded with special tools. SRI concepts and practices have continued to evolve as they are being adapted to rain-fed (un-irrigated) conditions and with transplanting being superseded by direct-seeding sometimes.

The central principles of SRI (Cornell University, New York)

- Rice field soils should be kept moist rather than continuously saturated, minimizing anaerobic conditions, as this improves root growth and supports the growth and diversity of aerobic soil organisms.
- Rice plants should be planted singly and spaced optimally widely to permit more growth of roots and canopy and to keep all leaves photosynthetically active.
- Rice seedlings should be transplanted when young, less than 15 days old with just two leaves, quickly, shallow and carefully, to avoid trauma to roots and to minimize transplant shock.

## SRI Principles

SRI methodology is based on **four main principles** that interact with each other:

- Early, quick and healthy plant establishment
- Reduced plant density
- Improved soil conditions through enrichment with organic matter
- Reduced and controlled water application.

Based on these *principles*, farmers can adapt recommended SRI *practices* to respond to their agro-ecological and socioeconomic conditions. Adaptations are often undertaken to accommodate changing weather patterns, soil conditions, labor availability, water control, access to organic inputs, and the decision whether to practice fully organic agriculture or not. The most common SRI practices for irrigated rice production are summarized in the following section.

In addition to irrigated rice, the SRI principles have been applied to rainfed rice and to other crops, such as wheat, sugarcane, finger millet, pulses, showing increased productivity over current conventional planting practices. When SRI principles are applied to other crops, we refer to it as the *System of Crop Intensification* or SCI.

Recommended SRI Management Practices for Irrigated Conditions

- ***Rice Plants*** › seedlings are transplanted:
- ***Very young*** › at the 2 leaf-stage, usually between 8 and 12 days old
- ***Carefully and quickly*** › protecting the seedlings' roots and minimizing the transplanting shock

- ❑ ***Singly*** › one plant per hill instead of 3-4 together to avoid root competition
- ❑ ***Widely spaced*** › to encourage greater root and canopy growth
- ❑ ***In a square grid pattern*** › 25x25 cm or wider in good quality soil
- ❑ ***Soil*** › The soil is enriched with organic matter to improve soil structure, nutrient and water holding capacity, and favor soil microbial development. Organic matter represents the base fertilization for the crop and is complemented if needed by fertilizer.
- ❑ ***Water*** › Only a minimum of water is applied during the vegetative growth period. A 1-2 cm layer of water is introduced into the paddy, followed by letting the plot dry until cracks become visible, at which time another thin layer of water is introduced. During flowering a thin layer of water is maintained, followed by alternate wetting and drying in the grain filling period, before draining the paddy 2-3 weeks before harvest. This method is called 'intermittent irrigation' or 'Alternative Wetting and Drying' (AWD). Some farmers irrigated their fields every evening, other leave their fields drying out over 3-8 days, depending on soil and climate conditions.
- ❑ ***Nutrients*** › As soils are improved through organic matter additions, many nutrients become available to the plant from the organic matter. Additionally the soil is also able to hold more nutrients in the rooting zone and release them when the plants need them. Depending on the yield level and on the farming system, some farmers use exclusive organic fertilization for their SRI plots. The majority of farmers complement the organic matter amendment with chemical fertilizers, most often urea, in order to achieve a balanced fertilization of the crop.
- ❑ ***Weeds*** › While avoiding flooded conditions in the rice fields, weeds grow more vigorously, and need ideally be kept under control at an early stage. A rotary hoe - a simple, inexpensive, mechanical push-weeder – is most often used starting at 10 days after transplanting, repeated ideally every 7-10 days until the canopy is closing (up to 4 times). The weeder has multiple functions and benefits. i) It incorporates the weeds into the soil, where they decompose and their nutrients can be recycled, ii) it provides a light superficial tillage and aerates the soil, ii) it stimulates root growth by root pruning, iii) it makes nutrients newly available to the plant by mixing water with organic matter enriched top soil. A re-greening effect of the plants can be observed 1-2 days after

weeding, and iv) it redistributes water across the plot, contributing to a continuous leveling of the plot and eliminating water patches in lower laying areas in the field that create anaerobic conditions for the plants. The use of the weeder contributes to homogeneous field conditions, creating a uniform crop stand and leading to increased yields.

## Why SRI is a Climate Smart Agriculture Practice?

Climate change is major environmental hazard which is affecting major crop production throughout the world. Therefore, techniques and methodologies need to be designed properly to combat climate change. In the present context SRI technology is major step ahead towards combating climate change. Under these changing scenario of climate change, climate smart agriculture practice includes some fundamental issues which needs to be implemented at field level throughout the world to maintain rice production under the changing scenario of climate change. Therefore, SRI includes certain vital characteristics such as reduced water requirements- higher crop water-use efficiency which benefits for both natural ecosystems and people in competion with agriculture for scare water supplies. Secondly, less use of inorganic fertilizer, reactive N is "the third major threat to our planet after biodiversity loss and climate change" - already returns are greatly diminishing. Thirdly, less dependency on agrochemicals for crop protectionwhich enhances the quality of both soil and water as well as buffering against the effects of climate change on drought, storms (resist lodging), cold temperature. Fourthly, some reduction in GHG such as $CH_4$ is reduced without producing offsetting NO emissions, also some reduction made in carbon footprint with less production, transportation and use of fertilizers.

## Approaches Towards Climate Smart Agriculture

SRI plants show improved resistance to drought, floods, storms, pests and diseases. SRI plants thrive with 30-50% less irrigation water per land area, due to deeper, larger, less senescing root systems, reduced competition among plants creates stronger plants above and below ground and organic matter-enriched soils able to store more water and furnish nutrients. SRI also promotes stronger and healthier plants as well as less humidity in the plant canopy. SRI contributes towards thicker tillers, deeper roots, wider spacing, increased uptake of silicon into leaves and tillers from soil that has aerobic conditions and reduced lodging (10%), against (55%) lodging under conventional cultivation methods.

From productivity perspectives SRI also refelects multidimensional benefits for the agricultural system. Rice yields are increased by 20-50% – sometimes >100-200%. Also SRI methods work for hybrids, HYVs, local and indigenous varieties. As a result reduction in water use due to irrigation purpose seems to be reduced by 30-50% per hectare on one side and on the other hand more grain output (30-100%). SRI also facilitates higher nutrient use efficiency for crop plants, less fertilizer and agrochemical inputs needed by 30-50%, and by 100% with organic SRI.

Mitigatory measures towards climate change by means of smart agricultural practices is an essential pre-requisites for sustainable crop production. In this context SRI is a positive step ahead. SRI enhances carbon sinks and lowers emissions that contribute to global warming potential. Firstly, SRI rice plants sequester more carbon – higher grain and straw yield, and more root biomass. Secondly, increased soil organic matter through SRI practices that improve the soil with more organic matter application and increased root exudates. Thirdly, *associated agro-ecological practices sequester carbon*, such as green manure production, integration with agroforestry, surface mulch applications. Fourthly, *reduced carbon footprint* due to less use of agrochemicals (including the manufacturing, and shipping of fertilizer). Fifthly, reduced greenhouse gas (GHG) emissions from paddy soils such as *Methane (CH4) emission is reduced* by between 22% and 64%, as soils are maintained under mostly aerobic conditions.

## Criticism

The productivity of SRI is under debate between supporters and critics of the system. Critics of SRI suggest that claims of yield increase in SRI are due to unscientific evaluations. They object that there is a lack of details on the methodology used in trials and a lack of publications in the peer-reviewed literature (McDonald *et al*, 2006 and 2008). Some critics have suggested that SRI success is unique to soil conditions in Madagascar (Surridge, 2004). While critics still claim there are over 700 articles and communications about SRI that are published in international scientific journals as of January 2016 that reveals the benefits of SRI system.

## Conclusion

Climate change is an alround phenomenon hampering ecological, economical and social perspectives of sustainable development. In the present scenario of the changing world, sustainable agriculture practices are boon for human civilization. Therefore, in agriculture sector we need to adopt and develop mechanisms that would lead to feed our growing population without hampering the sustainability of agro-ecosystem. It is due to the abuse of environment by mankind that has thrown human civilization into the era of climate change. It is a sort of stress which can be mitigated only by adopting suitable strategies towards smart agricultural practices. SRI technology is a part of sustainable agricultural practices which would help to maintain crop production by maintaining the overall sustainability of the agro-ecosystem in various dimensions.

## References

De Laulanié, H. (2011). Intensive Rice Farming in Madagascar. Tropicultura, 29(3): 183-187.

Journal Articles about the System of Rice Intensification(SRI) (2016). SRI International Network and Resources Center. Cornell University. Retrieved 18 January 2016.

Kabi, M.C. (2006). Organic farming: Status and prospects; seminar on organic farming: Resurgence of Traditional Agriculture. Souvenir, Bhubaneswar, P. 27-32.

Laulanié, H. (1993). Le système de riziculture intensive malgache. *Tropicultura,* 11: 110-114.

Laulanié, H. (2003). Le Riz à Madagascar: Un dèveloppement en dialogue avec les paysans, Editions Karthala, Paris.

McDonald, A.J., Hobbs, P.R. and Rihaa, S.J. (2008). Does the system of rice intensification outperform conventional best management?: A synopsis of the empirical record in Field Crops Research, 108(2): 188-191.

McDonald, A.J., Hobbs, P.R. and Rihaa, S.J. (2006). Field Crops Research Stubborn facts: Still no evidence that the System of Rice Intensification out-yields best management practices (BMPs) beyond Madagascar in Field Crops Research, 96(1): 31–36.

Mishra, A., Whitten, M., Ketelaar, J.W. and Salokhe, V.M. (2006). The System of Rice Intensification(SRI): A challenge for science, and an opportunity for farmer empowerment towards sustainable agriculture. International Journal of Agricultural Sustainability 4: 193-212.

Randriamiharisoa, R., Barison, J. and Uphoff, N. (2006). Soil biological contributions to the System of Rice Production. In: N. Uphoff et al. (eds.), Biological Approaches to Sustainable Soil Systems, 409424, CRC Press, Boca Raton, FL.

Stoop, W., Uphoff, N. and Kassam, A. (2002). A review of agricultural research issues raised by the System of Rice Intensification (SRI) from Madagascar: Opportunities for improving farming systems for resource-poor farmers, Agricultural Systems, 71: 249-274.

Subba Rao, N.S. (1979). Chemically and biologically fixed nitrogen-Potentials and Prospects , P-1-7. In: Recent Advances in Biological Nitrogen fixation, Ed, N.S, Subba Rao: Oxford & IBH Publishing Co. New Delhi.

Surridge, C. (2004). Rice cultivation: Feast or famine? *Nature,* 428: 360–361. doi:10.1038/428360a.

Uphoff, N. (2003). Higher yields with fewer external inputs? The System of Rice Intensification and potential contributions to agricultural sustainability, International Journal of Agricultural Sustainability, 1: 38-50.

Uphoff, N. (2007). Reducing the vulnerability of rural households through agroecological practice: Considering the System of Rice Intensification (SRI). Mondes en Développement, 35:4.

US NRC (2011). Food Production, Prices, and Hunger. In: Ch 5: Impacts in the Next Few Decades and Coming Centuries, p.161.

# 12

# Biofertilizers for Sustainable Agriculture Benefits, Types and Application

**Krishna Kumar Patel, Priyata Chaudhary and Upendra Singh**

## Introduction

The current population of our country is about 1.27 billion and the world population is predicted to grow from 6.9 billion in 2010 to 8.3 billion in 2030 and to 9.1 billion in 2050. By 2030, food demand is predicted to increase by 50% (70% by 2050). The main challenge facing the agricultural sector is not so much growing 70% more food in 40 years, but making 70% more food available on the plate (UNDESA, 2014). This rising food demand automatically put burden on the shoulder of formers especially for ASEAN countries such as India where population growth rate is high while resources are limited. India has about 141 million hectares area under cultivation out of its geographical area which 329 million hectares (Raghuwanshi, 2012; Parikh and James, 2012). Consequently, agriculture is, thus, one of the most important factors contributing to the economic growth of India (Kniivila, 2007). The production of the agricultural produce should be up to the mark to maintain this contribution. In order to enhance the crop production, fertilizers are needed because they supply the nutrients to the soil. By adding fertilizers, crop yields can often be doubled or even tripled.

Further, fertilizers ensure the most effective use of both land and water. Where rainfall is low or crops are irrigated, the yield per unit of water used may be more than doubled and the rooting depth of the crop increased through fertilizer application. Every plant nutrient, whether required in large or small amounts, has a specific role in plant growth and food production. One nutrient cannot be substituted for another (Anon 2013). The continued use of chemical fertilizers and its increasing trend since last 150 years have not only left soils degraded, polluted and less productive but have also posed severe health and environmental hazards. Thus, it is the great challenge before scientist, researchers, farmers, etc. to save fertile field, environment and also simultaneously to increase the crop yield to meet the growing demand of by developing new farming methods/technologies. Sustainable farming/or organic farming using of biofertilizers are among such methods able to solve these issues and make the ecosystem healthier to some extent. The current global market for organically raised agricultural products is valued at around US$ 30 billion with a growth rate of around 8 percent. Nearly 22 million hectares of land are now cultivated organically. Organic cultivation represents less than 1 percent of the world's conventional agricultural production and about 9 percent of the total agricultural area (Mazid and Khan 2014; Mishra 2014). Fertilisers can be roughly categorized into three types: chemical, organic and biofertilizer. Biofertilizer among these are supposed to be an ecofreindly way to replace chemical fertilizer. Figure 1 shows the global biofertilizer market revenue share by product segment in 2012.

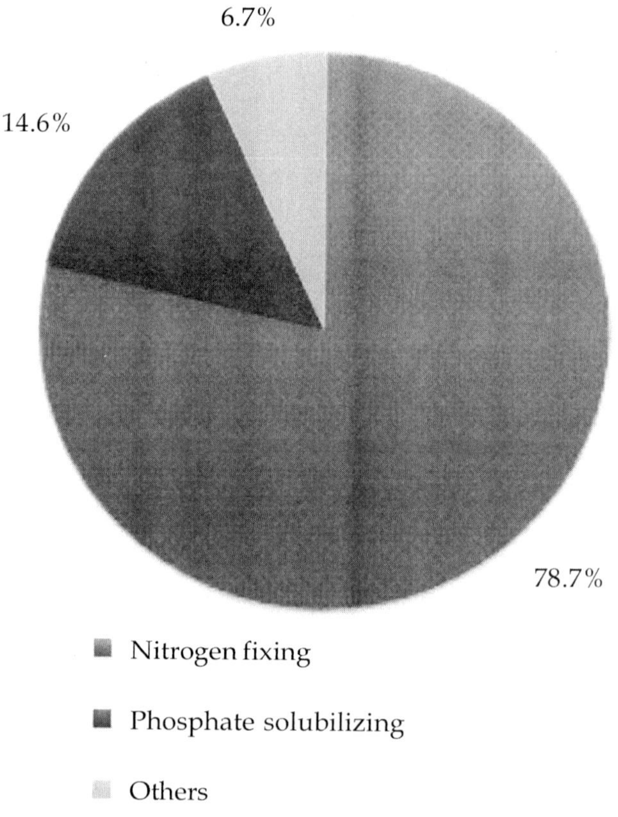


**Figure 1.** Global biofertilizer market revenue share by product segment in 2012 (*Source:* Novozyme, AgriLife, Primary Interviews, Transparency Market Research)

## What is Biofertilizer

A biofertilizer is an organic product containing a specific microorganism in concentrated form, which is derived either from the plant roots or from the soil of root zone. These bio-inputs, or bio-inoculants are the products containing living cells of different types of microorganisms that have the ability to mobilize nutritionally important elements from non-usable forms through biological stress and improve plant growth and yield. Biofertilizers, therefore, can be defined as substances contain living microorganisms or living cells or latent cells of efficient strains of microorganisms that help crop plants, for uptaking of nutrients by their interactions in the rhizosphere or that colonize in the rhizosphere, or the interior of the plant when applied through seed or soil. By increasing the availability of primary nutrients and/or growth stimulus to the target crop when applied to seed, plant surfaces, or soil, biofertilizers promote growth of the plant and accelerate certain microbial processes in the soil which augment the extent of availability of nutrients in a form easily assimilated by plants.

In addition, biofertilizers are cost effective, eco-friendly and have shown great potential as supplementary, renewable and environmental friendly sources of plant nutrients and are an important component of Integrated Nutrient Management (INM) and Integrated Plant Nutrition System (IPNS) (Raghuwanshi, 2012). By inoculation of biofertilizers farmers can enhance the crop yield up to 10-40% and fix nitrogen up to 40-50 Kg (Chavada, 2010; Youssef and Eissa, 2014).

Furthermore, biofertilizers improve soil texture, pH, and other properties of soil and continue use of these inputs in soil about 3-4 years need no application of as these parental inoculums are sufficient for growth and multiplication further. Biofertilizers could be applied to the field directly because its good moisture content about 75 % and it has been found as superior farmyard manure and other type of manure in terms of N (3.5% - 4%), P (2% - 2.5%), K (1.5%). In addition, plant growth promoting substances such as IAA amino acids, vitamins, etc. also process by this type of manure. In spite of this, the production of biofertilizers in 2010-11 in our country is about 37,997 mt while its production capacity is 98,000 mt/annum. India has about 225 biofertilizer production unit (NCOF, 2011) and average annual consumption of biofertilizers in the country is around 64g/ha (Mishra, 2014). Figure 2 represented the schematic mass production of bacterial biofertilizers.

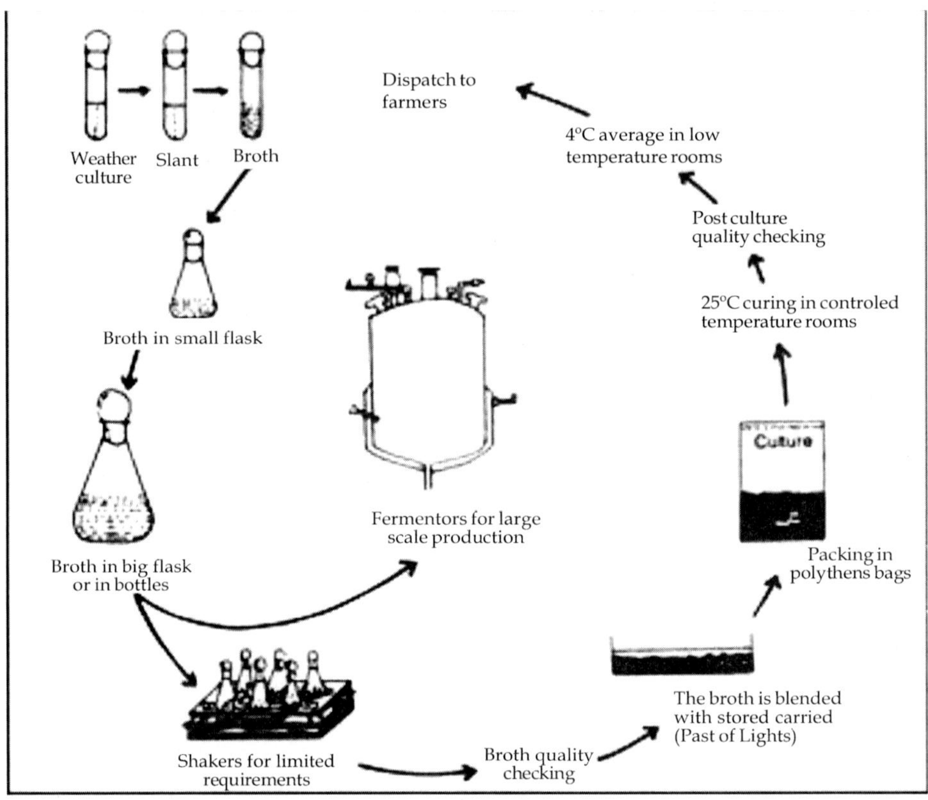


**Figure 2.** Schematic representation of mass production of bacterial biofertilizers (Anonymous, 2014)

Nevertheless, all type of fertilizers act as catalysts in providing nutrients to plants for their optimum growth and yield. They can be grouped in different ways based on their nature and function.

## Type of Biofertilizers

### Nitrogen Fixers

#### Azospirillum

*Azospirillum* (Fig. 3a) *lipoferum* and *A. brasilense* (*Spirillum lipoferum* in earlier literature) are primary inhabitants of soil, the rhizosphere and intercellular spaces of root cortex of graminaceous plants. They perform the associative symbiotic relation with the graminaceous plants. Azospirillum belongs to family *Spirilaceae*; it is heterotrophic and associative in nature and the bacteria of Genus Azospirillum are $N_2$ fixing organisms isolated from the root and above ground parts of a variety of crop plants. They are Gram negative, Vibrio or Spirillum having abundant accumulation of polybetahydroxybutyrate (70 %) in cytoplasm. In addition to their nitrogen fixing ability of about 20-40 kg/ha, they also produce growth-regulating substances. There are many species under this genus such as *A.amazonense, A.halopraeferens, A.brasilense, A.lipoferum, A.irakense,* etc. The benefits of inoculation of species mainly *A.lipoferum* and *A.brasilense*, however, have been seen worldwide. The *Azospirillum* form associative symbiosis with many plants particularly with those with the C4-dicarboxyliac pathway of photosynthesis (Hatch and Slack pathway), because they grow and fix nitrogen on salts of organic acids such as malic and aspartic acid. Thus it is mainly recommended for maize, sugarcane, sorghum, pearl millet, etc. Apart from nitrogen fixation, growth promoting substance production (IAA), disease resistance and drought tolerance are some of the additional benefits due to Azospirillum inoculation.

## Azotobacter

*Azotobacter* (Fig. 3b), belongs to family *Azotobacteriaceae*. It is aerobic, free living, and heterotrophic in nature, colonizing the roots not only remains on the root surface but also penetrates into the root tissues and lives in harmony with the plants (Anon, 2014). They do not, however, produce any visible nodules or outgrowth on root tissue. Of the several species of *Azotobacter*, *A. chroococcum* happens to be the dominant inhabitant in arable soils capable of fixing N2 (2-15 mg N2 fixed /g of carbon source) in culture media and the organism proliferates under both anaerobic and aerobic conditions but it is preferentially micro-aerophilic in the presence or absence of combined nitrogen in the medium. *Azotobacters* are present in neutral or alkaline soils and *A. chroococcum* is the most commonly occurring species in arable soils, while *A. vinelandii,A. beijerinckii, A. insignis* and *A. macrocytogenes* are also common species (Mishra, 2014). The bacterium produces abundant slime which helps in soil aggregation and it produces anti-fungal antibiotics that inhibit the growth of several pathogenic fungi in the root region and help prevent seedling mortality (Anon., 2014). The numbers of *A. chroococcum* in Indian soils rarely exceeds 105/g soil due to lack of organic matter and the presence of antagonistic microorganisms in soil. However, The population of *Azotobacter* is generally low in the rhizosphere of the crop plants and in uncultivated soils due to lack of organic matter and presence of antagonistic microorganisms in soil. The occurrence of this organism has been reported in the rhizosphere of a number of crop plants such as rice, maize, sugarcane, bajra, vegetables and plantation crops (Mishra *et al.*, 2013).

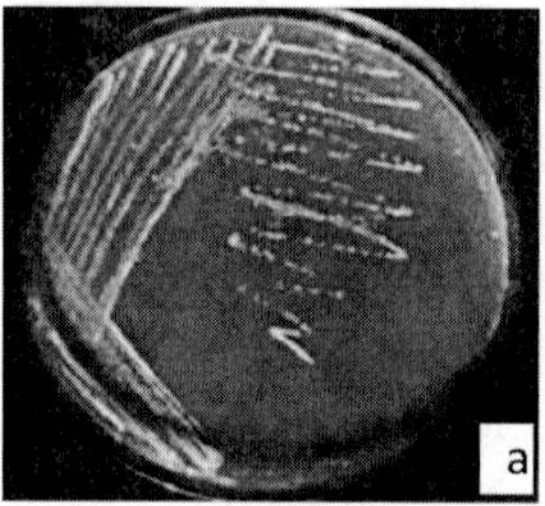

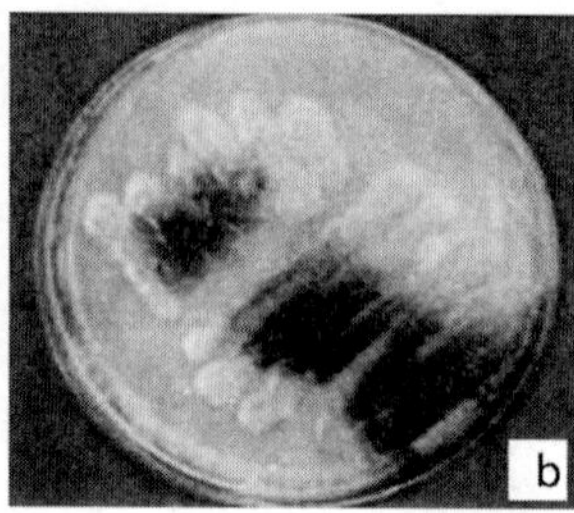

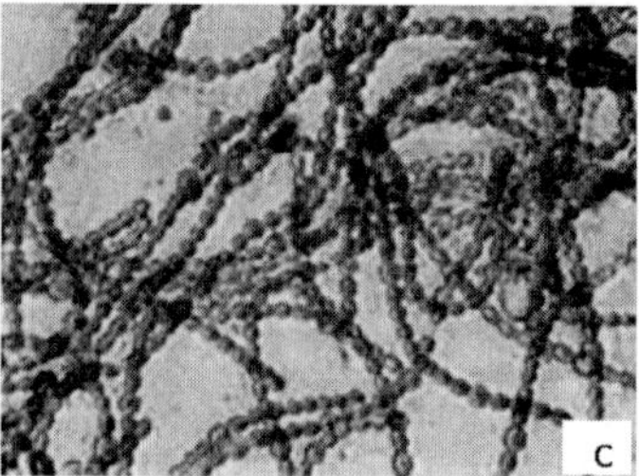


**Figure 3.** Strains of (a) Azospirillum, (b) Azotobacter and (c) Cyanobacteria (BGA)

### *Blue Green Algae (Cyanobacteria)*

Blue-green algae, which scientific name is cyanobacteria (Fig. 3c), are naturally found in most commonly in shallow, warm, and slow-moving or still water such freshwater ponds, freshwater ecosystem and wetlands. While some species are blue-green, it can vary in colour from olive and dark green to purple and even yellowish. Cyanobacteria is of concern because it can produce toxins under certain conditions. A combination of excess nutrients, sunlight and high temperatures can lead to a rapid increase in blue-green algae, called a "bloom. BGA are phototrophic in nature and produce auxin, indole acetic acid and gibberllic acid, and fix 20-30 kg nitrogen/ha in submerged rice fields and belong to eight different families (Mishra 2014). Both free-living as well as symbiotic cyanobacteria (blue green algae) have been harnessed in rice cultivation in India. The symbiotic associations form is capable of fixing nitrogen with fungi, liverworts, ferns and flowering plants. Soil nitrogen and Biological Nitrogen Fixation (BNF) by associated organisms are major sources of nitrogen that is key input required in large quantities for low land rice. According to Roger and Ladha (1992), 50-60% of total nitrogen need is supplied by the mineralization of organic nitrogen from the soil and BNF by free living and rice plant associated bacteria. Filamentous, consisting of chains of vegetative cells including specialized cells called heterocysts, which function as micronodules for synthesis and nitrogen-fixing machinery are the most nitrogen fixing BGA. In addition, researchers have reported that when a composite culture of BGA having heterocystous *Nostoc, Anabaena, Aulosira etc.* is given as primary inoculum in trays, polythene lined pots, later mass multiplied in the field for application as soil based flakes to the rice growing field at the rate of 10 kg/ha and found that the final product is not free from extraneous contaminants and also not very often monitored for checking the presence of desired algal flora.

So, it can be said that to achieve food security through sustainable agriculture, the requirement for fixed nitrogen must be increasingly met by BNF rather than by industrial nitrogen fixation (Kumar *et al.*, 2013; Mishra *et al.*, 2013; Mishra, 2014). Since, the benefits due to algalization could be to the extent of 20-30 kg N/ha under ideal conditions. However, there is a limitation in the preparation of BGA biofertilizer by the labour oriented methodology.

### *Azolla*

*A. pinnata* (Fig. 4), which can be propagated on commercial scale by vegetative means, is the most commonly occurring species in India. It is a small-leaf floating fern, which contains an endosymbiotic community living in the dorsal lobe cavity of the leaves. The presence in this cavity of a nitrogen fixing filamentous cyanobacteria - *Anabaena azollae* - turns this symbiotic association into the only fern-cyanobacteria association that presents agricultural interest by the nitrogen input that this plant could introduce in the fields (Carrapico *et al.*, 2012). *Azolla* is used as biofertilizer for rice crop and the important factors of *Azolla* are its quick and the efficient availability of its nitrogen to rice plants. Since, its fronds consist of sporophyte with a floating rhizome and small overlapping bi-lobed leaves and roots. *Azolla* is used as biofertilizer for wetland rice and it is known to contribute 40-60 kg N/ha per rice crop. For instance, rice growing areas have recently been evincing increased interest in the use of the symbiotic $N_2$ fixing water fern *Azolla* either as an alternate nitrogen sources or as a supplement to commercial nitrogen fertilizers. *Azolla* can be applied not only as green manure or biofertilizers, or biomanures for nitrogen fixation by incorporating it in the fields prior to rice planting but also it contribute significant amounts of phosphorous, potassium, sulfur, zinc, iron, molybdenum and other micronutrients.

**Figure 4.** *Azolla pinnata* sub sp. *africana* (Desv.) (Saunders and Fowler, 1992).

### *Rhizobium*

*Rhizobium* inoculant was first made in USA and commercialized by private enterprise in 1930s. *Rhizobium*, useful for legumes like chickpea, red-gram, pea, lentil, black gram, and oil-seed legumes like soybean and groundnut, and forage legumes like berseem and Lucerne, belongs to family *Rhizobiaceae*. It is a soil habitat bacterium and colonizes the roots of specific legumes to form tumor like growths called root nodules, which act as factories of ammonia production and is symbiotic in nature, fixes nitrogen 50-100 kg/ha with legumes only. The morphology and physiology of *Rhizobium* (Fig. 5a) will vary from free-living condition to the bacteroid of nodules. They are the most efficient biofertilizer as per the quantity of nitrogen fixed concerned. They have seven genera and highly specific to form nodule in legumes, referred as cross inoculation group.

In addition, it has the ability to fix atmospheric nitrogen in symbiotic association with legumes and certain non-legumes like *Parasponia* (Mishra, 2014). Generally, its population in the soil depends on the presence of legume crops in the field.

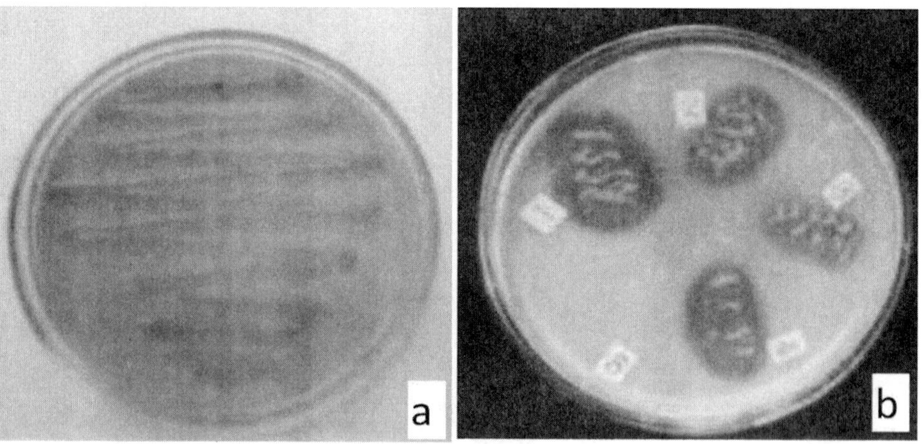


**Figure 5** *(a)* Rhizobium (b) Phosphobacteria

However, in absence of legumes the population found to be decreased and to form effective nodules, each legume requires a specific species of *Rhozobium*. Sometime to restore the population of effective strains of the *Rhizobium* near the rhizosphere to hasten N-fixation, an artificial seed inoculation is often needed (Mishra *et al.*, 2013).

## Phosphate Solubilizers

The species of *Pseudomonas, Bacillus, Penicillium, Aspergillus, Rhizobium, Burkholderia, Achromobacter, Agrobacterium, Microccocus, Aereobacter, Flavobacterium* and *Erwinia, etc.* among the various soil bacteria, fungi and plant rhizospheres are considerable population that secrete organic acids and lower the pH in their vicinity to bring about dissolution of bound phosphates in soil and consequently solubilize the phosphate. Phosphate solubilizing bacteria (Fig. 5b) includes both types such as aerobic and anaerobic strains, with a greater prevalence of aerobic strains in submerged soils. Since, the major microbiological means by which insoluble-phosphorous compounds are mobilized is by the production of organic acids, accompanied by acidification of the medium and thus the organic and inorganic acids convert tricalcium phosphate to diand-monobasic phosphates, resulting in an enhanced availability of the phosphorous to the plant. Various studies have reported that the ability of different bacterial species to solubilize insoluble inorganic phosphate compounds, such as tricalcium phosphate, dicalcium phosphate, hydroxyapatite, and rock phosphate. In spite of genera of *Pseudomonas* and *Bacillus* and *Fungi among* soil bacteria are more common, the phosphate solubilizing bacteria in the rhizosphere in comparison with non-rhizosphere soil commonly found higher in concentration.

## P Mobilizing Biofertilizers

Phosphate mobilizing or phosphorus solubilizing Biofertilizers/ microorganisms (bacteria, fungi, mycorrhiza etc.) converts insoluble soil phosphate into soluble forms by secreting several organic acids and under optimum conditions they can solubilize/mobilize about 30-50 kg $P_2O_5$/ ha due to which crop yield may increase by 10 to 20% (Anon 2011). Phosphorus mobilizing biofertilizers such as Arbuscular Mycorrhiza, AM fungi, etc. are the fungus that penetrates the cortical cells of the roots, increase surface area of roots and displace of absorption equilibrium of phosphate ions which increases the transfer of P ions and stimulate metabolic processes, and Arbuscles absorb these nutrients into the root system. Arbuscular mycorrhizal (AM) fungi facilitate phosphorus uptake in plants and also perform several other functions that are equally beneficial. Sikes (2010) studied the mechanism of mycorrhizal fungi protect plant roots from pathogens and reported that AM fungal colonization was not limited and simple roots are more mycorrhizal dependent.

## Biofertilizers for Micro nutrients (Silicate and Zinc Solubilizers)

Several reports confirm that *Bacillus* sp., isolated from the soil of granite crusher yard is capable of dissolving silicate minerals under *in vitro* condition. Different locations with varying degree of silicate solubilizing potential was chosen for the isolation of bacterium during the examination of anthrpogenic materials like cement, agro inputs (super phosphate and rock phosphate). For instance, Murali *et al.* (2005) isolated silicate solubilizers using modified Bunt and Rovira medium from soil samples collected from coconut palms. Majority of the silicate solubilizers were identified as *Bacillus*sp. and *Pseudomonas* sp.

Zinc, found in the earth's crust at a concentration of 0.008 percent, is of utmost importance as more than 50 percent of Indian soils exhibit zinc deficiency with content far below the critical level of 1.5 ppm of available zinc. The zinc fixation is, thus, important and it can be done by two mechanism first one operates in acidic soils and is closely related with cation exchange and other operates in alkaline conditions where fixation takes place by means of chemisorptions (chemisorptions of zinc on calcium carbonate form a solid-solution of $ZnCaCO_3$), and by complexation by organic legends (Mishra 2014). Therefore, the plant deficiencies in absorbing zinc from the soil can be overcome either by external application of soluble zinc sulfate ($ZnSO_4$) or by the host of microorganisms found in the soil that can be used as biofertilizers to supply zinc as well as other micronutrients like iron, copper, etc. However, nitrogen fixers like *Rhizobium, Azospirillum, Azotobacter, BGA* and phosphate solubilizing bacteria like *B. magaterium, Pseudomonas striata,* and phosphate mobilizing mycorrhiza have been widely accepted as bio-fertilizers and only supply major nutrients. The microorganisms viz., *B. subtilis, Thiobacillus thioxidans* and *Saccharomyces sp.* can be used to solubilize Zinc and thus these microorganisms can be used as biofertilizers for the solubilization of fixed micronutrients like zinc (Kumar *et al.,* 2013).

## Application of Biofertilizers

Each type has its own advantages and disadvantages. These advantages need to be integrated in order to achieve optimum performance by each type of fertilizer and to realize balanced nutrient management for crop growth. Use of biofertilizers is one of the important components of integrated nutrient management, as they are cost effective and renewable source of plant nutrients to supplement the chemical fertilizers for sustainable agriculture. Biofertilizer generally applied for seed treatment or seed inoculation, seedling root dip and main field application. In the given Table 1, the recommendations of biofertilizers in one packet of 200 g are represented.

**Table 1.** Biofertilizers recommendation (one packet - 200 g)

| Crop | Seed | Nursery | Seedling dip | Main field | Total requirement of packets per ha |
|---|---|---|---|---|---|
| Rice | 5 | 10 | 5 | 10 | 30 |
| Sorghum | 3 | - | - | 10 | 13 |
| Pearl millet | 3 | - | - | 10 | 13 |
| Ragi | 3 | - | 5 | 10 | 18 |
| Maize | 3 | - | - | 10 | 13 |
| Cotton | 3 | - | - | 10 | 13 |
| Sunflower | 3 | - | - | 10 | 13 |
| Castor | 3 | - | - | 10 | 13 |
| Sugarcane | 10 | - | - | 36 (3 split) | 46 |
| Turmeric | - | - | - | 24 (2 split) | 24 |
| Tobacco | 1 | 3 | - | 10 g/pit | 14 |
| Papaya | 2 | - | - | 10 | - |
| Mandarin Orange | 2 | - | - | 10 g/pit | - |
| Tomato | 1 | - | - | 10 | 14 |
| Banana | - | - | 5 | 10 g/pit | 130 |

Anon (2011)

For seed treatment or seed inoculation, one packet of the inoculants is mixed with 200 ml of rice kanji to make slurry. The seeds required for an acre are mixed in the slurry so as to have a uniform coating of the inoculants over the seeds and then shade dried for 30 minutes. The shade dried seeds should be sown within 24 hours. One packet of the inoculants (200 g) is sufficient to treat 10 kg of seeds.

Similarly, the seedling root dip method is used for transplanted crops. Two packets of the inoculants are mixed in 40 liters of water. The root portion of the seedlings required for an acre is dipped in the mixture for 5 to 10 minutes and then transplanted. However, four packets of the inoculants are mixed with 20 kgs of dried and powdered farm yard manure and then broadcasted in one acre of main field just before transplanting in main field application method. For instance, for all legumes *Rhizobium* is applied as seed inoculants and in the transplanted crops, *Azospirillum* is inoculated through seed, seedling root dip and soil application methods. However, for direct sown crops, *Azospirillum* is applied through seed treatment and soil application. Onther other hand, Phosphobacteria inoculated through seed, seedling root dip and soil application methods as in the case of *Azospirillum*. Bacterial biofertilizers can also be applied in combined application. The inoculants should be mixed in equal quantities. Phosphobacteria, therefore, can be mixed with *Azospirillum* and *Rhizobium*.

## Benefits and Drawbacks Biofertilizers

The nutrient supply is more balanced, which helps keep plants healthy. They enhance soil biological activity, which improves nutrient mobilization from organic and chemical sources and decomposition of toxic substances. They enhance the colonization of mycorrhizae, which improves phosphorous supply. They enhance soil structure, leading to better root growth. They increase the organic matter content of the soil, thereby improving the exchange capacity of nutrients, increasing soil water retention, promoting soil aggregates and buffering the soil against acidity, alkalinity, salinity, pesticides and toxic heavy metals. They release nutrients slowly and contribute to the residual pool of organic nitrogen and phosphorous in the soil, reducing nitrogen leaching loss and phosphorous fixation; they can also supply micronutrients. They encourage the growth of beneficial microorganisms and earthworms. They help to suppress certain soil-borne plant diseases and parasites. There by improving the exchange capacity of nutrients, increasing soil water retention, promoting soil aggregates and buffering the soil against acidity, alkalinity, salinity, pesticides and toxic heavy metals.

All biofertilizers have not only benefits but also several drawbacks such as short shelf life, lack of suitable carrier materials, susceptibility to high temperature, problems in transportation and storage, etc. that still need to be solved in order to promote the application of biofertilizers. In addition, application of a larger volume of biofertilizers is needed to provide enough nutrients for crop growth since these are comparatively low in nutrient content and also the nutrient release rate is too slow to meet crop requirements in a short time, hence some nutrient deficiency may occur. The nutrient composition of compost is highly variable and major plant nutrients may not exist in organic fertilizer in sufficient quantity to sustain maximum crop growth. Various benefits and drawbacks of biofertilizers are given in Table 2.

**Table 2.** Benefits and drawbacks of biofertilizers

| S.No. | Benefits |
|---|---|
| 1. | Balanced nutrient supply. |
| 2. | Suppress certain soil-borne plant diseases and parasites. |
| 3. | Enhance the colonization of mycorrhizae. |
| 4. | Enhance soil structure. |
| 5. | Increase the organic matter content of the soil. |
| 6. | Release nutrients slowly. |
| 7. | Provide protection against drought and some soil borne diseases. |
| 8. | Encourage the growth of beneficial microorganisms and earthworms, and stimulate plant growth. |
| 9. | Enhance biological activity of soil or activate the soil biologically |
| 10. | Increase Crop yield by 20-30%. |
| 11. | Restore natural soil fertility. |
| 12. | Ecofreindly |
| 13. | Reduces the costs towards fertilizers use, especially regarding nitrogen and phosphorus. |
| **Drawbacks** | |
| 1. | Short shelf life |
| 2. | Lacks of suitable carrier materials |
| 3. | Susceptible to high temperature |
| 4. | Problems in transportation and its storage |
| 5. | High cost in comparison to other type of fertilizer |
| 6. | Consists highly variable nutrient composition |
| 7. | Nutrient release is some too slow to nutrient deficiency |
| 8. | Sometimes it lacks major plant nutrients |

## Conclusion

Biofertilizer is a commercialized form of microbial inoculants that used in the agricultural field as a replacement to conventional (chemical) fertilizers which are not eco-friendly. In addition to this, chemical fertilizer are also responsible for polluting water, air and soil and spreading cancer causing agents. Researchers, therefore, have taken this challenge and developed a substance contains living cells of different types of microorganisms called as biofertilizer. Biofertilizer, thus, can be used to prevent pollution and to make this world healthy in a natural way. In India, it was started to produce in late seventies and its manufacturing is continue on a large scale, so far. In spite of this and despite good potentiality of biofertilizer usage, the actual utilization is very low at about 2% of its potential. Since, sustainability of productive soils is a major concern at the international level. The utilization of Biofertilizer results in several benefits to farmers, industries, environment and overall national economy since these are of low cost, renewable sources of plant nutrients and can be used either for seed treatment or soil application or liquid form. In addition, Biofertilizer generate plant primary nutrients and it also rejuvenates the degraded soils by protecting topsoil. So, the dwelling of fossil fuel reserves and the increasing costs of commercial nitrogen fertilizers implicate finding other alternatives, such as the use of biofertilizers.

## References

Anonymous (2011). Role of Biofertilizers in soil fertility and Agriculture. On line article by My Agriculture Information Bank. http://agriinfo.in, accessed date: 28.02.2015.

Anonymous (2013). Why use fertilizer. EroChem, http://eurochem. we are testing.co.uk/what-we-do/why-use-fertilizers. Accessed date: 26.02.2015.

Anonymous (2014). Organic Farming: Organic Inputs and Techniques. http://agritech.tnau.ac.in, accessed date: 26.02.2015.

Carrapico F, Teixeira G, Diniz MA (2002). Azolla as a Biofertilizer in Africa: A Challenge for the Future. Biotechnology of Biofertilizers, 277.

Chavada NB, Patel R, Vanpuria S, Raval BP, Thakkar PV (2010). A study on isolated diazotrophic (non-symbiotics) bacteria from saline desert soil as a biofertilizer. Int J Pharm Appl Sci, 1(1): 52-4.

Kniivila M (2007). Industrial development and economic growth: Implications for poverty reduction and income inequality. Industrial Development for the 21st Century: Sustainable Development Perspectives, 295-333.

Kumar K Shukla UN, Kumar D, Pant AK, Prasad Sk, Parasd K (2013). Bio-Fertilizers for Organic Agriculture. Popular Kheti, 1(4):91-96

Mazid M, Khan TA (2014). Future of Bio-fertilizers in Indian Agriculture: An Overview. International Journal of Agricultural and Food Research, 3(3): 10-23.

Mishra P (2014). Rejuvenation of Biofertilizer for Sustainable Agriculture and Economic Development. Consilience: The Journal of Sustainable Development, 11(1): 41-61.

Mishra D, Rajvir S, Mishra U, Kumar SS (2013). Role of bio-fertilizer in organic agriculture: a review. Research Journal of Recent Sciences ISSN, 2277, 2502.

Murali G, Gupta A, Nair RV (2005). Variations in hosting beneficial plant associated microorganisms by root (wilt) diseased and field tolerant coconut palms of west coast tall variety. Current Science, 89: 1922-1927.

NCOF (2011). Annual Report 2010-11, National Centre of Organic Farming, Department of Agriculture and Cooperation, Ghaziabad, UP, Publ June 2011.

NCOF (2012). Annual Report 2011-12, National Centre of Organic Farming, Department of Agriculture and Cooperation, Ghaziabad, UP, Publ June 2012.

Parikh SJ, James BR (2012). Soil: The Foundation of Agriculture. Nature Education Knowledge, 3(10): 2

Pikovskaya RI (1948). Mobilization of phosphorus in soil in connection with vital activity of some soil microbial species. Mikrobiologiya, 17:362–370

Raghuwanshi R (2012). Opportunities and challenges to sustainable agriculture in India, NEBIO, 3(2): 78-86.

Saunders RM K, Fowler, K. (1992). A morphological taxonomic revision of *Azolla* Lam. section *Rhizosperm* (Mey.) Mett. (Azollaceae). Botanical Journal of the Linnean Society. 109: 329-357.

UNDESA (United Nation Department of Economic and Social Affairs) (2014). Food and Food Security. http://www.un.org/waterforlifedecade/food_security.shtml, accessed date: 27.02.15.

Youssef, M.M.A., & Eissa, M.F.M. (2014). Biofertilizers and their role in management of plant parasitic nematodes–a review. J Biotechnol Pharm Res, 5, 1-6.

Sikes BA (2010). When do arbuscular mycorrhizal fungi protect plant roots from pathogens? Plant Signaling & Behavior, 5(6): 763–765.

# 13

# Role of Climate Change Under Rainfed Farming in Western Rajasthan, India

**M.L. Meena and Dheeraj Singh**

## Introduction

Climate is one of the main determinants of agricultural production. Throughout the world there is significant concern about the effects of climate change and its variability on agricultural production. Researchers and administrators are concerned with the potential damages and benefits that may arise in future from climate change impacts on agriculture, since these will affect domestic and international policies, trading pattern, resource use and food security. The climate change is any change in climate over time that is attributed directly or indirectly to human activity that alters the composition of the global atmosphere in addition to natural climate variability observed over comparable time periods (IPCC, 2007). Since climatic factors serve as direct inputs to agriculture, any change in climatic factors is bound to have a significant impact on crop yields and production. Studies have shown a significant effect of change in climatic factors on the average crop yield Dinar *et al.* (1998), Seo and Mendelsohn (2008), and Cline (2007). In developing countries, climate change will cause yield declines for most important crops and South Asia will be particularly hard hit (IFPRI, 2009). Many studies in the past have shown that India is likely to witness one of the highest agricultural productivity losses in the world in accordance with the climate change pattern observed and scenarios projected. Climate change projections made up to 2100 for India indicate an overall increase in temperature by 2-43$^{0}$ C with no substantial change in precipitation quantity (Kavikumar, 2010). In course of time where the industrial revolution occurred in western countries and usage of the fossil fuels increased rapidly, on the other side the natural buffering system for climate change forests, were destroyed indiscriminately for want of fuel, fodder and timbers in the developing countries. These factors were intensified by the human activities in the past 250 years, which had tremendous impact on the climate system.

According to the IPPCC the green house gas emission could cause the mean global temperature to rise by another 1.4°C to 5.8°C.

The uncertainties associated with climate change do not permit a precise estimation of its impact on agriculture and food production. However, what is happening already in terms of changing seasonal patterns and respective increases in temperature, moisture concentrations and $CO_2$ levels is likely to have diverse impacts on ecosystems—and therefore on crops, livestock, pests and pathogens. The physiological response of crops to changing climate is expected to be varied. Although some positive outcomes are expected, the new climatic conditions are more likely to have negative impacts such as a rise in the spread of diseases and pests, which will reduce yields. A meta-analysis of experiments on crop performance shows that a potential increase of photosynthetic activity of plants under higher $CO_2$ concentrations is only realistic if linked with optimum temperature and favourable rainfall patterns. But the temperature and rainfall patterns are expected to change in future climate scenarios in unpredictable ways. Globally 80 per cent of the agricultural land area is rainfed which generates 65 to 70 per cent staple foods but 70 per cent of the population inhabiting in these areas are poor due to low and variable productivity. India ranks first among the rainfed agricultural countries of the world in terms of both extent and value of produce. Rainfed agriculture is practiced in two-thirds of the total cropped area of 162 million hectares (66 per cent). Rainfed agriculture supports 40 per cent of the national food basket. The importance of rainfed agriculture is obvious from the fact that 80 per cent of seed spices, 91.00 per cent coarse grains, 90.00 per cent pulses, 85.00 per cent oilseeds and 65.00 per cent cotton are grown in rainfed areas. These areas receive an annual rainfall between 250 mm to 750 mm, which is unevenly distributed, highly uncertain and erratic. In certain areas, the total annual rainfall does not exceed 350 mm. As a result of low and erratic rainfall, a significant fall in food production is often noticed.

Within agriculture, it is the rainfed agriculture that will be most impacted by climate change. Temperature is an important weather parameter that will affect productivity of rainfed crops. The last three decades saw a sharp rise in all India means annual temperature. Though most rainfed crops tolerate high temperatures, rainfed crops grown during *rabi* are vulnerable to changes in minimum temperatures (Venkateswarlu and Rama Rao, 2010). As far as Rajasthan state is concerned 84.77 per cent of the net sown area was under rainfed condition during 2012-13. There are a number of studies which assessed the impact of climate change on Indian agriculture and vulnerability of the small and medium rainfed farmers in different aspects. But the perceptions on the climate change and the factors which drive the farmers to adopt or to follow coping mechanisms have not been studied in detail. The present study is a modest attempt to assess the perception of the rainfed farmers on climate change. The specific objectives of the study are to assess the impact of drought on the yield of rainfed crops, to identify the level of awareness on the climate change and to identify the factors influencing in decision making on the coping mechanism to mitigate the impact of climate change.

## Impact of climate change on agriculture

The climate is being modified at a faster rate due to henna farming of industries and high cattle population along with exploding human population. To meet the demand of the growing population, the natural resources such as forests are also being exploited at a faster rate. In spite of the rising demand for food and fodder, the climate change will further worsen the condition by reducing in the yield of dry land crops. Most of the studies projected that the decreased yield in rainfed wheat and mustard and loss in farm net revenue between 9 to 25 per cent for a temperature increase of 2 to 3.5° C. Sinha and Swaminathan (1991) showed that an increase of 2°C in temperature could decrease mustard yield by about 0.75 tons/ha in the high yield areas; and a 0.5° Celsius increase in winter temperature would reduce wheat yield by 0.45 ton/ha. Saseendran *et al.* (2008) showed that for every one degree rise in temperature the decline in mustard yield would be about 6 per cent. Major impacts of climate change will likely be on rainfed crops (other than mustard), which account for nearly 60 per cent of crop land area. In India, poorest farmers often practice rainfed agriculture. For the temperature rise of 2°C in mean temperature and 7 per cent increase in the mean precipitation would create a 12 per cent reduction in net revenues for the country as a whole (Dinar *et al.*, 1998).

## Growth and instability in the yield of major rainfed crops

Compound growth rates and instability index of major rainfed crops such as sorghum, maize, green gram, henna, cumin, mustard and cotton that are grown under rainfed condition in Pali district were worked out. The analysis was carried out during 2000-2001 to 2012-13. The results of growth and instability in yield of the crops are presented in Table 1. The results revealed that for the study period all the major crops registered negative growth in spite of the technologies such as new variety, fertilizers *etc.*, and the government to boost the rainfed agriculture, the yield could not be increased at significant level due to the vagaries in the monsoon and temperature in the district. The major factors which influence the yield of rainfed crop are the rainfall; the yield will vary according to rainfall level. The instability index revealed that the variation was very high for sorghum, cotton and wheat compared to other major rainfed crops.

**Table 1.** Growth and instability in the yield of major rainfed crops (N=240)

| S.No. | Crop | Compound growth rate (%) | Instability Index |
|---|---|---|---|
| 1 | Sorghum | -20.77 | 0.85 |
| 2 | Maize | -8.09 | 0.54 |
| 3 | Mungbean | -10.32 | 0.60 |
| 4 | Henna | -15.11 | 0.63 |
| 5 | Cumin | -18.08 | 0.74 |
| 6 | Mustard | -5.89 | 0.55 |
| 7 | Cotton | -12.37 | 0.83 |

**Table 2.** Effect of drought on rainfed crop yield in Pali district (N=240)

| S.No. | Name of crops | Per cent loss of normal yield |
|---|---|---|
| 1. | Sorghum | 43.0 |
| 2. | Maize | 14.09 |
| 3. | Mungbean | 28.23 |
| 4. | Henna | 34.09 |
| 5. | Cumin | 48.68 |
| 6. | Mustard | 29.56 |
| 7. | Cotton | 59.96 |

## Drought and its impact on rainfed crops yield

In Pali district out of the years (2000 to 2013) there were three chronic drought years which occurred during 2001, 2005 followed by 2010. During the severe drought year, the shortfall in rainfall was 43.67 per cent as compared to normal year in the district. The major effect of the drought reflected in the yield of the rainfed crops due to inadequate and poorly distributed rainfall. The drought need not be a lengthier one even a dry spell during the critical growth period as short drought can cause significant damage and harm local economy. Production loss which is often used as a measure of the cost of drought is only a part of the overall economic cost. The effect of drought on rainfed crop yield in Pali district is presented in Table 2. In maize only 10.22 per cent of yield reduction was registered. In all other rainfed crops yield reduction was very high to the extent of 53.66 per cent in cotton crop since it is a sensitive crop to drought, followed by cumin, sorghum and henna to the extent of 44.53 per cent, 38.93 per cent, 30.44 per cent. It is evident from Table 2 that if there was moderate deviation in precipitation there will be high reduction in the yield of rainfed crops. The findings confirm with the findings of Latha *et al.* (2012).

## Farmers' perception on the impact of climate change

The farmers' perception on the climate change was assessed using yes or no type questions and the results are presented in Table-3. Most of the farmers were not able to express their perception on climate change directly but they expressed through the effects or the changes that occurred compared to the earlier years or based on their elder's experiences. About 86.67 per cent of the sample respondents expressed that their net income was reduced over the years, 80.86 per cent of the farmers expressed that there was change in climate and rainfall patterns, 75.64 per cent farmers expressed reduction in yield, 70.69 per cent expressed that there was fast evaporation of soil moisture, 67.67 per cent farmers expressed that due to soil erosion and other factors day by day the land was degrading and it becomes unsuitable for cultivation, 51.52 per cent of the respondents expressed that the seasonal pattern is changing and 7.78 per cent of the respondents expressed that they have no idea on the changes in climate. From the Table 3 it is clear that the level of farmers' perception on the climate change was good.

**Table 3.** Farmers perception on the impact of climate change (N=240)

| S.No. | Factors | Small farmers | Medium farmers | Large farmers | Total farmers |
|---|---|---|---|---|---|
| 1. | Reduction yield | 89.32 | 80.54 | 57.08 | 75.64 |
| 2. | Reduction in net income | 94.56 | 84.36 | 77.70 | 85.54 |
| 3. | Pest and disease outbreak | 72.80 | 73.29 | 72.55 | 72.88 |
| 4. | Fast evaporation of soil moisture | 85.66 | 75.74 | 50.67 | 70.69 |
| 5. | Erratic rainfall | 100.00 | 79.10 | 81.12 | 86.74 |
| 6. | Crop failure | 97.90 | 97.90 | 64.88 | 86.89 |
| 7. | Shifting of seasons | 56.78 | 49.44 | 48.34 | 51.52 |
| 8. | Land is becoming unsuitable | 80.23 | 70.25 | 52.54 | 67.67 |
| 9. | Change in climatic and rainfall patterns | 84.28 | 80.31 | 78.00 | 80.86 |

**Table 4.** Reason for reduction in yield and net revenue (N=240)

| S. No. | Factors | Small farmers | Medium farmers | Large farmers | Total farmers |
|---|---|---|---|---|---|
| 1. | Change in temperature and seasonal pattern | 54.70 | 39.00 | 35.88 | 43.19 |
| 2. | Rainfall | 100.00 | 95.76 | 85.08 | 93.61 |
| 3. | Soil fertility and erosion | 62.23 | 56.34 | 27.87 | 48.81 |
| 4. | Pest and disease | 84.55 | 71.22 | 63.90 | 73.22 |

## Reason for reduction in yield and net revenue

The sample farmers were highly concerned about the reducing yield rate and net farm income since their livelihood and socio-economic status is determined by the net income. Almost 100 per cent of the small farmers and 95.76 per cent of the sample farmers reported that the reduction in the rainfall was the major reason for reduction in the yield levels over the period followed by the pest and disease to the extent of 73.22 per cent and changes in temperature and seasonal patterns were quoted as the reason for the reduction in the yield by 43.19 per cent of the sample respondents. About 48.81 per cent of the farmers expressed that the soil lost its vigour due to factors such as erosion, lack of organic manures etc., and 9.83 per cent expressed that they do not have any idea for the reduction in yield over the years. It is clear that the farmers know that the yield reduction occurring continuously and to some extent they have knowledge on the reason for yield reduction also.

## Coping mechanisms of dryland farmers to mitigate the impact of climate change

The coping mechanism was followed to mitigate the climate change through technologies as well as through the socio-economic aspect and the results are presented in Table 5. There are many coping mechanisms which were followed by the rainfed farmers of Pali district. The mixed and intercropping was the major coping mechanism which was adopted by 75.88 per cent, followed by integrated and mixed farming which was adopted by 73.06 per cent and change in cropping pattern to the extent of 44.99 per cent. When there is change in the climate in either rainfall or temperature from the normal condition, there would be reduction in yield and net income of the farmers. To mitigate the reduction in the net income, the farmers in the district have to adopt some socio-economic strategies to sustain their life. The major socio-economic coping is shifting the profession which is observed to the extent of 52.63 per cent followed by borrowing for consumption from private money lenders was 39.76 per cent, reduction in consumption expenditure was observed in small and marginal farmers and not in the case of large farmers. Nearly 14.23 per cent of the sample farmers sold land and livestock, where as 13.11 per cent of the respondents adopted crop insurance as a coping mechanism. The findings confirm with the findings of Maiti *et al.* (2014) and Bonsull *et al.* (2011).

**Table 5.** major coping mechanism adoption by rainfed farmers to mitigation the impact of climate change (N=240)

| S. No. | Coping mechanism | Small farmers | Medium farmers | Large farmers | Total farmers |
|---|---|---|---|---|---|
| a | **Technological mitigation** | | | | |
| | 1. Change in cropping pattern | 62.76 | 43.33 | 28.88 | 44.99 |
| | 2. Mixed/inter cropping | 94.67 | 74.90 | 58.09 | 75.88 |
| | 3. Cultivation tree crops | 00.00 | 12.45 | 79.08 | 30.51 |
| | 4. Soil organic matter enhancement | 47.98 | 48.98 | 18.44 | 38.46 |
| | 5. Drought resistance/tolerance crops | 15.67 | 46.21 | 21.08 | 27.65 |
| | 6. Integrated/mixed farming system | 94.23 | 79.06 | 45.90 | 73.06 |
| b | **Socio-economic factors** | | | | |
| | 1. Reduced consumption expenditure | 62.33 | 51.33 | 00.00 | 37.88 |
| | 2. Shifting to other profession | 81.78 | 51.33 | 24.77 | 52.63 |
| | 3. Borrowing | 89.05 | 17.45 | 12.79 | 39.76 |
| | 4. Crop insurance | 08.86 | 08.66 | 12.79 | 10.10 |
| | 5. Selling of land and livestock | 27.32 | 25.22 | 05.45 | 19.33 |

## Conclusion

Climatic variation as occurrence of drought have significant impact on the production of rainfed crops. The small and medium rainfed farmers were highly vulnerable to climate change and to a larger extent the small and medium rainfed farmers adopted coping mechanisms for climate change compared to large farmers. The farmers already act to the changes in the climatic changes both by adopting the technological coping mechanisms on the positive side and negatively through shifting to other professions. The study suggests that as the impact of climate change is intensifying day by day it should be addressed through policy perspective at the earliest to avoid short term effect such as yield and income loss and long-term effects such as quitting agricultural profession by the rainfed farmers.

## References

Aggarwal, PK (2008). "Global climate change and Indian agriculture: Impact, adaptation and mitigation", Indian Journal of Agricultural Sciences, 78, (11): 911-919.

Bonsull C, Macklin M.G, Anderson DE and Payton RW (2011). Climate change and the adoption of agriculture in north-west Europe. European Journal of Archaeology 5(1):9-13.

Cline WR (2007). Global warming and agriculture: Impact estimates by country, Peterson Institute of International Economics, NW, Washington, *D.C.*, U.S.A.

Dinar A, Mendelsonhn R, Evenson R, Parikh J, Sanghi A, Kumar K, McKinsey J, Longergan S (1998). (Eds.) Measuring the Impact of Climate Change on Indian Agriculture, World Bank Technical Paper 402, Washington, D.C., U.S.A.

IFPRI, (2009). Climate Change – Impact on Agriculture and Costs of Adaptation, Food Policy Report International Food Policy Research Institute, Washington, D.C., U.S.A., pp.30.

IPCC, (2007). Impacts, Adaptation and Vulnerability. The intergovernmental Panel on Climate Change, Cambridge University Press, U.K.

Kavikumar, KS (2010). Climate sensitivity of Indian Agriculture: Role of Technological Development and Information Diffusion, in Lead Papers-National Symposium on Climate Change and Rainfed Agriculture, Indian Society of Dryland Agriculture, Central Research Institute for Dryland Agriculture, Hyderabad, February 18-20, 2010, pp.1-18.

Latha A, Munisamy G and Bhat ARS (2012). Impact of climate change on rainfed agriculture in Indi: A vcase study of Dharwar. International Journal of Environment Science and Developmen*t*, 3(4):371.

Maiti S, Jha, SK, Garai S, Nag A, Chakarvarty, R.M Kadian, K.S, Chandel BS, Datta KK and Upadhyay, R.C. (2014). Adapting to climate change: Traditional coping mechanism followed by the Brokpa pastoral nomads of Arunachal Pradesh, India. Indian Journal of Traditional Knowledge 13(4):752-761.

Saseendran SA, Ma L, Malone RW, Heilman P, L.R. Ahuja LR, Kanwar RS, D.L. Karlen DL and Hoogenboom G (2008). Simulating management effects on crop production, tile drainage, and water quality using RZWQM-DSSAT. Geoderma 140:297–309.

Seo N and Mendelsohn R (2008). A Ricardian Analysis of the Impact of Climate Change on South American Farms, Chilean Journal of Agricultural Research, 68, (1):69-79.

Singh R, Saravan R, Feroze SM, Devarani L and Thelma RP (2012). A case study of drought and its impact on rural livelihood in Meghalya. Indian Journal of Dryland Agricultural Research and Development, 27 (1):90-94.

Sinha SK and Swaminathan MS (1991). Deforestation, Climate Change and Sustainable Nutrition Security: A Case Study of India. Climate Change 19: 201-209.

Venkateswarlu, B. and Rama Rao CA, (2010). Rainfed Agriculture: Challenges of Climate Change, Agriculture Today Yearbook: 43-45.

□□□

# 14

# Climate Change Scenario of Arunachal Pradesh and Role of Agro-met Advisory Services for Crop and Resource Planning

**K. Bhagawati and R. Bhagawati**

## Introduction

Arunachal Pradesh is the largest hill state of North Eastern Himalayan Region having the total geographical area of 83,743 Sq Km, stretching between 26°28′ to 29°28′ N latitude and 91°35′ to 97°27′ E longitude and is endowed with a unique geo-physiography. Topographically, Arunachal Pradesh is generally a hilly region, with entire state lies on the mighty Himalayan and Patkoi ranges. The elevation of the hills ranges from 60 meters to over 7000 meters. Mt. Kangto (7090 m), Mt. Nyegi Kangsang (7050 m) and Mt. Gorichen (6488 m) are three highest peaks of Arunachal Pradesh. The Himalayan mountain system divides the state mainly into five river valleys: the Kameng, the Subansiri, the Siang, the Lohit and the Tirap. These rivers, along with innumerable rivulets traversing through innumerable hill systems treading through the rugged terrains, steep hills, valleys and ultimately drain down to from two major river systems and valleys of the region– the Brahmaputra and the Barak. The state has rich biological as well as cultural diversity. It comes under "Indo-Myanmar Mega-Biodiversity Hot-Spots" and is the centre of origin of number of cultivated plants. The state is custodian of 23.52 per cent of total flowering plants of India including around 4,500 species of angiosperm and 550 species of orchids; and is regarded as nature's repository of medicinal plants (Haridasan, 1989). Arunachal Pradesh possesses India's second highest level of genetic resources (SAPCC, 2011). The fauna diversity includes 85 species of mammals and 760 species of birds (SAPCC, 2011). The region has been identified by Indian Council of Agricultural Research (ICAR) as the centre of rice germplasm, while National Bureau of Plant Genetic Resource (NBPGR) has highlighted the region being rich in wild relatives of crop plants. It is the centre of origin of citrus crops. Owing to its rich biological diversity, the state is regarded as the '*Paradise of the Botanists*.' Also, Arunachal Pradesh is the home of 26 major tribes and 110 sub-tribes, with about two third of population comes under schedule tribes that are indigenous to the state.

## Climate Change Scenario

Arunachal Pradesh is the land of climatic diversity and agro-climatically the state is divided into five zones viz. Alpine Zone (18.5 %), Temperate Zone (27.6%), Sub-Tropical Hill Zone (25.7%), Mid-Tropical Hill Zone (7.6%) and Mid-Tropical Plain Zone (20.6%). Arunachal Pradesh is one of the highest rainfall recipient states of the country with more than 3500 mm in a year, which is distributed over the period of 8 to 9 months excepting in winter, however, most of rainfall is between May and September. Higher regions experience snowfall during winter. The average annual rainfall is 1000 mm in the higher elevations and 5750 mm in the foothill areas. This climatic and topographic diversity accounts for wide diversity in vegetation and agriculture. The normal temperature during winter months varies between -5°C to 21°C, while it varies from 22°C to 33°C during monsoon months (SAPCC, 2011). In the summer months, the temperature sometime reaches to around 40°C in foothills of the state.

The long term analysis for trends in observed seasonal precipitation and temperature over Arunachal Pradesh using India Meteorological Department (IMD) gridded and temperature at daily times scale shows that the rise in temperature is appreciable with more significant in case of mean minimum temperature trends compared to maximum temperature (Table 1). Although no much variation was observed in yearly total amount of rainfall, the intensity of rainfall increases in the recent years due to fall in number of rainy days from the normal of 144 days to average of 102 days in the last 5 years as recorded in Meteorological observatory at ICAR, Basar. The distribution of rainfall also observed very erratic in the recent decade. Overall analysis indicates that Eastern Himalaya in general and Arunachal Pradesh in particular are experiencing widespread warming generally 0.01° to 0.04°C per year (Sharma *et al.*, 2009).

According to the PRECIS regional climate model, annual rainfall is projected to decrease by 5 to 15 per cent in the 2030's as compared to base line and increase by 25 to 35 per cent towards 2080. Decrease in rainfall is projected for all seasons except pre-monsoon for 2030. Maximum temperature is projected to increase by 2.2°C to 2.8°C during 2030 as compared to baseline and towards 2080s the increase is projected by 3.4° C to 5°C. Minimum temperature is projected to increase by 1°C to 2.6°C during 2030 and by 2.8°C to 5°C during 2080 (SAPCC, 2011).

**Table 1**. Climatic conditions (long term averages)[1]

| | | | | | | | | | | | | |
|---|---|---|---|---|---|---|---|---|---|---|---|---|
| TEMPERATURE TRENDS | | | | | | | | | | | | |
| Mean Maximum Temperature Trends in °C per year | | | | | | | | | | | | |
| | Annual | | | Winter | | Summer | | Monsoon | | | Post Monsoon | |
| | +0.02* | | | +0.02* | | No trend | | No trend | | | +0.02* | |
| MEAN MINIMUM TEMPERATURE TRENDS IN °C PER YEAR | | | | | | | | | | | | |
| | Annual | | | Winter | | Summer | | Monsoon | | | Post Monsoon | |
| | +0.02* | | | +0.02* | | +0.02* | | +0.01* | | | +0.02* | |
| MONTHLY MEAN TEMPERATURE TRENDS IN °C PER YEAR | | | | | | | | | | | | |
| Months | Jan | Feb | Mar | Apr | May | Jun | Jul | Aug | Sep | Oct | Nov | Dec |
| | +0.02 | +0.03* | -0.02 | -0.03* | +0.03* | NT | -0.01 | NT | NT | +0.01 | +0.02* | +0.02* |
| RAINFALL TRENDS | | | | | | | | | | | | |
| | | Annual | | Winter | | Summer | | Monsoon | | | Post Monsoon | |
| | | -3.63 | | -0.10 | | No trend | | -2.30 | | | -0.83 | |
| MONTHLY RAINFALL TRENDS IN MM PER YEAR | | | | | | | | | | | | |
| Months | Jan | Feb | Mar | Apr | May | Jun | Jul | Aug | Sep | Oct | Nov | Dec |
| | -0.29 | -0.05 | +0.49 | +0.86 | -2.19 | -0.82 | +0.06 | -3.29 | +0.26 | -0.88 | -0.09 | -0.13 |

Increasing (+) and decreasing (-) trends significant at 95% level of significance are shown in bold and marked with '*' sign.

[1]Agro-climatic classifications (Trends: 1950-2014, IMD)

## Impact of Climate Change on Agriculture

Rainfall play an important role in determining the fate of agriculture in Arunachal Pradesh as only 19 per cent of gross cropped area is under irrigation (SAPCC, 2011). Due to climate change, the reliability of agriculture on rain has reduced in the recent years. Seasonal rainfall has been marked by delayed onsets, declining number of rainy days and increased intensities altering farming/cropping calendars with negative effects on the yields. This variable nature of rainfall imposes a delicate balance between the onset, duration and amount of rain and the timing of agricultural activities. Climatic manifestations over the last decade have made it difficult to maintain this fragile balance. The persistent low rainfall during the months of April, May and June leads to the delayed sowing of *Kahrif* crops. Also, the consistent mid-season dry spell during month of August adversely affects the standing crops. The excessive rainfall during the end of September or beginning of October leads to crop loss. During the last decade, climate-related crop failures in the eastern part of Arunachal Pradesh attracted much attention (Arunachal Times, 2009; 2012). Among many manifestations of climatic instability and change have been floods that devastated large areas of near-ready grain fields, several episodes of late rains during planting seasons, and persistent droughts in large portions of the region. The frequency of extreme weather events like floods, landslides and drought increased in the recent decade (Shrestha, 2004). About 80 per cent of the soils are acidic in nature with ferrous toxicity in valleys and aluminum toxicity in the hills/uplands (Ngachan, 2013). Acidity of soil is major problem in Arunachal Pradesh, which is further aggravated by increasing intensity of rainfall with decreasing number of rainy days.

In terms of food crops production, the state still have minus 6 per cent deficit, producing around 2,50,400 tones of grains against the requirement of 2,68,100 tones (Ngachan, 2013). Uncertain and uneven rain due to climate change further increases the pressure on the sector. The net cereal production in the region is projected to decline by at least 4 to 10 per cent by the end of this century, under most conservative climate change scenario (Lal, 2005).

Temperate crops are another important venture in the state with around one third of the geographical area of the state is under temperate climate. Rise in minimum temperature have significant impact on temperate fruits of Arunachal Pradesh, especially apple and kiwi, because of their chilling temperature requirement. During growing season, the rise in minimum temperature leads to shift in apple and kiwi cultivation to higher elevation due to non-fulfillment of chilling temperature requirement, which is in accordance to the finding in Himachal Pradesh (Gautam *et al.* 2014). Consequently, the established orchards in Shergoan and Dirang are declining.

## Disease Dynamics

Weeds, insects and diseases are benefitted by warming environment. The climate change is a concern to researchers and farmers because it would lead to change in insect and disease dynamics i.e. the distribution, abundance and management of insects and pathogens. Climate change will probably alter the geographical and temporal distribution of pests and insects. New diseases may arise in certain regions, and other diseases may cease to be economically important, especially if the host plant migrates into new areas (Coakley *et al.*, 1999). More aggressive strains of pathogen with broad host range, such as *Rhizoctonia, Sclerotinia, Sclerotium* and other necrotrophic pathogens can migrate from agroecosystems to natural vegetation, and less aggressive pathogens from natural plant communities can start causing damage in monocultures of nearby regions (Chakraborty, 2000). Regarding unspecialized necrotrophs, the range of hosts can be extended due to crop migration. The rice blast disease (*Pyricularia oryzae*) which is much prevalent in the region and main cause behind hindrance of achieving full potential of rice cultivars is favored by rise in temperature in the state and its effect is observed up to 2000 m altitude. At mid hill condition, rice blast first appears on foliage by the end of April or beginning of May. From the recent study it is observed that quick build-up of the disease on foliage occurred in the beginning of June and attained the proportion of epiphytotic by the end of the month. Studies indicated that initiation of leaf infection occurred when temperature-humidity index reached to 66. Rainfall being one of important factor behind the initiation of the disease, the erratic rainfall in the recent years makes the disease very much unpredictable. The simulation model shows that the risk of blast epidemic is likely to increase under current trend of temperature rise in Arunachal Pradesh (Luo *et al.*, 1995). Survey revealed that the intensity of false smut disease (*Ustilaginoidea virens*) of rice was highest in shady, fertile and water stagnant fields. It was found that the locally grown low land rice varieties like Lemuk, Leli, Mungmai and Kaying are resistant to the disease.

In maize, Turcicum leaf blight disease mainly appear by middle of June when the temperature reach to 24°C and relative humidity to 86 per cent. The disease build-up steadily continued to increase and reached maximum (68%) by 7$^{th}$ of August. White rust, alternaria leaf spot and downy mildew were major diseases of mustard in the state. Bacterial stalk rot of Indian mustard was recorded as a new threat for mustard production system in the state. On an average, about 10 to 78% of plants were affected by the disease. The first symptom of disease was noticed during the 2$^{nd}$ week of November. Turnip mosaic virus (TuMV) in broad-leaved mustard (*Brassica juncea* var. *rugosa*) was confirmed by symptomatology, transmission electron microscopy, reverse transcription-polymerase chain reaction and partial characterization of cytoplasmic inclusion protein and coat protein (CP) domains. Phylogenetic analysis of the partial CP sequences of the new TuMV isolates from Indian mustard, broad leaved mustard and broccoli revealed their closest relationship with members of the World-B genogroup of TuMV. This is the first molecular evidence of TuMV infection in *Brassica* spp. from India (Singh *et al.*, 2015).

Tikka diseases (*Cercospora personata* and *C. arachidicola*), *Sclerotium* rot (*Sclerotium rolfsi*), rust (*Puccinia arachidis*) were found to be the most serious threat to groundnut cultivation under mid-hill conditions. It was observed that tikka disease mostly appear on 23$^{rd}$ June when the average temperature reached to 24°C and relative humidity above 85 per cent. The rise in temperature increases the disease incidence and severity up to 62 per cent. Appearance of *Sclerotium* rot was noticed during the month of July when the temperature is above 24°C. In potato, the prominent diseases are Late Blight, Early Blight, Seed Tuber Rot, Brown Rot and Black Scurf. Bacterial wilt of solanaceous crops caused by Ralstonia solanacearum continuously occurred during rainy season starting from the month of May and last up to September. The cultivation of colocasia, which is one of the important crops of the state, is severely hampered by blight disease (*Phytophthora colocasiae*), with incidence reached its peak stage during 3rd week of September favored by temperature range of 21°C to 26°C and humidity of 80 to 90 per cent.

In citrus, fruit drop, die-back, sooty mould, scab and lichens increased with the increase in slope (Bag, 2002), with die-back and scab were most severe in sandy soil. Lichens and mosses were found to be most detrimental factor leading to decline in the productivity of mandarin orange throughout the state that is favored by rising temperature. Recent field inspection revealed that citrus greening and citrus tristeza virus disease is emerging as the major factor behind decline of khasi mandarin orchards. The population and infestation of trunk borer is found to be increased in the recent years. Banana bunchy top disease was recorded in entire area having 0.1-75 per cent disease incidence of different varieties of banana. In Large Cardamom, disease like chirke and foorkey were observed in different districts of the state.

Apple scab disease is most prevalent in the state with highest severity in Red Gold followed by Royal Delicious, Maharajji, while Maling-9, Red June are found to have mild severity. While varieties like Fanny, Prima, Priscilla, Liberty are almost resistant to the disease. The severity of scab also increases with age and altitude. In bamboo, the common diseases are Culm rot (*Fusarium* sp.), steam blight (*Curvularia* sp.), leaf blight (*Rhizoctonia solani*), Helminthosporium leaf spot, Curvularia leaf spot, Rusty leaf spot and sooty mould.

## Livestock and Fisheries

Livestock sector plays multiple roles in the livelihood of people in agriculturally underdeveloped states like Arunachal Pradesh; especially the poor. The sector is negatively affected by a decline in forage quality, heat stress and disease in the recent years. In case of poultry, the mortality due to chronic heat stress is found to be significant in foot hills (Jini *et al.*, 2015). The rise in temperature associated with high humidity make environment congenial for diseases like *Ranikhet, Fowl Pox* and *Coccidiosis*. Temperature also found to have detrimental effect on the laying capacity of the hens and lowers the quality of egg in Arunachal Pradesh (Jini *et al.*, 2015). The performance in terms of body weight gain and egg production is also reduced in the higher altitudes during winter season due to cold stress (Jini *et al.,* 2015). In pig, the breeds Hampshire and Ghungroo performed well in the state in terms of body weight gain, while local breeds are more resistant to disease.

The impacts of climate change on aquaculture are more complex than those on terrestrial agriculture owing to the much wider variety of species produced.The small-scale fish farmers in developing countries practice a low input, low-output form of aquaculture depending heavily on ecosystem services and naturally available feed to support their fish. Rice–fish systems at *Apatanis* of Arunachal Pradesh are often depend upon wild fish entering paddy fields; their production system is in turn affected by reduced wild stocks. Changes in rainfall distribution and amount cause a spectrum of changes in water availability ranging from droughts and shortages to floods. Rising temperatures, especially in lower altitudes, reduce levels of dissolved oxygen and increase metabolic rates of fish that leads to increases in fish deaths, declines in production or increases in feed requirements and increasing the risk and spread of disease. While the rise in temperature is found to be beneficial for the fish cultivation in the higher altitude, as rising temperatures results in increased growth rates and food conversion efficiencies, longer growing seasons, reduced cold water mortality and expansion of areas suitable for aquaculture.

Thus, climate change has the potential to act as a 'risk multiplier' in Arunachal Pradesh, where agriculture and other natural resource-based system are already failing to keep pace with the demand on them.

## Role of Agro-Met Advisory Services

Directly or indirectly, weather contributes to approximately three-quarters of annual losses in farm production. Around 65 per cent of the Indian agriculture is heavily dependent on natural factors, particularly rainfall and it account for more than 50 per cent of variability in crop yields. When specifically tailored weather support is available as per the need of the farmers, it contribute greatly towards making short-term adjustment in the daily farm operations, which minimize input losses and improve the quality and quantity of farm produce. The seasonal weather outlook also provides guidelines for long-rang or seasonal planning and selection of crops and varieties most suited to anticipated weather conditions.

Weather forecast based Agro-Met Advisories thus become vital to stabilize yield through management of resources and inputs like irrigation, fertilizer and pesticides. The information help farmers and the planners to exploit the potential of good weather and minimize the impact of bad weather. Besides, it also helps farmers to avoid the adverse effect of weather events like heavy rainfall, dry spell, high wind speed which may influence the crop productivity.

## Objective of Agro-met Advisory Services (AAS)

The main emphasis of the existing AAS system is to collect and organize climate/weather, soil and crop information, and to amalgamate them with weather forecast to assist farmers in taking management decisions.

Weather based advisories as supported system has been organized after characterization of agro-climate, including length of crop growing period, moisture availability period, distribution of rainfall and evaporative demand of the regions, weather requirements of cultivars and weather sensitivity of farm input applications.

## Agro-meteorological Information Requirements

1. Rainfall: Information on quantum of precipitation at various crop phenophases and at different probability level; and information on the probability of occurrence of dry/wet spells,
2. Soil Moisture: Help in scheduling the irrigation
3. Temperature: Occurrence of high/low temperature during crop season due to heat/cold waves significantly affects the crop growth and its yield. The performance of livestock and poultry is also very much dependent on the ambient temperature.
4. Wind: Pesticide/insecticide spray scheduling
5. Pest/disease: Pest/disease incidences are influenced by prevailing weather conditions

## Process of Agro-met Advisory Services

The detail flowchart of the Agro-met Advisory process/system is depicted in Figure 1.

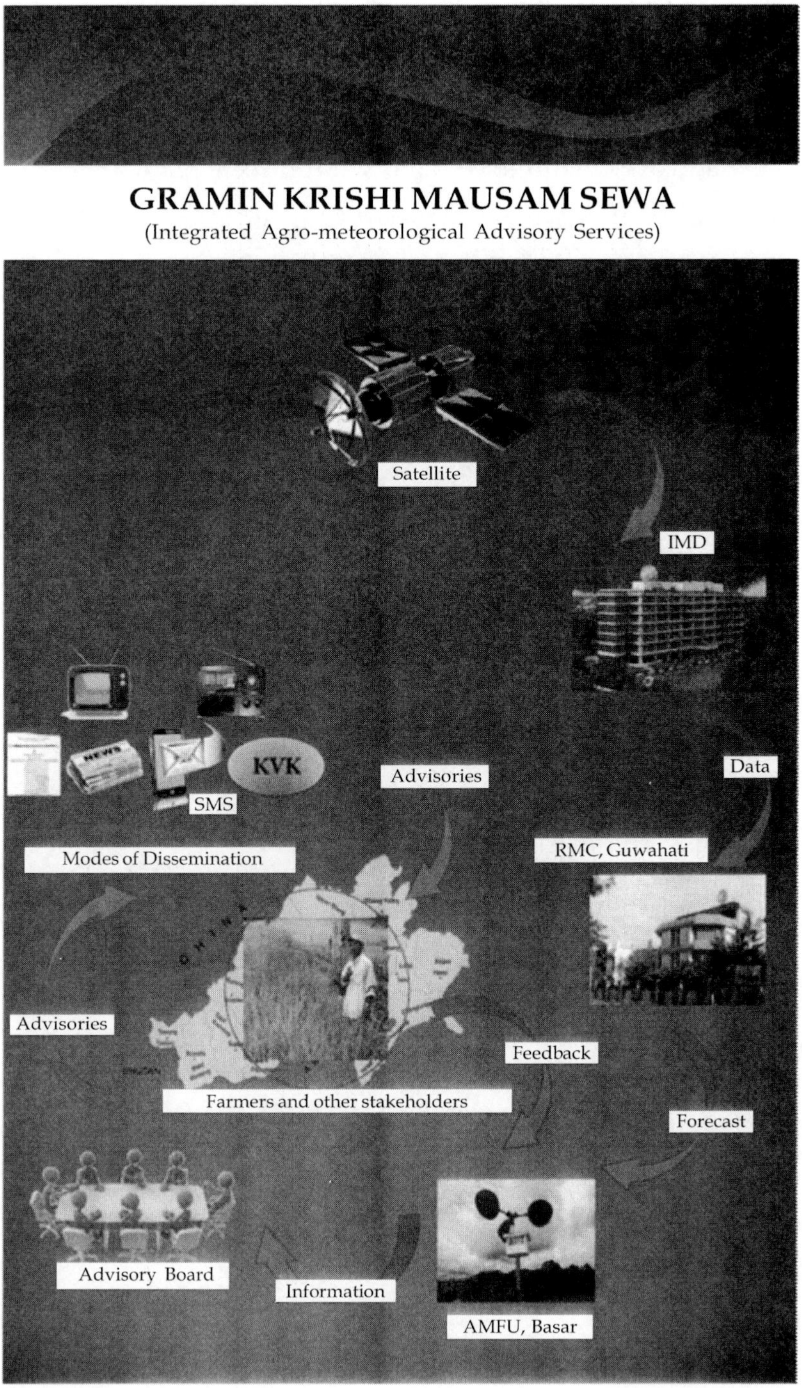

## GRAMIN KRISHI MAUSAM SEWA

(Integrated Agro-meteorological Advisory Services)



**Figure 1.** The Agro-met-Advisory Services System and Process

1. Weather forecast by IMD (input)
    - Twice in a week (Tuesday & Friday), IMD communicates the weather forecast for five days and for each districts.
    - Variables: Rainfall, minimum & maximum temperature, humidity, cloud cover, wind speed and wind direction.
2. Preparation of Agro-met Advisory Bulletin
    - The expert team of AMFUs consisting of scientists from different disciplines prepare the agro-met advisories for each districts by analyzing the weather forecast and realizing the previous week weather.
3. Dissemination to stakeholders
    - The bulletins/advisories are disseminated through different media viz. website, mobile sms, newspaper, radio, television etc.
    - It is also disseminated through different KVKs and NGOs.
    - The bulletins are also put on the website of IMD and universities for wide circulation.
4. Feedback from farmers and stakeholders
    - The feedbacks were collected and analyzed to refine the advisories/forecast.

## Media for dissemination of advisories

1. All India Radio and Television
2. Newspaper
3. Mobile Phone/SMS
4. Internet
5. Kisan Call Centres
6. KVKs and NGOs
7. Bulletins

## Component of Agro-met Advisory Bulletins

The bulletin mainly contains weather information (current & past week), crop information and weather based advisories:

1. AMFU Details: Agro-climatic region, state, district, mandal/block/ sub, date of forecast and validity period of forecast,
2. Weather Deviation Summary: Weather deviation summary of the preceding weeks of realized weather or since last bulletin (till forecast date) and predicted weather of next 5 days,
3. Weather-based Agro-Advisory
    - Crop planning: How the weather influenced the choice of crop. Information on crop planning, selection of sowing/harvesting time etc.
    - Crop management advice: The phenophase of crop is identified. The weather-based agroment advisory is prepared by considering low the change in weather could influence the field preparation, irrigation scheduling, fertilizer application etc,
    - Crop management under adverse/malevolent weather: The advisories contain possible mitigation steps for extreme weather events such as extreme temperatures, heavy rains, floods, strong winds etc.
4. Weather-based livestock/poultry advisory
    - Seasonal livestock/poultry management strategies
    - Weather-based feeding and housing management
    - Advisory for routine vaccination and disease management
    - Other location-specific and need-based cultural practices

## Conclusion

Variability of climate acts as a multiplier of uncertainties and complexities of agriculture and allied sector. The agricultural system need to adapt and behave according to the parallel change in environment and climate, i.e. the system need to be self-adaptive. This demands multidisciplinary holistic approach in a very sustainable way. Climate change will place an additional burden on efforts to meet long-term development goals of Arunachal Pradesh in particular. To cope with the current crisis, the ongoing development initiatives need to be strengthened to reduce vulnerability to climate change by adopting suitable policies and technologies. The agriculture and the farming system of the state must make necessary adjustment and readjustment with the changing climate to enhance the resilience of the sector. Reliable and well timed weather forecast/information is of vital importance for appropriate, foresighted and up-to-date planning of agricultural activities. Weather that dictates the success and productivity of crops and livestock, is most critical input that aids sustainable resource utilization and conservation. Agro-met Advisory Services aids the farmers and other stake holders in proper decision making is very crucial in the current era of climatic variability.

## References

Arunachal Times. (2009). Paddy Crop Failure, Farmers Apprehensive. 15th July'2009. http://www.arunachaltimes.in/archives/jul09%2015.html.

Arunachal Times. (2012). Flash flood causes extensive damage to crops, roads; local MLA 'unconcerned'. 14th September, 2012. http: //www. arunachaltimes.in /sep12%2014.html.

Bag, T.K. (2002). Plant Protection Research in Arunachal Pradesh. In: Singh, A.K. (Eds), Resource Management Perspective of Arunachal Pradesh. ICAR Research Complex for NEH Region, Arunachal Pradesh Centre, Basar.

Chakraborty, S., Tiedemann, A.V. and Teng, P.S. (2000). Climate change: potential impact on plant diseases. Environmental Pollution, 108: 317-326.

Coakley, S.M., Scherm, H. and Chakraborty, S. (1999). Climate change and plant disease management. Annual Review of Phytopathology, 37:399-426.

Gautam, H.R., Sharma, I.M. and Kumar, R.A. (2014). Climate change is affecting apple cultivation in Himachal Pradesh. Current Science, 106(4):498-499.

Haridasan, K., Shukla, G.P. and Benewal, B.S. (1989). Medicinal Plants of Arunachal Pradesh. SFRI Information Bulletin, No.5. State Forest Research Institute, Itanagar.

Hegde, S.N. (2002). Arunachal Pradesh State Biodiveristy Stretagy & Action Plan – Final Report. State Forest Research Institute, Itanagar.

Jini, D., Bhagawati, K., Bhagawati, R. and Rajkhowa, D.J. (2015). Identification of Critical Periods Environmentally Sensitive to Normal Performance of Vanaraja Poultry Breed in Climatically Different Locations. International Letters of Natural Sciences, 46:76-83.

Lal, M. (2005). 'Implications of climate change on agricultural productivity and food security in South Asia'. In Key vulnerable regions and climate change - Identifying thresholds for impacts and adaptation in relation to Article 2 of the UNFCCC. European Climate Forum.

Luo, Y., Tebeest, D.O., Teng, P.S. and Fabellar, N.G. (1995). Simulation studies on risk analysis of rice leaf blast epidemics associated with global climate change in several Asian countries. Journal of Biogeography, 22: 673-678.

Ngachan, S.V. (2013). Building Climate Resilience for Food Security and Livelihood. In: Bhagawati, R., Choudhary, V.K., Bhagawati, K., Rajkhowa, D.J. and Ngachan, S.V. (eds) Climate Change: Its Impact and Mitigation Strategies for Hill Ecosystem. ICAR Research Complex for NEH Region, Arunachal Pradesh Centre, Basar.

SAPCC. (2011). Arunachal Pradesh State Action Plan on Climate Change. Department of Environment and Forest, Government of Arunachal Pradesh. Itanagar.

Sharma, E., Chettri, N., Tse-ring, K., Shrestha, A.B., Jing, F., Mool, P. and Eriksson, M. (2009). Climate Change Impacts and Vulnerability in the Eastern Himalayas. ICMOD, Kathmandu, Nepal.

Shrestha, A.B. (2004). 'Climate change in Nepal and its impact on Himalayan glaciers'. In Hare, W.L., Battaglini, A., Cramer, W., Schaeffer, M., Jaeger, C. (eds) Climate hotspots: Key vulnerable regions, climate change and limits to warming, Proceedings of the European Climate Change Forum Symposium. Potsdam: Potsdam Institute for Climate Impact Research.

❑❑❑

# Section 3

## Climate Change and Pest Scenario

# 15

# Encountering Climate Change in Agricultural Crops: Are we Ready with Tools for Pest Management?

**S. Bhagat, Amrender Kumar, B.K. Bhattacharya, A. Birah, P.D. Meena, V. Kumar and C. Chattopadhyay**

## Introduction

Climate change is of great concern to India in view of food and nutritional security of its ever growing population. The impacts of climate change are global, but Indian subcontinent is more vulnerable due to dependence of its large population on agriculture. It can cause millions of deaths a year worldwide and costing more than a trillion of US dollar. Climate change has direct effect on agriculture due to 0.74°C average global increase in temperature in the last 100 years and atmospheric $CO_2$ concentration having increased from 280 ppm (1750) to 400 ppm (2013). These changes have direct effect on growth and multiplication/fecundity, spread and severity/infestation of many plant pathogens/insect pests, which in turn are affecting the pattern of incidence of pests (including diseases). Changing pest scenario due to climate change has warranted the need for future studies on such models which can predict the severity of important pathogens of major crops in real-field conditions. Simultaneously, pest management strategies may be reoriented to cope up with changing scenario for sustainable food production.

According to Intergovernmental Panel on Climate Change (IPCC), the climate change refers to 'a change in the state of the climate that can be identified (e.g. using statistical tests) by changes in the mean and/or the variability of its properties that persists for an extended period, typically for decades or longer. It refers to any change in climate over time, whether due to natural variability or as a result of human activity' (IPCC, 2007). Increased emission of carbon dioxide ($CO_2$) and other green house gases, predominantly methane ($CH_4$) and nitrous oxide ($N_2O$) have been identified as the main agents causing increase in global temperature. According to IPCC, cold days and cooler nights have become less frequent and hot days, hotter nights, heat waves are more common. The IPCC reports that the frequency of heavy precipitation has increased over most land areas, which is consistent with warming and increase of atmospheric water vapour.

Effect of climate change on insect-pests and diseases of agricultural crops is multidimensional. Magnitude of this impact could vary with the type of species and their growth patterns. It may be assumed that the vegetation tolerating high temperature, salinity and having high $CO_2$-use efficiency could be better than other species. IPCC in its report of 1995 predicted that doubling in the level of $CO_2$ could possibly increase yield in several crops by 30%. The increased production could be off-set partly or entirely by the insect pest, pathogens or weeds. It is, therefore, important to consider all the biotic components under the changing pattern of climate. There is also thought about shorter winters, which may affect the oil yields of the rapeseed-mustard crops.

Development of diseases and pests is strongly dependent upon the temperature and humidity. Any change in them, depending upon their base value, can significantly alter the scenario, which ultimately may result in yield loss. Any small change in temperature can result in changed virulence as well as appearance of new pests in a region. Likewise, crop-weed competition may be affected, depending upon their growth behaviour. The following scenarios can be visualized regarding impact of climate change on pest dynamics in agriculture (Khan *et al.*, 2009). With an increase in concentration of carbon dioxide, the nutritional status of crop will change, and the net effect on agricultural production will depend upon interaction between pests and crops. Gradual climate warming will lead to changes in the composition of pest fauna in different areas. The high population growth rate of many species will ensure changes in pest distribution. If the rise in winter temperature takes place, the duration of hibernation of pests may decrease, thus increasing their activity. Uncongenial areas for pests due to low temperature at present may become suitable due to rise in temperature. However, we should not forget that insects could adapt to slow changes in the environment and with increase in temperature, their favourable range of temperature may also shift.

## Climate change and disease scenario

In India, there limited efforts have been made in this area for any insect-pest or disease of any crop (Subba Rao *et al.*, 2007; Chattopadhyay and Huda, 2009). However, at the genomic level, advances in technologies for the high-throughput analysis of gene expression have made it possible to begin discriminating responses to different biotic, abiotic stresses and potential trade-offs in responses. Ecologists are now addressing the role of plant disease in ecosystem processes and the challenge of scaling-up from individual infection probabilities to epidemics and broader impacts (Garrett *et al.,* 2006). Swaminathan (1986) indicated that the number of diseases on the same crops were much higher in tropics than under temperate conditions to indicate how rising temperatures could impact occurrence of plant diseases on agricultural crops. Presently, most of the work related to climate change vis-à-vis plant diseases is going on in rice (blast, bacterial leaf blight), wheat (*Puccinia, Septoria*) and horticultural (*Meloidogyne*) crops. The trend indicates that severity of majority of diseases is found to be higher with elevated $CO_2$ levels (Chakraborty, 2008), an off-shoot of climate change. It is also being opined that climate change could lead to a changed profile (variants) of pathogen, insect-pest ("climate change can activate 'sleeper' pathogens, whilst others may cease to be of economic importance" - Bergot *et al.,* 2004). The facultative pathogens with broad host range may survive better. There is also possibility of broadening of host range of the facultative pathogens. The need for further work in this area has been highlighted in adaptation experiments using twice-ambient $CO_2$, which increased the aggressiveness (Chakraborty and Datta, 2003) and fecundity (Chakraborty *et al.,* 2000) of *Colletotrichum gloeosporioides,* which causes anthracnose of tropical legumes. Elevated $CO_2$ may modify pathogen aggressiveness and/or host susceptibility and affect the initial establishment of the pathogen, especially fungi, on the host (Plessl *et al.,* 2005; Matros *et al.,* 2006). The host resistance has increased, possibly due to changes in host morphology, physiology and composition. Increased fecundity and growth of some fungal pathogens under elevated $CO_2$ has also been reported (Chakraborty *et al.,* 2000; Kumar *et al.,* 2013, 2014). However, it has been reported that greater plant canopy size, especially in combination with humidity and increased host abundance, can increase pathogen load (Chakraborty and Datta, 2003; Mitchell *et al.,* 2003; Pangga *et al.,* 2004). Sporulation by the pathogenic fungi may be 15-20 folds higher, leading to massive increase in the pathogen (Mitchell *et al.,* 2003). New strains may emerge, with adaptation occurring faster and their evolution may get accelerated. Elevated $CO_2$ may modify pathogen aggressiveness and/or host susceptibility and affect the initial establishment of the pathogen, especially fungi, on the host (Coakley *et al.,* 1999; Plessl *et al.,* 2005; Matros *et al.,* 2006). In most examples, host resistance has increased, possibly due to changes in host morphology, physiology and composition. Increased fecundity and growth of some fungal pathogens under elevated $CO_2$ has also been reported (Coakley *et al.*, 1999; Chakraborty *et al.*, 2000). Among the 27 diseases examined under elevated $CO_2$ levels, 13 caused higher crop losses than expected. Ten of the diseases had a reduced impact, and four had the same effect as they do now (NSW DPI, 2007). With increased plant biomass at high $CO_2$ the stubble-borne fungal pathogen *Fusarium pseudograminearum* causing crown rot of wheat may become more severe (Melloy *et al.*, 2010). *Xanthomonas axonopodis* pv. *citri*, was predicted to extend further south to major Australian citrus growing regions with a 1-5°C temperature increase (van Rijswijk *et al.* unpublished). Analysis of archive samples from the Rothamsted long-term (since 1850s till 2010) wheat production and fertilizer experiment shows that historical records of $SO_2$ emissions are well correlated with the ratio of two pathogens (*Phaeosphaeria nodorum* / *Mycosphaerella graminicola*) (Fitt *et al.*, 2011). Low solar radiation and short-day periodicity could result in higher infections by *Fusarium, Sclerotinia* and *Verticillium* (Nagarajan and Muralidharan, 1995). Root rot is an emerging threat for rapeseed-mustard production system, recently reported from the farmers' field in some pockets of the country (Meena *et al.*, 2010), which was initially identified as stand-alone bacterial or fungal incidence or in combinations (*Erwinia carotovora* pv. *carotovora, Fusarium, Rhizoctonia solani* and *Sclerotium rolfsii*). In view of the fact that some isolates of *Alternaria brassicae* sporulated at 35°C and several isolates had increased fecundity under higher RH, it seems that as per recent changes towards warmer and humid winters, existence of such isolates could pose more danger to the oilseed Brassicas due to Alternaria blight in times to come. The immense variation available among only thirteen representative isolates of *A. brassicae* also indicates their ability to adapt to varied climatic situations (Goyal *et al.*, 2011). In Germany, rapeseed-mustard pathogens such as *A. brassicae, Sclerotinia sclerotiorum*, etc. are predicted to be favoured by average warmer temperatures (Siebold and von Tiedemann, 2012), which matches our observations.

In India, yellow rust of wheat severity reached up to 100% with incidence being 30-40% in the terai and northern hills in 2010-13. Spot blotch (*Septoria* sp.) has also become very important on wheat in recent times, which has been attributed to climate change. It seems that there is severe occurrence of Indian Cassava Mosaic Virus in Kerala due to shift in climatic conditions; a new report of African Cassava Mosaic Virus, Sri Lankan Cassava Mosaic Virus is attributed to increase in temperature and $CO_2$ levels. The Groundnut Bud Necrosis Virus (GBNV) has been on the rise on several field and horticultural crops across the country. The rise in temperature at Kanpur may have been beyond the tolerance limit of the mite vectoring Sterility Mosaic virus of pigeonpea, which could have influenced decline in the disease there. On the other hand, the weather factors might have shifted in favour of the vector at Bangalore that may have resulted in rise of the disease on the crop there. The climate variability may have also influenced Phytophthora blight incidence at Kanpur and Pantnagar in mutually opposite directions (Kumar *et al.* 2012). Since report of Stemphylium blight on chickpea from Bangladesh in 1987, the disease has gradually moved to Nepal, India and now has been found infesting both lentil, chickpea in the North-Eastern Plain Zone, Terai region viz., Pantnagar and Pusa (Bihar) (Ghosh *et al.* 2012). Contrastingly, Ascochyta blight, which used to be a major problem in chickpea in the North-Western Plains of India, has almost vanished from the area, which could also be due to substantial reduction in the area under the crop in the region and shift in the same towards southern India. Phyllody is being noticed on mungbean with up to 8% incidence (Mohapatra and Chattopadhyay, 2012).

**Table 1.** List of plant diseases likely to occur in climate changed scenario

| Component of climate change | Pathogen/Disease | Impact |
|---|---|---|
| **Elevated $CO_2$ level (550 ppm $CO_2$)** | | |
| | Biotrophic fungi such as rust | Promotes the growth and development due to enhanced carbohydrate |
| | Downy mildew (*Peronospora manshurica*), brown spots (*Septoria glycines*) and sudden death syndrome (*Fusarium virguliforme*) of soybean | Altered expression of diseases |
| | Powdery mildew in barley (*Blumeria graminis*) | Increased resistance to powdery mildew |
| | Blast (*Pyricularia oryzae*) and sheath blight (*Rhizoctonia solani*) | More susceptible injury |
| | Powdery mildew and anthracnose (*Colletotrichum gloeosporioides*) | Increased reproduction of pathogen |
| | Rice blast (*Magnaporthe grisea*), wheat scab (*Fusarium* spp.), stripe rust (*Puccinia striiformis*) and powdery mildew (*Blumeria graminis*) | Increased infestation |

| | | |
|---|---|---|
| | *Rhizoctonia solani, Sclerotium rolfsii, Fusarium oxysporum* f. sp. *ricini, Macrophomina phaseolina, Alternaria tenuissima* | Increased fungal biomass with increased sclerotial bodies (*R. solani*) Increased cellulase activity (*Trichoderma viride, T. harzianum, F. o.* f. sp. *ricini* and *M. phaseolina*) |
| **Effect of increase in temperature** | | |
| | Stripe rust isolates (*Puccinia striiformis*) | Increased aggressiveness |
| | Common bunt (*Tilletia caries*) and Karnal bunt (*Tilletia indica*) in wheat | Increased infection |
| | Dry root rot (*Rhizoctonia bataticola*) of chickpea | Increased infestation |
| | Phytophthora blight (*Phytophthora drechsleri* f. sp. *cajani*) of pigeon pea | Increased infestation |
| | Needle blight (*Dothistromaseptosporum*) | Increasing northward |
| | leaf rust (*Puccinia recondita*) in wheat,broomrape (*Orobanche cumana*) in sunflower, blackshank (*Phytophthora nicotianae*) in tobacco and bacterialb light (*Xanthomonas oryzae* pv. *oryzae*) in rice | Temperature sensitivity |

*Ref:* Gautam *et al.*, 2013

Due to rise in maximum temperature in post-rainy (*rabi*) season and shortening of cold period, powdery mildew incidence on chickpea (*Leveillula taurica*) in parts of southern India viz., northern Karnataka (Ghosh *et al.*, 2012) and on oilseeds *Brassica* (*Erysiphe cruciferarum*) has been observed appearing earlier (December) than usual time (late Jan, Feb) quite frequently in non-traditional crop growing areas of Madhya Pradesh, Haryana, central Uttar Pradesh, parts of Rajasthan and Bihar (Chattopadhyay *et al.*, 2012) possibly due to shortening of cold spell during the crop period. There are indications for increase in sheath blight, sheath rot, false smut diseases of rice, fruit rot and die back of chilli, bitter gourd and potato, including post-harvest losses (Biswas, 2011). In 2011 there has already been an upsurge in false smut incidence on rice crops in West Bengal (Ghosh *et al.*, 2012); it reached very high severity in 2013 with additional areas of Jharkhand, Orissa, Bihar and is expected to cause serious problem in rice production in coming years.

Free Air $CO_2$ Enrichment (FACE) facility has been utilized in India to study pathogen biology and host–pathogen interactions for the biotrophic stripe rust (*Puccinia striiformis*), necrotrophic crown rot (*Fusarium pseudograminearum*) and barley yellow dwarf virus (BYDV). In addition, naturally occurring levels of soil-borne pathogens, *Heterodera avenae, Rhizoctonia solani, Gaeumannomyces graminis* var. *tritici, Pratylenchus neglectus* and *P. thornei, F. pseudograminearum* and *F. culmorum* and *Bipolaris* spp. were quantified from the soil using PCR-based diagnostics from each FACE ring both prior to sowing and post-harvesting (Chakraborty *et al.*, 2008). Using the 'Rice FACE' facility in northern Japan, Kobayashi *et al.* (2006 a, b) studied the effect of 200–280 ppm above-ambient $CO_2$ on rice blast and sheath blight diseases for three seasons. Severity of leaf blast (*Magnaporthe oryzae*) was consistently higher at the elevated $CO_2$ levels in all the three years assessed at two different stages of rice growth. However, severity of panicle blast caused by the same pathogen was not consistently higher at higher $CO_2$ levels. The incidence of rice plants naturally infected with sheath blight (*R. solani*) was generally higher at the elevated $CO_2$ concentrations under high nitrogen levels. But this trend was not apparent for sheath blight severity.

## Climate change and insect-pest scenario

The climate has profound effects on the populations of invertebrate pests like insects, mites, etc. and affects their population build-up and dispersal systems. Climate change may influence the physiology, abundance, phenology and distribution of the insect pests (Lastvka, 2009), and the major factors including temperature, $CO_2$ concentration, precipitation, natural enemies and their host plant. Extreme weather events such as intense rainstorms, high wind or elevated temperatures also affect their survival. The climate change impacts on pests may include shifts in species distributions with species shifting their ranges to higher latitudes and elevations, changes in phenology with life cycles beginning earlier in spring and continuing later in autumn, increase in population growth rates and number of generations, change in migratory behaviour, alterations in crop-pest synchrony and natural enemy-pest interaction, and changes in inter-specific interactions. Changes in community structure and extinction of some species are also expected (Thomas *et al.*, 2004). Many insects have large population sizes and short generation times, and their phenology, fecundity, survival, selection and habitat use can respond rapidly to the climate change. These changes to insect life-history may in turn produce rapid changes in their abundance and distribution. Due to recent climate changes, widespread, generalist species at their cool range margins have expanded their distribution ranges, whereas the ranges of localized, habitat-specialist species and those at their warm margins have narrowed (Hill *et al.*, 2002; Konvicka *et al.*, 2003). In India, Bihar hairy caterpillar (*Spilarctia obliqua*) surprisingly on mustard has been noted to be on the rise. Oilseeds Brassicas have been affected a lot by the painted bug (*Bagrada cruciferarum*) in the western and by saw fly (*Athalia proxima*) in the eastern India (Chattopadhyay *et al.*, 2012).

Gao *et al.* (2009) used OTCs to examine interactions across three trophic levels, cotton (*Gossypium hirsutum*), aphid herbivore (*Aphis gossypii*) and its coccinellid predator (*Propylaea japonica*), as affected by elevated $CO_2$ concentrations and crop cultivars. Two levels of $CO_2$, viz. ambient (375 ppm) and double the ambient (750 ppm) were used. Plant carbon: nitrogen (C:N) ratios, condensed tannin, and gossypol content were significantly higher while nitrogen-content was significantly lower in the plants exposed to elevated $CO_2$ levels compared to those exposed to ambient $CO_2$ concentrations. No significant difference in the survival and lifetime fecundity of *P. japonica* was observed between cultivars and $CO_2$ concentration treatments. However, stage-specific larval durations of the lady beetle were significantly longer when fed on aphids from the elevated $CO_2$ concentrations. It was speculated that *A. gossypii* may become a more serious pest under an environment with elevated $CO_2$ concentrations because of increased survivorship of aphid and longer development time of lady beetle.

**Table 2.** Impact of climate change on insect pest

| Particulars of climate change | Name of pests/ pathogens | Impact | References |
|---|---|---|---|
| Elevated temperature | *Spodoptera litura* | Developmental duration decreased with increase in temperature for immature stages of *S. litura* till 30°C, but increased at 35°C in egg, $3^{rd}$ & 4th larval instars and pupal stage indicating a nonlinear response at extreme temperatures | CRIDA, 2014 |

| | | | |
|---|---|---|---|
| Temperature dependent | Maize aphid | Nymphal development and adu lt longevity showed a linear decreasing trend till 25°C and a nonlinear response above this temperature Fecundity was highest at 20°C and was reduced by >50% at 30°C | |
| Elevated $CO_2$ | Peanut aphids (*Aphis craccivora*) | The number of nymphs laid per female was significantly higher at e-$CO_2$ than ambient. The increased mean fecundity and significant reduction in longevity and development time | |
| | Many foliage feeding Lepidopterans in agricultural and forest species | Enhanced feeding by insects in order to obtain sufficient nitrogen for their metabolism. Slower development increased length of life stages. More foliage feeding than the normal | Fand *et al.*, 2012 |
| | Gypsy moth *Lymantria dispar* in Mapple | Reduced larval weight gain. Increased larval feeding. Prolonged development | |
| High temperature and water stress | Midge *Stenodiplosis sorghicola* (Coq.) and Spotted stem borer *Chilo partellus* Swinhoe in sorghum | Breakdown of resistance against target insect-pests. Heavy loss in yield due to increased pest damage | |
| | Bollworms *Heliothis virescens* (F.) *Helico verpa armigera* (Hubner) | Negative impacts on transgene expression in *Bt* cotton. Reduced | |

| | | | |
|---|---|---|---|
| | and *Helicoverpa punctigera* (Wallen) in cotton | production of *Bt* toxins Enhanced susceptibility of the crops to insect-pests | |
| Recent abnormal weather patterns | sugarcane woolly aphid *Ceratovacuna lanigera* | 30% yield losses Reduced cane recovery | Joshi, and Viraktamath, 2004; Srikanth, 2007 |
| | Rice plant hoppers *Nilaparvata lugens* (Stal) and *Sogatella furcifera* | Crop failure over more than 33,000 ha paddy area | IARI News, 2008 IARI News, 2009 |
| Recent abnormal weather patterns, Changed cropping environment (introduction of *Bt* cotton) | Mealybug, *Phenacoccus solenopsis* in Cotton, vegetables and ornamentals | Heavy yield (30-40%) loss to the cotton. Increased cost of crop protection due to overuse of chemical pesticides | Dhawan *et al.*, 2007 |
| | Papaya mealy bug *Paracoccus marginatus* | Significant yield loss to the papaya growers | Tanwar *et al.*, 2010 |
| Efficacy of insect biocontrol by fungi | Leaf webber of Mango (*Orthaga exvinaceae*) | Increased infestation | Rao *et al.*, 2014 |
| Reliability of economic thrust hold levels | Leaf miner of mango (*Acrocercops syngramma*) | Increased infestation | |
| Insect diversity in ecosystems | Stem borer (*Batocera rufomaculata*) | Increased infestation | |
| Parasitism | Fruit borer (*Deanolis albizonalis / Citripestis eutraphera*) | Increased infestation | |
| | Early shoot borer (*Chlumetia transversa*) | Increased infestation | |
| | Scale insects (*Aspidiotus destructor*) | Increased infestation | |
| | Mealy scale (*Chloropulvinaria polygonata*) | Increased infestation | |
| | Snow scale (*Aulocaspis tubercula*) | Increased infestation | |

| | | |
|---|---|---|
| Mealy bug (*Rastrococcus iceriyodes, Ferrisia virgata*) | Increased infestation | |
| Thrips (*Scirtothrips dorsalis*) | Increased infestation | |
| Corn earworms *Heliothis zea* (Boddie) and *Helicoverpa armigera* in maize | Altitudes wise range expansion and increased overwintering survival in USA | Fand *et al.*, 2012 |
| European corn borer *Ostrinia nubilalis* | Northward shifts in the potential distribution up to 1220 km are estimated to occur an additional generation per season | |
| Old world Bollworm *Helicoverpa armigera* | Phenomenal increase in the United Kingdom from 1969-2004 and outbreaks at the northern edge of its range in Europe | |
| Cottony cushion scale *Icerya purchasi* | Cottony cushion scale *Icerya purchasi* | |
| Oak processionary moth *Thaumetopoea processionea* | Northward range extension from central and southern Europe into Belgium, Netherlands and Denmark | |
| Cottony camellia scale *Chloropulvinaria floccifera* | More abundant in the United Kingdom, extending its range northwards in England and increasing its host range in the last decade | |
| Cotton bollworm/ Pulse pod borer *Helicoverpa armigera* Cotton, Pulses, vegetables | Expansion of geographic range in Northern India. Adult flights/ migratory behaviour | |

Studies under NICRA (ICAR) since 2011 has indicated that rise in ambient temperature due to climate change could result in 2-4 additional generations of lepidopteran pests (viz., *Helicoverpa armigera*) with shortened life cycle by 5-6 days across locations (Rao *et al.*, 2014), which could possibly lead to increased crop losses (Sharma *et al.*, 2010). There are indications of increased predatory activity on pests (Karuppaiah and Sujayanand, 2012) while climate change has been ascribed as among important factors having influenced high infestation of *Nilaparvata lugens* on rice crops in different parts of Asia in recent times (Ali *et al.*, 2014). Apprehensions have been expressed towards adverse impact of climate change on pollinators (Srivastava *et al.*, 2014) and lac productivity (Mohanasundaram *et al.*, 2014).

Global warming resulting in elevated carbon dioxide ($CO_2$) and temperature in the atmosphere could influence plant parasitic nematodes directly by interfering with their developmental rate, survival strategies and indirectly by altering host physiology. It may also influence free-living (microbial feeding) nematodes due to changes in quality and availability of food under enriched $CO_2$ conditions. Available information on effect of global warming on soil nematodes though limited, indicate that abundance of soil nematodes in general is either increased or unaffected by elevated $CO_2$ levels while individual species and trophic groups differ considerably in their response to climate change. Herbivorous nematodes showed neutral or positive response to $CO_2$ enrichment effects with some species showing the potential to build up rapidly and interfere with plant's response to global warming. Studies have also demonstrated that the geographical distribution range of plant parasitic nematodes may expand with global warming spreading nematode problems to newer areas (Somasekhar *et al.*, 2010). There are also reports of upsurge in infestation by *Rotylenchulus* and *Pratylenchus* on several crops viz., chickpea, vegetables, etc.

Presently, the India Meteorological Department (IMD, Govt. of India) and the National Centre for Medium Range Weather Forecasting (NCMRWF) in collaboration with Indian Council of Agricultural Research (ICAR) are regularly issuing location-specific weather forecast and agrometeorological advisory as per different climatic conditions and cropping systems. ICAR launched National Initiative on Climate Resilient Agriculture (NICRA) in February, 2011 to boost research on the impact of climate change and its mitigation at national level. The project aims to enhance resilience of Indian agriculture to climate change, climate vulnerability through strategic research and technology demonstration. Research on adaptation and mitigation covers crops, livestock, fisheries and natural resource management. It also demonstrates site-specific technology packages on farmers' fields for adapting to current climate risks. This will certainly enhance the capacity of scientists and other stakeholders in climate resilient agricultural research and its application (http://www.icar.org.in). The mitigation of the adverse effect of climate is challenging. Acquaintances between pragmatic and modelling studies could prop-up swift advancement in perception, prediction of climate change effects.

There are several insect-pest and disease forecasting networks across the globe viz., maize (EPICORN – for Southern corn leaf blight), tomatoes and potatoes (EPIDEM, TOMCAST, BLITECAST – for early and late blights), apple (Maryblight, EPIVEN – for fire blight and scab) etc. Further research advancement led to development of weather-based location-specific forewarning models based on multiple or stepwise regression, discriminant function analysis, artificial neural network for diseases (Desai *et al.,* 2004; Chattopadhyay *et al.*, 2005, 2011; Laxmi and Kumar 2011; Kumar *et al.,* 2013). These have been validated with success, even with issue of agro-advisories for public use.

At present, remote sensing (ISRO) data are also being used in generating weather forecasts, providing crop estimate in terms of net sown area and yield, issued in operational mode for the last few years with reasonable accuracy for rice, wheat, mustard and potato. Through an ICAR-ISRO (SAC) collaboration under MoA (DAC)-funded project FASAL (Forecasting Agricultural output using Space, Agrometeorological and Land-based observations), crop production has been forecasted at national level using multi-date temporal AWiFS (Advanced Wide Field Sensor on IRS-P6; 56 m x 56 m) and RADAR (RAdio Detection and Ranging) data. Two forecasts are made during the season at different crop growth stages. Encouraged by these successes, the IMD envisages implementation of FASAL initially at 46 centres, which is likely to be extended to 130 stations in due course (IMD, 2013). Use of remote sensing (RS) and Geographic Information System (GIS) could be explored for analysis of satellite-based agro-met data products, mapping geographical distribution of diseases and delineating the hotspot zones. Super-imposition with causative abiotic and biotic factors on visual pest maps can be useful for disease forecasting. Since, diseased plants increase reflectance particularly in chlorophyll absorption band (0.5-0.7 m) and water absorption bands (1.45-1.95m), forecasting plant disease is possible by remote sensing. Though information on this aspect is scanty, disease severity and yield loss estimation using changed reflectance pattern of diseased plants can be attempted. Remote sensing (greenness vegetation index derived from LANDSAT MSS digital data, four bands) has been successfully used to distinguish the healthy wheat in India from diseased wheat in Pakistan. Favourable weather from January to April and a sudden rise in temperature at mid-April are the main causes for yellow rust of wheat. Routine monitoring at surface for weather and by remote sensing could help predict epidemic well before first appearance of the disease on the crop, giving a positive edge to make accurate decision related to disease management. It has also been possible to detect Sclerotinia rot-affected mustard using remote-sensing technology (Dutta *et al.*, 2006; Bhattacharya *et al.*, 2007; Bhattacharya and Chattopadhyay, 2013). These successful experiences could certainly be effective boosters for any future endeavour. But the potential benefits of short-to-medium range weather forecast from numerical weather prediction (NWP) models or future climate projections have been least harnessed in India for regional crop protection services. Recent momentum to assimilate more updated satellite-based spatio-temporal atmospheric and land surface products from Indian geostationary satellites (Kalpana-1, INSAT 3A) for high resolution (5-15 km) weather forecasts from advanced NWP model such as WRF (Weather Research and Forecasters) is encouraging. Such regular high-resolution forecast products are available for the registered users (http://www.mosdac.gov.in). The Mahalanobis National Crop Forecasting Centre (MNCFC) at IARI, New Delhi under the behest of MoA (DAC) provides crop masks, crop forecasts. Therefore, an *Integrated Decision Support System (IDSS) for Crop Protection Services* (Ghosh *et al.*, 2012) can be imagined with its evolution in a phased manner, which could have the following three components: (A) Operational focus (with periodic production of alarm zones encompassing 127 agro-climatic zones through well-tested models, forecast weather, high-resolution remote sensing data and operational crop map in the GIS framework), (B) Research priorities [(i) Development of forecast models for major insect-pests and diseases with large-scale applicability, validation in farmers' fields and model refinements, (ii) Evaluation and improvement in quality of well-validated satellite-based products, improved data assimilation approaches, (iii) field-to-satellite-based remote sensing with high-resolution observations to differentiate among crops, among phenological stages within crop growth period, biotic stresses from abiotic stresses (moisture and nutrients), normal health and (iv) development of models for minor pests and diseases in view of climate change scenario], (C) Human Resources renewal [(i) creation of experts on handling of spatial data, who could be brave enough to think differently, bold enough to believe that as a team they could bring a positive change in the present practices of pest management and talented enough to do it, (ii) familiarization of policy makers with more of digital products for interpretation and (iii) regular feedback mechanism from farmers through Village Resources Centre (VRC) network using satellite communication].

On-line decision support systems to forecast different pests (including insect-pests and diseases) are in use across the globe viz., tan spot, septoria leaf blotch, leaf rust and Fusarium head blight /scab diseases of wheat (http://www.ag.ndsu.edu/ndsuag/features/small-grain-disease-forecasting), canola light leaf spot (*Pyrenopeziza brassicae* on http://www.rothamsted.bbsrc.ac.uk/Research/Centres/Content. php? Section=Leafspot&Page=llsforecast) and Phoma stem canker (*Leptosphaeria maculans, L. biglobosa* on http://www.rothamsted. bbsrc.ac.uk/Research/Centres/Content.php?Section= Leafspot&Page= phomaforecast) or for the *Lypaphis erysimi* on oilseeds Brassica (http://www.drmr.res.in/aphidforecast/). Models of plant disease have now been developed to incorporate more sophisticated climate predictions.

In view of changing climate, the devised and to-be-born models need to be oriented to dynamic mode. The models already developed are based on some observations on meteorological parameters, pest infestation levels recorded in the past and hence are based on previous pest-weather correlation. However, with change in climate, the pest-weather relationship is also bound to change apart from behaviour of the hosts, newer varieties, cropping practices, etc. (Chakraborty *et al.*, 2008). Importance of decision support systems have been highlighted by several workers (Rossi *et al.*, 2012) with need to project future crop enemies as a proactive effort (Ghosh *et al.*, 2012; Kumar *et al.*, 2013; Juroszek and Tiedemann, 2015). Multiple, interconnected processes will require interdisciplinary science, long-term funding and the increased use of meta-analysis (Kozlov and Zvereva, 2011). At the same time, there is a need for catching up with the latest scientific developments viz., networking in epidemiology/epizoology, pest forecasting, digital pest diagnostics and severity estimation to improve dissemination of knowledge in plant health (Bock *et al.*, 2010; Norton and Taylor, 2010). Many reviews available on the topic of plant pests and climate change agree that there is a need for more empirical data on the subject (Chakraborty and Newton, 2011). The assumption is that more and better data will make prediction more accurate and/or reliable (Shaw and Osborne, 2011). Dynamic models incorporate the recorded data of each crop season for a particular pest to suitably revise itself and thus remain stable, relevant enough to continue providing accurate forecast.

But due to inadequacies in hard-core data, the vital questions still haunt - what are the possible impacts of climate change on diseases and insect-pests of major crops or any shift in pathogen status with change of climate in the agro-ecological region growing major crops in India? We may have seen some changes in scenario of insect-pests, pathogens and ascribed them to changed climatic patterns. However, establishing such correlation through research remains to be done in India. There is no clear indication on injury profile on attainable yield and changes thereof in course of any change in pest scenario and hence we are not sure about the fitness status of any pest *vis-á-vis* the changed climatic situation apart from the status of our readiness to encounter the same.

The technique in System of Rice Intensification is reported to have helped 55-70% reduction in major diseases of the crop in Vietnam during 2005-06 (Uphoff, 2007), which seems an effective strategy to cope with the effects of climate change. Similar efforts would be warranted to adapt and manage the effects of climate change on pests and diseases of different crops. Prediction and management of climate change effects on plant health are complicated by indirect effects and the interactions with global change drivers. Uncertainty in models of plant disease development under climate change calls for a diversity of management strategies, from more participatory approaches to interdisciplinary science (Pautasso *et al.*, 2012). Accordingly, there could be a pro-active approach to breed for resistance to pests, pathogens and their variants likely to be dangerous apart from modelling the future possible pest and disease scenario in major crops. Monitoring of variability in major insect-pests, pathogens and nematodes affecting agricultural crops to keep track of upheaval of 'sleeper races' becoming severe under changed climate, modelling future enemies of plants and designing crops with resistance to such would-be menaces as also modifying IPM strategies suitably to enable them be more climate-resilient and pest smart may need emphasis. One among them is farming in cluster approach nursery onwards as practiced by farmers of Tripura state. The government in the state has enabled farmers to register more than 2000 clusters of farmers, who not only share their land for farming but also all inputs and benefits, which provides an insurance cover against adverse impact on individual plot / farmer.

## Novel Plant Protection Approaches

Changing pattern of disease and insect-pest scenario due to climate change has warranted the need for improved novel agricultural practices and use of eco-friendly approach for sustainable crop production. Due to change in climate, there is shift in seasons, cropping pattern, insect-pest and disease scenario, etc. Hence, the choice of crop management practices based on the real time situation is very important. In such scenarios, weather-based insect-pest and disease monitoring, inoculums/ population monitoring, and rapid diagnostics/systematic would play a significant role. There is need to adopt novel approaches to counter the resurgence of diseases/ insect-pests under changed climatic scenario. Educating farmers towards maintaining records on farming practices undertaken date-wise may enable experts to guide on decision making as per need of the hour. Integrated pest management strategies would undoubtedly be the only solution to cope with complex insect-pest / disease scenario. It covers multiple approaches including the use of healthy disease free seeds with innate forms of broad and durable disease resistance and various types of cropping systems that promote the conservation of natural/native bio-control agents. In addition, monitoring and early warning systems for forecasting disease / insect-pest epidemics / epizootics should be developed for important host–pathogens / insect-pests, which have a direct link on the crop production, income generation of the farmers and food security. Use of botanical pesticides and plant-derived soil amendments such as neem oil, neem cake and karanja seed extract also help in mitigation of climate change because it helps in the reduction of nitrous oxide emission by nitrification inhibitors such as nitrapyrin and dicyandiamide.

## Conclusion

Though the effect of climate change is very vast, only limited research on impact of climate change on plant diseases / insect-pests has been done under real field conditions. Most of the recent plant protection technologies have been developed either in the laboratory or artificial condition (which are likely to occur in the changed climate scenario). However, some assessments are now being done in few countries, regions, crops and particular pathogens / insect-pests under field condition to counter the current as well as upcoming problems of crop insect-pests / diseases. Now, emphasis must shift from impact assessment to developing adaptation and mitigation strategies and options. We have to systematically evaluate the efficacy of current physical, chemical and biological control tactics, including disease-resistant cultivars under climate change, and to include future climate scenarios in all research aimed at developing new tools and tactics. Disease risk analyses based on host–pathogen / pests interactions should be done, and research on host response and adaptation should be conducted to understand how an imminent change in the climate could affect plant diseases. India is fortunate enough to have such a diverse climate suitable to grow various types of crop plants with diverse pest population, which can help to counter the pest/disease problems in the changed climate scenario.

Today's challenge is to 'produce more from less'. In the era of climate change, diagnostics of pests and diseases and capacity building of farmers, extension and even research personnel for adaptation to changed pest scenario under future climates assumes significance, wherein Integrated Crop management in Private Public Partnership mode could be very effective. Farmers' decisions are of vital importance for good yields of crops. Forecasted weather products and area-wide weather networks are becoming more prevalent. Now, the challenge is to bring continuous improvement in productivity, profitability, stability and sustainability of major farming systems, wherein scientific management of plant pests holds a pivotal role (Swaminathan, 1995). Crop loss models, representing a dynamic interaction between pests and host, are essential for forecasting losses thereof. Accurate information concerning possible yield losses due to occurrence of a pest is needed by growers or plant protection specialists to decide on cost-effective control measures. With an increasing concern for cleaner environment and discouragement for pesticide use, there is need to approach pest management through knowledge on their dynamics as an art of living with them (Zadoks, 1985). Thus, future research and education in Plant Pathology in India does need to address the issue of future climates in pest management, for which fund requirement would certainly be lesser than many ambitious ones.

## References

Ali, MP; Huang, D; Nachman, G; Ahmed, N; Begum, MA and Rabbi, MF. (2014). Will climate change affect outbreak patterns of planthoppers in Bangladesh? *PLoS ONE* 9(3): e91678. doi: 10.1371/journal.pone.0091678.

Bergot, M., Cloppet, E., Pérarnaud, V., Déqué, M., Marçais, B. and Desprez-Loustau, M. L. (2004). Simulation of potential range expansion of oak disease caused by *Phytophthora cinnamomi* under climate change. Global Change Biology 10: 1-14.

Bhattacharya BK and Chattopadhyay C. (2013). A multi-stage tracking for mustard rot disease combining surface meteorology and satellite remote sensing. Computers and Electronics in Agriculture 90: 35-44.

Bhattacharya, B.K., Chattopadhyay, C., Dutta, S., Meena, R.L., Roy, S., Parihar, J.S. and Patel, N.K. (2007). Detection of Sclerotinia rot incidence in Indian mustard from polar orbiting satellite platform. In: *Proceedings, 12th International Rapeseed Congress, Vol. IV* (Plant Protection), Wuhan, China, 26-30 Mar 2007, Fu Tingdong and Guan Chunyun (eds.), Science Press USA Inc., pp 79-83.

Biswas A (2011). Plant diseases in the perspective of climate change. SATSA Mukhopatra–Annual Tech Issue15: 84-86.

Bock, C.H.; Poole, G.H.; Parker, P.E. and Gottwald, T.R. (2010). Plant disease severity estimated visually, by digital photography and image analysis, and by hyperspectral imaging. Critical Reviews in Plant Sciences 29: 59–107.

Central Research Institute for Dryland Agriculture (CRIDA). (2014). Annual Report. CRIDA, Santoshnagar, Hyderabad, India.

Chakraborty, S. (2008). Meeting Sukumar Chakraborty. ICPP Newslr 4: 2.

Chakraborty, S. and Datta, S. (2003). How will plant pathogens adapt to host plant resistance at elevated $CO_2$ under a changing climate? New Pathologist 159: 733-742.

Chakraborty, S. and Newton, A.C. (2011). Climate change, plant diseases and food security: an overview. Plant Pathology 60: 2–14.

Chakraborty, S., Luck, J., Hollaway, G., Freeman, A., Norton, R., Garrett, K.A., Percy, K., Hopkins, A., Davis, C. and Karnosky, D.F. (2008). Impacts of global change on diseases of agricultural crops and forest trees. *CAB Reviews: Perspectives in Agriculture, Veterinary Sciences,* Nutrition and Natural Resources 3: 1-15.

Chakraborty, S., Pangga, I. B., Lupton, J., Hart,L., Room, P. M. and Yates, D. (2000). Production and dispersal of *Colletotrichum gloeosporiodies* spores on *Stylosanthes scabra* under elevated $CO_2$. Environmental Pollution 108: 381-387.

Chattopadhyay, C., Bhattacharya, B.K., Kumar, V., Kumar, A. and Meena, P.D. (2012). Impact of climate change on pests and diseases of oilseeds *Brassica* – the scenario unfolding in India. J. Oilseed Brassica 2: 48-55.

Chattopadhyay, C. and Huda, A.K.S. (2009). Changing climate forcing alteration in cropping pattern to trigger new disease scenario in oilseeds and pulses in Indian sub-continent. *Abstracts,* National Symposium on Climate Change, Plant Protection and Food Security Interface, Kalyani, 17-19 Dec 2009. pp. 70.

Chattopadhyay, C., Agrawal, R., Kumar, A., Bhar, L.M., Meena, P.D., Meena, R.L., Khan, S.A., Chattopadhyay, A.K., Awasthi, R.P., Singh, S.N., Chakravarthy, N.V.K., Kumar, A., Singh, R.B. and Bhunia, C.K. (2005). Epidemiology and forecasting of Alternaria blight of oilseed *Brassica* in India – a case study. Zeitschrift für Pflanzenkrankheiten und Pflanzenschutz 112: 351-365.

Chattopadhyay, C., Agrawal, R., Kumar, A., Meena, R.L., Faujdar, K., Chakravarthy, N.V.K., Kumar, A., Goyal, P., Meena, P. D. and Shekhar, C. (2011). Epidemiology and development of forecasting models for white rust of *Brassica juncea* in India. Archives of Phytopathology and Plant Protection 44: 751-763.

Coakley, S.M., Scherm, H. and Chakraborty, S. (1999). Climate change and plant disease management. Annu. Rev. Phytopathol. 37: 399–426.

Desai, A.G., Chattopadhyay, C., Agrawal, R., Kumar, A., Meena, R.L., Meena, P.D., Sharma, K.C., Rao, M.S., Prasad, Y.G. and Ramakrishna, Y.S. (2004).

*Brassica juncea* powdery mildew epidemiology and weather based forecasting models for India – a case study. Zeitschrift für Pflanzenkrankheiten und Pflanzenschutz. 111: 429-438.

Dhawan, A.K., Singh, K., Saini, S., Mohindru, B., Kaur, A., Singh, G. and Singh, S. (2007). Incidence and damage potential of mealybug, *Phenacoccus solenopsis* Tinsley, on cotton in Punjab. Indian J. Ecol. 34: 110-116.

Dutta, S., Bhattacharya, B.K., Rajak, D.R., Chattopadhyay, C., Patel, N.K. and Parihar, J.S. (2006). Disease detection in mustard crop using EO-1 hyperion satellite data. J. Indian Soc. Remote Sensing (Photonirvachak) 34: 325-329.

Fand B B. Kamble A L. and Kumar M. (2012). Will climate change pose serious threat to crop pest management-A critical review. International Journal of Scientific and Research Publications 2: 1-14.

Fitt, B.D.L.; Fraaije, B.A.; Chandramohan, P and Shaw, M.W. (2011). Impacts of changing air composition on severity of arable crop disease epidemics. Plant Pathology 60: 44–53.

Gao F, Zhu S, Sun Y, Du L, Parajulee M, Kang L and Ge F. (2009). Interactive effects of elevated $CO_2$ and cotton cultivar on tri-trophic interaction of Gossypium hirsutum, Aphis gossyppii and Propylaea japonica. Environ. Entomol. 37: 29-37.

Garrett, K.A., Dendy, S.P., Frank, E.E., Rouse, M.N. and Travers, S.E. (2006). Climate change effects on plant disease: Genomes to ecosystems. Annu. Rev. Phytopathol. 44: 489–509.

Gautam, H.R., Bhardwaj, M.L. and Kumar, R. (2013). Climate change and its impact on plant diseases. Current Science 105: 1685-91.

Ghosh, A., Das, A., Bhattacharya, B.K., Kumar, V., Kumar, A., Meena, P.D. and Chattopadhyay, C. (2012). Impact of climate change on pests and diseases - the scenario unfolding in Indian agriculture. SATSA Mukhopatra Annual Technical Issue 16: 15- 29.

Goyal, P., Chahar, M., Mathur, A.P., Kumar, A. and Chattopadhyay, C. (2011). Morphological and cultural variation indifferent oilseed *Brassica* isolates of *Alternaria brassicae* from different geographical regions of India. The Indian Journal of Agricultural Sciences 81: 1052-1058.

Hill, J.K., Thomas, C.D., Fox, R., Telfer, M.G., Willis, S.G., Asher, J. and Huntley, B. (2002). Responses of butterflies to twentieth century climate warming: implications for future ranges. – Proceeding of the Royal Society of London B 269: 2163–2171.

IARI (2008). Brown plant hopper outbreak in rice. News 24.

IARI. (2009). Brown plant hopper outbreak in rice. News 25.

IMD (India Meteorological Department). (2013). Future projections in AAS. Retrieved from http://www.imd.gov.in/services/agri_new/Activities.html. on 15 Feb 2013 at 1242 hrs IST. Informatics in Agricultural Research, organised by ISAS at New Delhi, 18-20 Dec 2012, pp 100-101

IPCC (2007). Fourth Assessment Report: Climate Change. Available at: http://www.ipcc.ch (Accessed on Dec. 27, 2011).

IPCC. (2007). Climate Change: Impacts, Adaptation and Vulnerability. In: (eds.: Parry, M.L., Canziani, O.F., Palutikof, J.P., van der Linden, P.J., Hanson, C.E.), Cambridge University Press, Cambridge, UK, pp. 976.

Joshi, S. and Viraktamath, C.A. (2004). The sugarcane woolly aphid, *Ceratovacuna lanigera* Zehntner (Hemiptera: Aphididae): its biology, pest status and control. *Curr. Sci.* 87: 307-316.

Juroszek, P. and Tiedemann, A.V. (2015). Use of disease and crop models linked to regional climate projections to predict future disease risks in agricultural crops – a review. Journal of Plant Diseases and Protection (in press).

Karuppaiah, V., Sujayanand, G.K. (2012). Impact of climate change on population dynamics of insect pests. World J Agri Sci 8: 240–246.

Khan, S.A., Kumar, S., Hussain, M.J., and Kalra, N. (2009). Climate Change, Climate Variability and Indian Agriculture: Impacts Vulnerability and Adaptation Strategies. S.N. Singh (ed.), *Climate Change and Crops,* Environmental Science and Engineering. DOI 10.1007/978-3-540-88246-6 2, C-Springer-Verlag Berlin Heidelberg.

Kobayashi, T., Ishiguro, K., Nakajima, T., Kim, H.Y., Okada, M. and Kobayashi, K. (2006a). Effects of elevated atmospheric $CO_2$ concentration on the infection of rice blast and sheath blight. Phytopathology 96: 425-431.

Kobayashi, K., Okada, M., Kim, H.Y., Lieffering, M., Miura, S., Hasegawa, T. (2006b). Paddy rice responses to free-air [$CO_2$] enrichment. In: Nösberger J, Long SP, Norby RJ, Stitt M, Hendrey GR, Blum H, eds. Managed ecosystems and $CO_2$: case studies, processes, and perspectives. Berlin: Springer-Verlag, pp. 87–104.

Konvicka, M., Maradova, M., Benes, J., Fric, Z. and Kepka, P. (2003). Uphill shifts in distribution of butterflies in the Czech Republic: effects of changing climate detected on a regional scale. Global Ecol. Biogeogr., 12: 403–410.

Kozlov, M. V. and Zvereva, E. L. (2011). A second life for old data: global patterns in pollution ecology revealed from published observational studies. Environmental Pollution 159: 1067–1075.

Kumar, A., Chattopadhyay, C., Singh, K.N., Vennila, S. and Rao, V.U.M. (2012). Trend analysis of climate variables for pigeonpea growing area. In: Abstract, International Conference on Statistics and Computation: 145-148.

Kumar, A., Chattopadhyay, C., Singh, N.K., Venilla, S. and Rao, V. (2014). Trend analysis of climatic variables in pigeonpea growing regions in India. Mausam 65: 161-170.

Kumar, A., Kumar, V., Bhattacharya, B.K., Singh, N. and Chattopadhyay, C. (2013). Integrated disease management: need for climate-resilient technologies. J. Mycol Pl Pathol 43: 28-36.

Lastuvka, Z. (2009). Climate change and its possible influence on the occurrence and importance of insect pests. Plant Protect. Sci. 45: S53-S62

Laxmi, R. R. and Kumar, A. (2011). Forecasting of powdery mildew in mustard (*Brassica juncea*) crop using artificial neural networks approach. Indian J. Agric. Sci. 81: 855–860.

Manning, W. and Tiedemann, A. (1995). Climate change: Effects of increased atmospheric carbon dioxide ($CO_2$), ozone ($O_3$), and ultraviolet-B (UV-B) radiation on plant diseases. Environmental Pollution 88: 219-245.

Matros, A., Amme, S., Kettig, B., Buck-Sorlin, G. H., Sonnewald, U. and Mock, H. P. (2006). Growth at elevated $CO_2$ concentrations leads to modified profiles of secondary metabolites in tobacco cv. SamsunNN and to increased resistance against infection with potato virus Y. Plant, Cell and Environment 29: 126-137.

Meena, P.D., Awasthi, R.P., Chattopadhyay, C., Kolte, S.J. and Kumar, A. (2010). Alternaria blight: a chronic disease in rapeseed-mustard. Journal of Oilseed Brassica 1: 1-11.

Melloy, P., Hollaway, G., Luck, J., Norton, R., Aitken, E. and Chakraborty, S. (2010). Production and fitness of *Fusarium pseudograminearum* inoculum at elevated carbon dioxide in FACE. Global Change Biology 16: 3363-3373.

Mitchell, C.E., Reich, P.B., Tilman, D. and Groth, J.V. (2003). Effects of elevated $CO_2$, nitrogen deposition, and decreased species diversity on foliar fungal plant disease. Global Change Biology 9: 438-451.

Mohapatra, S. D. and Chattopadhyay, C. (2012). Occurrence of mungbean phyllody and its Management. *Abstracts*. 3rd Global Conference on Plant Pathology for Food Security, Org. by ISMPP and MPUAT, Udaipur, Jan. 10-13, 2012. pp. 259.

Mohanasundaram, A., Monobrullah, M., Sharma, K.K., Singh, R.K., Meena, S.C. and Verma, S. (2014). Effect of weather parameters on production of summer rangeeni lac crop at Ranchi, Jharkhand. In: *Compendium-cum-Abstract*, International Symposium on New-Dimensions in Agrometeorology for Sustainable Agriculture, 16-18 Oct 2014, GBPUAT, Pantnagar (UA), Association Agrometeorologists, pp. 329.

Nagarajan, S. and Muralidharan, K. (1995). Dynamics of Plant Diseases, Allied Publ. Ltd., New Delhi, 247 p.

Norton, G. and Taylor, M. (2010). What pest is that? Recent developments in digital pest diagnostics. Outlooks on Pest Management 21: 236–238.

NSW Department of Primary Industries. (2007). Editor's Note: Science Alert. 02 March 2007 Available at: http://www. sciencealert.com.au/news/20070203-13999.html.

Pangga, I.B., Chakraborty, S. and Yates, D. (2004). Canopy size and induced resistance in *Stylosanthes scabra* determine anthracnose severity at high $CO_2$. Phytopathology 94: 221-227.

Pautasso, M., Döring, T.F., Garbelotto, M., Pellis, L. and Jeger, M.J. (2012). Impacts of climate change on plant diseases—opinions and trends. European J. Plant Pathol. DOI 10.1007/s10658-012-9936-1

Plessl, M., Heller, W., Payer, H.-D., Elstner, E. F., Habermeyer, J. and Heiser, I. (2005). Growth parameters and resistance against *Drechslera teres* of spring barley (*Hordeum vulgare* L. cv. Scarlett) grown at elevated ozone and carbon dioxide concentrations. Plant Biology 7: 694-705.

Rao, M.S., Manimanjari, D., Aruna, D., Desai, S., Rao, V.U.M. and Maheswari, M. (2014). Prediction of number of generations of *Helicoverpa armigera* (Hub.) on pigeonpea in India during climate change period. In: *Compendium-cum-Abstract, International* Symposium on New-Dimensions in Agrometeorology for Sustainable Agriculture, 16-18 Oct 2014, GBPUAT, Pantnagar (UA), Association Agrometeorologists, pp. 306.

Rao, T.M., Jayanthi, P.D.K., Verghese, A., Arthikirubha, A., Sowmya, B.R. and Bhatt, R.M. (2014). Emerging pests of mango under climate change scenario. Technical Bulletin no.18, IIHR, Bengaluru.

Root, T.L. and Hughes, L. (2005). Present and future phe-nological changes in wild plants and animals. In: T. Lovejoy & L. Hannah (Eds.), Climate Change and Biodiversity. Yale University Press, New Haven. pp 61–69.

Rossi, V., Caffi, T. and Salinari, F. (2012). Helping farmers face the increasing complexity of decision-making for crop protection. Phytopathologia Mediterranea 51: 457"479.

Sharma, H.C., Srivastava, C.P., Durairaj, C. and Gowda, C.L.L. (2010). Pest management in grain legumes and climate change. In: *Climate Change and Management o Cool Season Grain Legume Crops* (Yadav, SS; McNeil, DL; Redden, R and Patil, SA eds.), Dordrecht, The Netherlands: Springer Science & Business media. pp. 115-140.

Shaw, M.W. and Osborne, T.M. (2011). Geographic distribution of plant pathogens in response to climate change. Plant Pathology 60: 31–43.

Siebold, M. and von Tiedemann, A. (2012). Potential effects of global warming on oilseed rape pathogens in Northern Germany. Fungal Ecology 5: 62–72.

Somasekhar, N., Prasad, J.S. and Ganguly, A.K. (2010). Impact of Climate Change on Soil Nematodes - Implications for Sustainable Agriculture. Indian J. Nematology 40: 125-134.

Srikanth, J. (2007). World and Indian scenario of sugarcane woolly aphid. In: Woolly Aphid Management in Sugarcane (eds. Mukunthan, N., Srikanth, J., Singaravelu, B., Rajula Shanthy, T., Thiagarajan, R. and Puthira Prathap, D.), Extension Publication, Sugarcane Breeding Institute, Coimbatore. 154: 1-12.

Srivastava, P., Chauhan, D. and Khan, M.S. (2014). Climate changes: a threat to insect pollinators in litchi crop at Pantnagar. In: *Compendium-cum-Abstract, International* Symposium on New-Dimensions in Agrometeorology for Sustainable Agriculture, 16-18 Oct 2014, GBPUAT, Pantnagar (UA), Association Agrometeorologists, pp. 323-324.

Subba Rao, A. V. M., Agarwal, P. K., Huda, A.K.S. and Chattopadhyay, C. (2007). Using InfoCrop – A user friendly crop simulation model for mustard. In: Climate and Crop Disease Risk Management – An International Initiative in Asia Pacific Region. CRIDA, Hyderabad. pp. 16.

Swaminathan, M. S. (1986). Changing paradigms in Indian agriculture – the way ahead. The Hindu. pp. 22-46.

Swaminathan, M.S. (1995). Foreword. In: *Dynamics of Plant Diseases,* S. Nagarajan and K. Muralidharan, Allied Publ. Ltd., New Delhi, pp.

Tanwar, R.K., Jeyakumar P. and Vennila, S. (2010). Papaya mealybug and its management strategies. Technical Bulletin 22. National Centre for Integrated Pest Management, New Delhi.

Thomas, C.D., Cameron, A., Green, R.E., Bakkenes, M., Beaumont, L.J., Collingham, Y.C., Erasmus, B.F.N., Ferreira de Siqueira, M., Grainger, A., Hannah, L., Hughes, L., Huntley, B., A.S. van Jaarsveld, Midg-ley, G.F., Miles, L, Ortega-Huerta, M.A., Peterson, A.T., Phillips O.L. and Williams, S.E. (2004). Extinction risk from climate change. Nature 427: 145–148.

Uphoff, N. (2007). Reducing the vulnerability of rural households through agroecological practice: considering the System of Rice Intensification (SRI). Modern Development 35: 85-100.

Zadoks, J.C. (1985). On the conceptual basis of crop loss assessment: the threshold theory. Ann. Rev. Phytopathol. 23: 455-473.

❑❑❑

# 16

# Climate Change and Changing Pest Scenario

C.P. Viji, K. Phanikumar and V. Sudhavani

## Introduction

Climate is a measure of the average pattern of variation in temperature, humidity, atmospheric pressure, wind, precipitation and other meteorological variables in a given region over long period of time. Climatic change refers to the shifts in the mean states of the climate or in all its variability, persisting for an extended period (decade or longer). The expected climatic change includes an increase in average temperature, an increase in $CO_2$ concentration and varied rainfall regimes due to increased ultra violet-B radiation. Fossil fuel burning and deforestation have emerged as principal anthropogenic sources of rising atmospheric $CO_2$ and other greenhouse gases and consequent global warming and climatic change.

Climate change affects agriculture in a number of ways as agriculture is highly dependent on specific climate conditions. Changes in average temperatures, rainfall, and climate extremes (e.g., heat waves), changes in sea level, changes in atmospheric carbon dioxide and ground-level ozone concentrations, changes in pests and diseases, changes in the nutritional quality of some foods, the frequency and severity of droughts and floods could pose challenges for farmers. Current speculation about global climatic change is that, most agricultural land will experience more extreme environmental fluctuations. Climatic change also modifies plant growth with changes in crop architecture, size, density, microclimate and quality of susceptible tissue thereby affecting the physiology of host-pest interactions.

Insects are the world's most diverse class of organisms and as such they play a major role in the succession, functioning and carbon cycling that occurs in most natural and human made ecosystems. Effects of global warming and climatic change in insects are expected to trigger major changes in geographical distribution of insect pests, changes in population growth rates, severity of pests, species extinction etc. Climatic change is also likely to change the activity and abundance of natural enemies, efficacy of crop protection technologies and biosecurity measures to be taken up. The effect of climatic change on crop plants, insect pests, their natural enemies and their interactions are discussed below.

## Influence of Climatic change on insect pests

The effects of climate change on insect pest can be both direct (temperature) and indirect ($CO_2$) on different life history traits of insects. Temperature cause direct effects like survival, growth and development, voltinism and dispersal. $CO_2$ causes indirect effect by altering the host physiology and it has both positive and negative effects.

### A. Temperature

Multiple lines of scientific evidence shows that the climate system is warming. Average global air temperatures are already 1.5$^{o}$ higher than that they were at the start of 20$^{th}$ century and has been risen about 1 degree over just the last 30 years with an average of 0.13°C per decade. According to the latest report from the Intergovernmental Panel on Climate change (IPCC) global climate is likely to rise by 2-3°C in 2100.

Effect of temperature on agricultural crops and insect pests is multidimensional. The change in temperature will have direct effects on insects by affecting their development, and also indirect effects via their host plants and natural enemies.

Insects are able to function faster and more efficiently at higher temperatures as they can feed, develop, reproduce and disperse more rapidly when the climate is warm. Temperature directly influences the rate of biochemical reactions and has a strong influence on growth rate and development of insects. The total heat energy required to complete a certain stage of development in the life cycle of a species is constant called 'Thermal constant'. Increased temperatures may allow faster population growth of pests as the thermal requirement is obtained in a short period of time. Increase in temperature will have a greater effect on insects than rising $CO_2$ concentration.

#### *I. Geographical distribution*

It is expected that the range of many insects will expand or change, and new combinations of pests may emerge when natural ecosystems respond to altered temperature and precipitation profiles.

Poleward expansion of species distribution has been attributed to climate warming. The ranges of species are likely to shift to the north and many species will extend their northern boundary. Eg. The potential distribution of *Bactrocera zonata* will expand poleward into areas which are currently too cold. Mid-latitude populations (tropical species) are more susceptible to climatic change. Due to global warming scenario, tropical and widespread *Drosophila* sp. will face a proportional reduction in distribution range.

Models predicted a shift in the ranges of pest species to higher altitudes. The rising temperatures will encourage the uphill spread of insects. Change in cloud cover due to increasing temperatures also resulted in increasing the elevation extent in insects. Eg. Stonefly, *Acroneuria abnormis* shifted uphill 60 to 250 m in 30 years, butterfly *Erebia epiphron* shifted 98 m uphill in 19 years.

Distribution of insect pests will also be influenced by the changes in cropping patterns triggered by climate change. Major insect pests, such as cereal stem borers (*Chilo*, *Sesamia*, and *Scirpophaga*), pod borers (*Helicoverpa*, *Maruca*, and *Spodoptera*), aphids and whiteflies may move to temperate regions leading to greater damage in cereals, legumes, vegetables, and fruit crops.

Increase in temperature favours the distribution of vectors. Eg. Increase in population of *Coraebus florentinus* responsible for oak decline. 10% increase of winged forms of cereal aphid (*Rhopalosiphum padi*) is noted which could result in greater spread and incidence of barley yellow dwarf virus for which this aphid is a vector.

### *II. Pest status*

Climatic change widens the 'invasion niche' or the set of environmental conditions under which the pest can successfully invade. Climate change may also trigger organisms to find with new, vulnerable hosts. Eg. Egyptian cottony cushion scale *Icerya aegyptiaca*, scale *Aulacaspis* sp. feeding on kusum, *Schleichera oleosa* in which it was not previously recorded. New pests may become able to invade previously uninhabitable areas and there is a likely widening of pest's habitats. Eg. Colorado potato beetle (*Leptinotarsa decemlineata*) and the European corn borer (*Ostrinia nubilalis*) will have newly established areas.

Migrant pests are expected to respond more quickly to climate change and may be able to colonize newly available crops/habitats. Significantly predicted inter annual abundance of migrant monarchs were observed due to lengthening of the breeding season. Higher temperatures are also associated with increased migrant numbers. Movement of non native pest species occurs and increased problems of many insect pests are expected as they develop more rapidly in response to rising temperatures.

High summer temperature increased the outbreak of sawfly, *Diprion pini*. There will be a rapid spread of pests to new areas due to favourable climate. Eg. Spread of coconut mite, *Aceria guerreronis* rapidly to all coconut growing states of India. Pests may become established in areas where their presence was previously only sporadic. Eg. Slug caterpillar, *Macroplecta nararia* a sporadic pest occurring in epidemic form in Andhra Pradesh.

Increased temperatures may cause a slight increase of non-indigenous invasive insect species and migratory pests. Eg. Risk of leaf mining flies of the genus *Liriomyza*. It has also been observed that, those species which were not recognised as economically important have now become major pests. New record of pests is also noticed. Eg. Indian rose beetle, *Adoretus versutus* and black leaf chaffer beetle, *Apogonia blanchardi* in cocoa.

Climatic change has also attributed to local extinction or displacement of species. Eg. Butterflies, *Coenonympha tullia* and *Aricia artaxerxes* were locally extinct at 50% of sites due to habitat loss as a result of climatic change in Britain.

Distribution of pests is also expected to change. Global warming may be responsible for the decline in abundance of *Plutella xylostella* and increase in *Helicoverpa armigera* and *Trichoplusia ni* in Japan. Modelling works also suggests that under warmer climate, exotic pest may establish their populations. Eg. Southern pine beetle could establish their populations in Europe. Natural ecosystems will be affected due to the change in insect diversity.

Increased temperatures may allow faster population growth of pests and therefore the propensity for increased mutation rates and evolution, potentially leading to more virulent or pesticide resistant populations.

### *III. Biological Parameters*

Insect development, survival, reproduction and life table parameters are also influenced by the kind of host plant and quality of the diet. The Parameters influences the insect development is described below:

#### a. Voltinism/ Number of generations

Acceleration of pests development due to the faster achieving of number of degree-days is likely to increase of the number of generations of some pests. Eg. In Ethiopia, it was too cold for coffee berry borer, *Hypothenemus hampei* to complete even one generation per year before 1984, but thereafter, because of rising temperatures in the area, 1-2 generations per year/coffee season could be completed.

Rapid generation increase was observed in pests like Colorado potato beetle (*Leptinotarsa decemlineata*) and European corn borer (*Ostrinia nubilalis*). Generation numbers increased from 10-18 to 14-24 for two-spotted spider mite, from 9-14 to 14-20 for European red mite, and from 2-4 to up to 5 for navel orangeworm. A significant decrease in the length of overlap of first and second generation larval emergence was identified in *Cydia pomonella* due to climatic change. It is estimated that in aphids, a $2^{o}$ temperature increase causes 1-5 additional life cycles per season.

b. Population size

Overall pest pressure is expected to increase substantially due to the increase in number of generations of some pests and higher population density in the consequence of prolonged growing season and the period favourable for reproduction. Eg. Increase in temperature may increase scale insect abundance by three orders of magnitude.

Global climatic change will play a critical role in the ability of insect pests to overwinter. Predicted rise in temperature will favour winter survival in some insects. Eg. High winter survival in green spruce aphid (*Elatobium abietinum*) leads to more intense and frequent tree defoliation resulting in decline in productivity of Sitka spruce (*Picea sitchensis*) in Europe. Higher winter survival rates, leading to higher initial pest counts in spring, or of extended pest development times in the summer are factors that are likely to exacerbate future pest pressure. Earlier timing of spring events also resulted in earlier timing of population peaks.

Global warming will lead to earlier beginning and prolongation of growing season in temperate region and will have pronounced effect on phenology and life history adaptation in many insects. A two week shift in phenology with prolonged second generation and an additional third generation was observed in Codling moth, *Cydia pomonella* due to climatic change.

In contrast, in mid-latitudes, higher incidence of heat death in summer may constrain pest population sizes. Studies showed the declined survival rates of brown plant hopper, *Nilaparvatha lugens* and rice leaf folder, *Cnaphalocrocis medinalis* due to rising temperatures in rice ecosystem.

#### c. Longevity

The reduction of longevity or duration of an insect with increase in temperature occurs due to the accelerated use of energy. Decrease in longevity enables the insect to complete more generations in a shorter time. A decrease in development period with an increase in temperatures was reported in case of lepidopteran insect pests like *Cnaphalocrocis medinalis, Elasmopalpus lignosellus, Spodoptera litura.* Aphid *Myzus persicae* had reduced developmental time with decreased adult and progeny weights.

Climatic change will bring long term survival of many northern and mountain species due to increased temperatures and reducing the winter kill. Winter mortality of adults of *Nezara viridula* and *Halyomorpha halys* is predicted to be reduced by 15% by each rise of 1°C.

#### d. Other biological parameters

Temperature was negatively correlated with adult body size. Eg. Whitefly *Bemesia tabaci*, aphid *Myzus persicae.* Temperature above the specific optimum range lead to decreased growth rates, reduced fecundity and increased rates of mortality for a multitude of species.

### B. Carbon dioxide ($CO_2$)

Global warming is caused by the emission of greenhouse gases. 72% of the totally emitted greenhouse gases is carbon dioxide ($CO_2$), 18% methane and 9% nitrous oxide (NOx). $CO_2$ is inevitably created by burning fuels like oil, natural gas, diesel, petrol, ethanol. The emissions of $CO_2$ have been dramatically increased within the last 50 years and are still increasing by 3% each year. The amount of $CO_2$ in the atmosphere increased by about 40% compared with pre-industrial levels. Rough estimates on the effect of doubling of atmospheric $CO_2$ levels show a major redistribution of Earth's vegetation.

Enhanced $CO_2$ ($eCO_2$) levels increase photosynthesis, growth, yield and C:N ratios in most plant species which may affect the quality and quantity of food available to insect herbivores which are indirectly affected by those changes in their host plants. The modified host physiology and canopy microclimate at elevated carbon dioxide influences production, dispersal and survival and feeding behaviour of insect pests. Insect-host plant interactions will also alter with due to the change in the nutritional quality and secondary metabolites of the host plants.

When pest feed on less nutritious host plant under $eCO_2$ conditions, larval developmental time may be lengthened resulting in more crop damage. The size and weight of the larvae was also increased when fed on $eCO_2$ foliage. Insects feed more to meet their nutritional requirements resulting in the development of aggressive traits. However, individual species responses to $eCO_2$ vary. Some experiments showed that the mean development time (days) of each stage egg, larva, pupa, pre-oviposition and total life span is decreased on $eCO_2$ foliage. It is also reported that elevated $CO_2$ reduced the abundance of sucking insects but increased the abundance of chewing insects.

Predicted increases in atmospheric carbon dioxide ($CO_2$) concentrations could modify crop resistance to insect pests by altering plant quality. The short generation times of aphids may allow them to exploit such changes and colonise previously resistant plant genotypes. Eg. Pea aphid, *Acyrthosiphon pisum* on aphid resistant lucerne (*Medicago sativa*) genotypes.

Decreased N content, lower water content and rising concentration of secondary metabolites in plants due to change in atmospheric $CO_2$ stimulates plant defence and resistance to the colonisation of phytophagous insects.

## C. Rainfall

An increase in global temperatures will lead to an intensification of the hydrological cycle. Increase in surface air temperature causes an increase in evaporation leading to higher levels of water vapour in the atmosphere. Moreover, warmer atmosphere is capable of holding more water vapour. This leads to increase in heavy precipitation events in winter, drying of soils and vegetation due to increased evaporation in the summer resulting in widespread droughts and wildfires. It is estimated that there is a 7% increase in extreme rainfall intensity for every degree increase in global atmospheric temperature.

Altered precipitation affects canopy architecture through either drought or flooding stress with corresponding effects on pests. Dry seasonal rainfall results in leaf flushing of host plant *Callichlamys latifolia* which was the primary driver for population outbreaks for butterfly, *Aphrissa statira*. In Sub-Saharan Africa, changes in rainfall patterns are driving migratory patterns of the desert locust, *Schistocerca gregaria*.

Enhanced summer rainfall and drought conditions on soil dwelling *Agriotes lineatus* (wireworms) promote rapid increase in the population of wireworms in the upper soil. If several very rainy days coincide with the presence of the youngest cutworms, they will suffer high mortality.

Global climate change scenarios predict more frequent and extended droughts, especially in the mid-latitudes. Drought is likely to cause local or regional extinctions of insects/ plants with potentially severe consequences on different ecosystems. Drought affects directly the crop physiology and growth and indirectly by the activity of insect pests, heat waves and fire. Drought and heat waves cause extensive forest damage since the activity of woodborers are positively influenced by prolonged water stress. Defoliators are also profited better from the increased nitrogen in plant tissues linked to moderate or intermittent water stress. The palatability of the leaves is also expected to change due to drought stress leading to high pest incidence.

Water stress conditions can alter the composition of secondary plant metabolites. Eg. Drought stress and water logging changes the level of glucosinolates (GS) in broccoli, *Brassica oleracea* L. and significantly higher populations of *Myzus persicae* aphids were detected on plants with limited water supply.

## Influence of Climatic change on host plant

### A. Crop growth and development

Rising temperature, carbon dioxide and altered precipitation modifies plant growth and development with concomitant changes in canopy architecture, size, density, microclimate and the quantity of susceptible tissue. Estimates show that there is an increase in crop yield 10-20% for $C_3$ crops and 0-10% for $C_4$ crops. Eg. An increase of one degree centigrade in the temperature during the crop growth period increased the mustard yield by around 140 Kg $ha^{-1}$.

Rise in temperature increases the geographical suitability of growing crops at higher latitudes in tropical and sub-tropical areas. Eg. Cotton production areas increases. With the crop, mealy bug, *Pseudococcus solenopsis* increases with additional 2.0 generations per year and 4.0 fold increase in population abundance.

Climatic change is responsible for northward and uphill distribution of many European plant species. Forests are likely to be contracted in the south and expanded in the north which inturn will cause alteration in the distribution of insect pests and species diversity.

It is reported that under $eCO_2$ conditions, soluble sugars are increased and starch is decreased by elevated temperature whereas photosynthesis was decreased in plant due to global warming. The reduction in protein content and increase of C:N ratio due to dilution of nitrogen in leaves also implies a reduction in food quality. Secondary metabolites like polyphenols, phenolics and polyphenolics concentrations in green leaves under $eCO_2$ increased. Crop biomass is predicted to increase in response to elevated $CO_2$ concentrations. The life table parameters of insect pests are influenced by the kind of host plant and quality of the diet.

### B. Crop damage

The production of agricultural commodities faces increased risk of pests, diseases and other stresses due to climate change and variability. Prolonged growing season and period favourable for reproduction increased the number of generations of some pests and high population density consequently resulting in high yield losses.

Changes in geographical range of insects and insect abundance will increase the extent of crop losses. Cosmopolitan pest species may be resilient to climatic change. Specialists may be forced to move poleward concurrently with their host species or go extinct.

Elevated temperature accelerates plant growth and developmental rates to modify canopy architecture and microclimate. We can expect increased problems with many pest insects as they develop more rapidly in response to rising temperatures. Crop yield loss or damage is predicted to be high as the pest load will increase. Crop will also have yield losses associated with increased frequency of high temperature stress, inadequate winter chill period for optimum fruiting in spring and extended droughts.

Elevated $CO_2$ is reported to cause significant increase in total biomass of crops but nutritional quality of leaves declines substantially due to dilution of nitrogen by 10-30%. Many insects will alter how much they eat in response to changing plant nutrition. Developmental period of insects is predicted to increase due to less nutritious host plants causing more damage and thereby decreasing the yield.

## Influence of Climatic change on pest control

Global warming and climatic change reduces the effectiveness of host plant resistance, transgenic plants, natural enemies, biopesticides, and synthetic chemicals for pest management.

### A. Natural enemies

Any increase in the frequency or severity of extreme weather events due to climatic change including droughts, heat waves, windstorms, floods could disrupt the pest-predator relationship that normally keeps pest population in check. The effectiveness of natural enemies in controlling pests will decrease if pest distributions shift into regions outside the distribution of their natural enemies. Pests no longer have natural predators or environmental variables to control their population size. Pests establish in areas where natural enemies are not present which may lead to outbreak.

The fitness of natural enemies can be altered in response to changes in herbivore quality and size induced by temperature and $CO_2$ effects on plants. The susceptibility of the pests to predation and parasitism could be decreased through the production of additional plant foliage or altered timing of pest life cycles in response to plant phonological changes. Also, mismatch between pests and natural enemies in space and time decreases their effectiveness for bio-control.

In contrast, it is also reported that $CO_2$ increased the number of parasitoids. Drought also alters the rate of parasitism.

### B. Biosecurity

Biosecurity authorities should consider the effects of climate change when undertaking pest risk assessments. Major insect pests, such as cereal stem borers (*Chilo*, *Sesamia*, and *Scirpophaga*), pod borers (*Helicoverpa*, *Maruca*, and *Spodoptera*), aphids, and whiteflies may move to temperate regions, leading to greater damage in cereals, grain legumes, vegetables, and fruit crops. To prevent the introduction and spread of pests enhanced quarantine and monitoring measures should be implemented in areas that are projected to be suitable for the establishment of the pest under current and future climatic conditions. Eg. Poleward spread of *Bactrocera zonata*.

Global trafficking of timber and wood products and climate change are likely to result in exotic pests such as the Asian longhorn beetle (*Anoplophora glabripennis*) becoming more prevalent, it is therefore essential that vigilance is maintained in reporting new pests and altered patterns of damage.

## Influence of Climatic change on pollinators

Plant responses to global warming include altered flower, nectar, and pollen production, could modify floral resource availability and reproductive output of pollinating insects. Similarly, pollinator responses, such as altered foraging activity, body size, and life span, could affect patterns of pollen flow and pollination success of flowering plants.

Climatic change may result in phonological shift of bee flight periods. The pollen season is expected to start on an average of 10 days earlier which may bring a mismatch between flowering and activity of bees. Studies show a significant shift towards an earlier start to activity 6-10 days per decade.

Sustainable food security is attained by increasing the agricultural production to meet the demand of growing population. A major challenge to sustainability comes from climate change. The production of agricultural commodities faces increased risk of pests, diseases and other stresses due to global warming and climate change. Care should be taken that crop production technologies and management strategies adopted by farmers cope with climatic change.

## References


Andrew, N. R. and Hughes, L. (2004). Species diversity and structure of phytophagous beetle assemblages along a latitudinal gradient: predicting the potential impacts of climate change. Ecological Entomology, 29(5): 527-542.

Baker, R. H. A., Cannon, R. J. C. and Walters, K. F. A. (1996). An assessment of the risks posed by selected non-indigenous pests to UK crops under climate change. Aspects of Applied Biology, (45):323-330.

Blumel, S. (2012). Climate change and plant health - increasing importance of biocontrol options for risk management of quarantine pests. Proceedings of the IOBC/WPRS Working Group "Biological Control of Fungal and Bacterial Plant Pathogens", Graz, Austria, 7-10 June, 2010. IOBC/WPRS Bulletin, 78:11-14.

Boulton, A. J. and Lake, P. S. (2008). Effects of drought on stream insects and its ecological consequences. Aquatic insects: challenges to populations, Pp: 81-102.

Cannon, R. J. C. (1998). The implications of predicted climate change for insect pests in the UK, with emphasis on non-indigenous species. Global Change Biology, 4(7):785-796.

Cheke, R.A. and Tratalos, J.A. (2007). Migration, patchiness and population processes illustrated by two migrant pests. Bioscience, 57: 145-154.

Dale, A. G. and Frank, S. D. (2014). Urban warming trumps natural enemy regulation of herbivorous pests. Ecological Applications, 24(7):1596-1607.

Emmanuel, N., Kiran Patro, T.S.K.K., Chalapathi Rao, N.B.V. and Sujatha, A. (2014). Impact of climactic variations on the outbreak of pests in horticultural ecosystems. In: Proceedings: National Conference on

Emerging Challenges and Opportunities in Biotic and Abiotic Stress Management (ECOBASM-2014), December 13-14, 2014. Pp: 63-66.

Evans, H., Straw, N. and Watt, A. (2002). Climatic change: implications for insect pests. (Climate change: impacts on UK forests) Forestry Commission Bulletin, 125:99-118.

Fabre, F., Plantegenest, M., Mieuzet, L., Dedryver, C. A., Leterrier, J. L. and Jacquot, E. (2005). Effects of climate and land use on the occurrence of viruliferous aphids and the epidemiology of barley yellow dwarf disease. Agriculture, Ecosystems & Environment, 106(1):49-55.

Finlay, K. J. and Luck, J. E. (2011). Response of the bird cherry-oat aphid (*Rhopalosiphum padi*) to climate change in relation to its pest status, vectoring potential and function in a crop-vector-virus pathosystem. Agriculture, Ecosystems & Environment, 144(1):405-421.

Franco, A. M. A., Hill, J. K., Kitschke, C., Collingham, Y. C., Roy, D. B., Fox, R., Huntley, B. and Thomas, C. D. (2006). Impacts of climate warming and habitat loss on extinctions at species' low-latitude range boundaries. Global Change Biology, 12(8):1545-1553.

Ghosh, A., Das, A., Bhattacharya, B. K., Kumar, V., Kumar, A., Meena, P. D. and Chattopadhyay, C. (2012). Impact of climate change on insect pests and diseases - the scenario unfolding in Indian agriculture. (Climate change: its impact, adaptation and mitigation in agriculture). SATSA Mukhaptra Annual Technical Issue, 16:15-29.

Hardy, P. B., Sparks, T. H. and Dennis, R. L. H. (2014). The impact of climatic change on butterfly geography: does climatic change produce coincident trends in populations, distributions and ranges? Biodiversity and Conservation, 23(4):855-876.

Hillstrom, M. L. and Lindroth, R. L. (2008). Elevated atmospheric carbon dioxide and ozone alter forest insect abundance and community composition. Insect Conservation and Diversity, 1(4):233-241.

Jaramillo, J., Chabi-Olaye, A., Kamonjo, C., Jaramillo, A., Vega, F.E., Poehling, H.M. and Borgemeister, C. (2009). Thermal tolerance of the coffee berry borer, *Hypothenemus hampei*: predictions of climate change impact on a tropical insect pest. PLoS ONE, (August):e6487.

Johnson, S.N., Ryalls, J.M.W. and Karley, A.J. (2014). Global climate change and crop resistance to aphids: contrasting responses of lucerne genotypes to elevated atmospheric carbon dioxide. Annals of Applied Biology, 165(1):62-72.

Khan, M. A. M., Ulrichs, C. and Mewis, I. (2011). Drought stress - impact on glucosinolate profile and performance of phloem feeding cruciferous insects. Acta Horticulturae, 917:111-117.

Kiritani, K. (2006). Predicting impacts of global warming on population dynamics and distribution of arthropods in Japan. Population Ecology, 48: 5-12.

Kocmankova, E., Trnka, M., Juroch, J., Dubrovsky, M., Semeradova, D., Mozny, M. and Zalud, Z. (2009). Impact of climate change on the occurrence and activity of harmful organisms. Plant Protection Science, 45(Special Issue):S48-S52.

Luedeling, E. Steinmann, K. P., Zhang, M. H., Brown, P. H., Grant, J. and Girvetz, E. H. (2011). Climate change effects on walnut pests in California. Global Change Biology,17(1):228-238.

Marchioro, C.A. and Foerster, L.A. (2011). Development and survival of the diamond back moth, *Plutella xylostella* (L.) (Lepidoptera: Yponomeutidae) as a function of temperature: effect on the number of generations in tropical and subtropical regions. Neotropical Entomology, 40: 533-541.

Mathukumalli Srinivasa Rao, Dammu Manimanjari, Anantha Chitiprolu Rama Rao, Pettem Swathi and Maheswari, M. (2014). Effect of climatic change on *Spodoptera litura* Fab. on peanut: A life table approach. Crop protection, 66: 98-106.

Ni, W. L., Li, Z. H., Chen, H. J., Wan, F. H., Qu, W. W., Zhang, Z. and Kriticos, D. J. (2012). Including climate change in pest risk assessment: the peach fruit fly, *Bactrocera zonata* (Diptera: Tephritidae). Bulletin of Entomological Research, 102(2):173-183.

Overgaard, J., Kearney, M.R., Hoffmann, A.A. (2014). Sensitivity to thermal extremes in Australian *Drosophila* implies similar impacts of climate change on the distribution of widespread and tropical species. Global Change Biology, 20(6):1738-1750.

Pangga, Hanan, I.B. and Chakraborty, J.S. (2013). Climate change impacts on plant canopy architecture: implications for pest and pathogen management. European Journal of Plant Pathology, 135(3):595-610.

Rouault, G., Candau, J.N., Lieutier, F., Nageleisen, L.M., Martin, J.C. and Warzee, N. (2006). Effects of drought and heat on forest insect populations in relation to the 2003 drought in Western Europe. Annals of Forest Science, 63(6): 613-624.

Salle, A., Nageleisen, L. M. and Lieutier, F. (2014). Bark and wood boring insects involved in oak declines in Europe: current knowledge and future prospects in a context of climate change. Forest Ecology and Management, 328: 79-93.

Scaven, V.L. and Rafferty, N.E. (2013). Physiological effects of climate warming on flowering plants and insect pollinators and potential consequences for their interactions. Current Zoology, 59(3):418-426.

Sharma, H.C. (2014). Climate change effects on insects: implications for crop protection and food security. Journal of Crop Improvement, 28(2):229-259.

Sheldon, A.L. (2012). Possible climate-induced shift of stoneflies in a southern Appalachian catchment. Freshwater Science, 31(3):765-774.

Srygley, R.B., Dudley, R., Oliveira, E.G., Aizprua, R., Pelaez, N. Z. and Riveros, A. J. (2010). Global Change Biology, 16(3):936-945.

Staley, J.T., Hodgson, C.J., Mortimer, S.R., Morecroft, M.D., Masters, G.J., Brown V.K. and Taylor, M.E. (2007). Effects of summer rainfall manipulations on the abundance and vertical distribution of herbivorous soil macro-invertebrates. European Journal of Soil Biology, 43: 189-198.

Stoeckli, S., Hirschi, M., Spirig, C., Calanca, P., Rotach, M.W. and Samietz, J. (2012). Impact of climate change on voltinism and prospective diapause induction of a global pest insect - *Cydia pomonella* (L.). PLoS ONE, 7(4):e35723.

Suseelendra Desai and Rao, M. S. (2011). Impact of climate change on insect pests, pathogens and their natural enemies. Natural Rubber Research, 24(1): 174-186.

Svobodova, E., Trnka, M., Dubrovsky, M., Semeradova, D., Eitzinger, J., Stepanek, P. and Zalud, Z. (2014). Determination of areas with the most significant shift in persistence of pests in Europe under climate change. Pest Management Science, 70(5):708-715.

Thomson, L.J., Macfadyen, S. and Hoffmann, A. A. (2010). Predicting the effects of climate change on natural enemies of agricultural pests. Biological Control, 52(3):296-306.

Toth, I.K., Torrance, L., Fenton, B. and Lees, A. K. (2008). Impact of climate change on pests and diseases of potatoes in Scotland: risks and recommendations. The Dundee Conference. Crop Protection in Northern Britain, 2008, Dundee, UK, 26-27 February, 2008. Pp. 187-192.

Trumble, J.T., and Butler, C.D. (2009). Climate change will exacerbate California's insect pest problems. California Agriculture, 63(2):73-78.

Wolfe, D.W., Ziska, L., Petzoldt, C., Seaman, A., Chase, L. and Hayhoe, K. (2008). Projected change in climate thresholds in the Northeastern U.S.: implications for crops, pests, livestock, and farmers. (Special Issue: Assessment of climate change, impacts, and solutions in the Northeast United States.) Mitigation and Adaptation Strategies for Global Change, 13(5/6):555-575.

Yamamura, K. and K. Kiritani, (1998). A simple method to estimate the potential increase in the number of generations under global warming in temperate zones. Applied Entomology and Zoology, 33: 289-298.

❑❑❑

# 17

# Impact of Climate Change on Insect Pests and Their Management Strategies

**Abhishek Pareek, B.M. Meena, Sitaram Sharma, M.L. Tetarwal R.K. Kalyan and B.L. Meena**

## Introduction

Climate change is an important determinant of abundance and distribution of species. It is concerned with everyone since it posess potential threat to environment, and agricultural productivity and production throughout the world. It has implications for livelihood and survival of human beings. According to Intergovernmental Panel on Climate Change (IPCC, 2001), it is defined as "Change in climate over time, either due to natural variability or as a result of human activity." As described by IPCC, the most of global warming observed over last 50 years is attributed to the human activities. The rise in global climate temperature is the result of the enhanced green house effect that is caused due to the increased levels of green house gasses (GHG) like Carbon-dioxide ($CO_2$), Chlorofluorocarbon (CFC), Methane ($CH_4$) and Nitrous oxide ($N_2O$) in the atmosphere. Over past hundred years, $CO_2$ concentration in the atmosphere has increased drastically from 280 ppm to 370 ppm and is likely to be doubled in 2100. Likewise, the global temperature has increased by 0.6 + 0.2°C and is expected to reach 1.1-5.4°C by the end of next century (IPCC, 2007). According to IPCC, if temperatures rise by about 2°C over the next 100 years, negative effects of global warming would begin to extend to most regions of the world and directly affect most of the organisms on the earth. The climatic variability, together with increase in atmospheric temperature and carbon dioxide, change in precipitation pattern, extended period of drought do have lot of implication in agriculture sector. These climatic variables interact with plants in numerous ways with diverse mechanisms and affect directly in terms of tissue and organ-specific photosynthetic allocation. Such changes in climate also profoundly affect the population dynamics and the status of insect pests (Woiwod, 1997). These effects could either be direct, through the influence of weather on the insects' physiology and behaviour (Samways, 2005, Parmesan, 2007 and Merrill *et al.,* 2008), or may be mediated by host plants, competitors or natural enemies (Harrington *et al.,* 2001 and Bale *et al.,* 2002). In addition, the impacts include changes in phenology, distribution and community composition of ecosystem that finally leads to extinction of species (Walther *et al.,* 2002).

For species to survive in the changing climates, they must either adapt *in situ* to new conditions or shift their distributions in pursuit of more favourable ones. Many insects have large population sizes and short generation times, and their phenology, fecundity, survival, selection and habitat use can respond rapidly to the climate change. These changes to insect life-history may in turn produce rapid changes in their abundance and distribution. Increased temperature will cause insect pests to be more abundant and almost all insects will be affected by changes in temperature (Bale *et al.,* 2002). Porter *et al.* (1991) listed various effects of temperature on insects, including: changes in geographical range, overwintering, population growth rates, number of generations per annum, crop–pest synchronization, dispersal and migration, and availability of host plants and refugia. In-season effects of warming include the potential for increased levels of feeding and growth, including the possibility of additional generations in a given year (Cannon, 1998). This will alter the crop yield, and also influence the effectiveness of insect-pest management practices. Increased global temperature will also influence the phenology of insects including early arrival of insect pests in their agricultural habitats and emergence time of a range of insect pests (Dewar and Watt, 1992; Whittaker and Tribe, 1996, 1998). This will require early and more frequent application of insecticides to reduce the pest damage. Increased temperatures will also increase the pest population, and water stressed plants at times may result in increased insect populations and pest outbreaks. This will affect the crop yield and availability of food grains and threaten food security. The climatic change impacts on pests may include:

- Changes in diversity and abundance of insect pests
- Changes in geographical distribution of insect pests
- Increased overwintering insects
- Rapid population growth and no. of generations
- Changes in synchrony between insect pests and their host crops
- Introduction of alternative hosts plants

- Changes in host plant resistance
- Changes in insect biotypes
- Changes in tritrophic interactions
- Impact on extinction of species
- Changes in activity and relative abundance of natural enemies
- Increased risk of invasive pest species
- Reduced efficacy of crop protection technologies.

Climate change will also result in increased problems of insect transmitted diseases. These changes will have major implications for crop protection and food security, particularly in the developing countries, where the need to increase and sustain food production is most urgent. Long-term monitoring of population levels and insect behaviour, particularly in identifiably sensitive regions, may provide some of the first indications of a biological response to climate change. The impact of climate change will vary across regions, crops and species. A large number of models and protocols have been designed to measure the effects of climate change for different species and in different disciplines. There is a need for interdisciplinary cooperation to measure the effects of climate change on the environment and food security. It will be important to keep ahead of undesirable pest adaptations, and consider global warming and climate change for planning research and development efforts for integrated pest management (IPM) in the future (Sharma, 2010).

## Climate Change and Indian Agriculture

The Indian climate has undergone significant changes showing increasing trends in annual temperature with an average of 0.56°C rise over last 100 years (IPCC, 2007; Rao *et al.* 2009; IMD, 2010). Warming was more pronounced during post monsoon and winter season with increase in number of hotter days in a year (IMD, 2010). Even though, there was slight increase in total rainfall received, number of rainy days decreased. The rainfed zone of the country has shown significant negative trends in annual rainfall (De and Mukhopadhyay, 1998; Lal, 2003, Rao *et al.,* 2009). The semi arid regions of the country had maximum probability of prevalence of droughts of varying magnitudes (20-30%), leading to sharp decline in water tables and crop failures (Lal, 2003; Rao *et al.,* 2009;). By the end of next century (2100), the temperature in India is likely to increase by 1-5°C (De and Mukhopadhyay, 1998; Lal, 2003; IPCC, 2007; IMD, 2010). According to the estimates of NATCOM (2004), there will be 15-40% increase in rainfall with high degree of variation in its distribution. Apart from this, the country is likely to experience frequently occurring extreme events like heat and cold waves, heavy tropical cyclones, frosts, droughts and floods (NATCOM, 2004; IPCC, 2007).

Being a tropical country, India is more challenged with impacts of looming climate change (Chahal *et al.,* 2008). Already, the productivity of Indian agriculture is limited by its high dependency on monsoon rainfall which is most often erratic and inadequate in its distribution (Chand and Raju, 2009). The country is experiencing declining trend of agricultural productivity due to fluctuating temperatures (Samra and Singh, 2004, Aggarwal, 2008; Joshi and Viraktamath, 2004), frequently occurring droughts and floods, problem soils, and increased outbreaks of insect-pests (Joshi and Viraktamath, 2004; Srikanth, 2007; Dhawan *et al.,* 2007; IARI News, 2008; IRRI News, 2009) and diseases. These problems are likely to be aggravated further by changing climate which put forth major challenge to attain a goal of food security.

## Impact of Climate Change and Insects-Pests

### Rising Temperature

Temperature is identified as dominant abiotic factor directly affects herbivorous insects. Insects being poikilothermic, have temperature of their bodies is approximately the same as that of the environment. Therefore, the developmental rates of their life stages are strongly dependent on temperature. Almost all the insects will be affected to some degrees by changes in temperature and there may be multiple effects upon insect life histories. Laboratory and modeling experiments support the notion that the biology of insect pests are likely to respond to increased temperatures (Fye and McAda, 1972, Cammell and Knight, 1991 and Fleming & Volney, 1995). With every degree rise in global temperature, the life cycle of insect will be shorter. The quicker the life cycle, the higher will be the population of pests. In temperate regions, most insects have their growth period during the warmer part of the year because of which, species whose niche space is defined by climatic regime, will respond more predictably to climate change while those in which the niche is limited by other abiotic or biotic factors will be less predictable (Bale *et al.,* 2002). In the first case, the general prediction is that if global temperatures increase, the species will shift their geographical ranges closer to the poles or to higher elevations and increase their population size (Sutherst, 2000, Harrington *et al.,* 2001, Bale *et al.,* 2002 and Samways, 2005).

The increase in temperature associated with climatic change, would impact crop pest insect populations in several complex ways like (a) extension of geographical range (b) increased over-wintering (c) changes in population growth rate (d) increased number of generations (e) extension of development season (f) changes in crop pest synchrony (g) changes in interspecific interactions (h) increased risks of invasions by migrant pests and (i) introduction of alternative hosts and over-wintering hosts.

## i) Changes in insect- Pests diversity

Increased realization of the deterioration of biological systems over the last couple of decades has led to a better appreciation of the loss of genes, species and ecosystems. Insects comprise the largest group of animal kingdom and play vital role in providing various ecosystem services (Kremen *et al.* 1993; Kannan and James, 2009). The insect diversity in a habitat indicates the health status of an ecosystem as they are very good indicators of environmental change (Gregory *et al.,* 2009), play an important role in food chains. About 6.83% of world insect species are inhabitant in India (Alfred, 1998). The climate change may affect the relative abundance of different insect species and the species unable to adapt the changes may be lost in the due course of time (Thomas *et al.,* 2004). The Western Ghats in India is the only habitat to many rare, endemic and exotic species of colourful butterflies in the world (Hampson, 1908; Anand and Pereira, 2008). In the present day scenario, many butterfly species are under a real threat due to depletion of the natural vegetation for various anthropogenic developmental activities (Costanza *et al.* 1987; Sachs, 2008; Sidhu and Mehta, 2008). There is a need to increase functional diversity in agro-ecosystems vulnerable to climate change to improve system resilience, and decrease the extent of losses due to insect pests (Newton *et al.,* 2009). However, changes in cropping patterns as a result of climate change will drastically affect the balance between insect pests and their natural enemies. Main effects of climate change on insect pests and natural enemies communities result in decreased abundance of decomposers and predators, and increased herbivory, which may have negative consequences for structure and services of the entire ecosystems. Responses of insect pests and natural enemies depend on both temperature and precipitation, and ecosystem-wide adverse effects are likely to increase under predicted climate change (Zvereva and Kozlov, 2010). Consequences of temperature increases of 1 to 2°C will be comparable in magnitude to the currently seen climate change in the Antarctic region (Bokhorst *et al.,* 2008). Increase in rainfall in the Pampas region of Argentina will largely affect the species with poor dispersal capabilities, which will limit their ability to expand their home range. The most affected among the beetle species are the habitat specialists

(Xannepuccia *et al.,* 2009). Large scale changes in rainfall due to climate change will have a major effect on the abundance and diversity of both insect pests and natural enemies. In case of extreme drought due to climate change are likely to decrease multi-trophic diversity and change the composition of ecosystem (environment and organisms). Habitat fragmentation results in the subdivision of habitats into smaller units resulting in their increased insularity as well as loss of total habitat area. Such fragmentation changes the microenvironment at the fragmented edges and edges effects include microclimate, changes in light, temperature and humidity and each of these can have a significant impact on the vitality and composition of the species in the fragment. Speciation takes between 100 and 1,000,000 years, providing between 10 and 10,000 new species per annum. Nearly 99.9% of all species that ever existed have become extinct. Due to climate change the current extinction rates are 100 to 1,000 times greater than what has happened earlier, and nearly 45 to 275 species are becoming extinct everyday (Sharma, 2010).

### ii) Expansion of geographical ranges

The geographic distribution and abundance of plants and animals in nature is determined by species specific climate requirements essential for their growth, survival and reproduction. Any increase in temperature is bound to influence the distribution of insects. It is predicted that a 1°C rise in temperature would enable speed 200 km northwards (in northern hemisphere) or 40 m upward (in altitude). The areas that are not favorable at present due to low temperature may become favorable with rise in temperature. Minimum temperature rather than maximum temperature plays an important role in determining the global distribution of insect species, hence any increase in temperature will result in a greater ability to overwinter at higher altitudes, ultimately causing a shift of pest intensity from south to north. Many insect species have geographic ranges that are not directly limited by vegetation, but instead are restricted by temperature. Earlier researches have shown that with rise in temperature, the insect-pests are expected to extend their geographic range from tropics and subtropics to temperate regions at higher altitudes along with shifts in cultivation areas of their host plants (Hill and Dymock, 1989; Kuchlein *et al.*, 1997; Parmesan and Yohe, 2003; Logan *et al.*, 2003; Elphinstone and Toth 2008; Sharma *et al.*, 2005; 2010). This may lead to increased abundance of tropical insect species (Cannon, 1998; Patterson *et al.* 1999; Bale *et al.* 2002; Diffenbaugh *et al.* 2008) and sudden outbreaks of insect-pests can wipe out certain crop species, entirely (Kannan and James, 2009). At the same time; warming in temperate region may lead to decrease in relative abundance of temperature sensitive insect population (Petzoldt and Seaman, 2010: Sharma, 2005; 2010). Mostly the Polar Regions are constrained from the insect outbreaks due to low temperature and frequently occurring frosts (Volney and Fleming, 2000). In future, projected climate warming (Carroll *et al.,* 2004) and increased drought incidence (Logan *et al.,* 2003) is expected to cause more frequent insect outbreaks in temperate regions also.

As the species richness of insect tends to increase with temperature, it is presumed that with an increase in temperature more species will be gained than lost. Overwintering survival and timing of the commencement spring are important at higher latitudes, leading to population build-up of, for example, global warming results altitude wise range expansion and increased overwintering survival of corn earworms *Heliothis zea* (Boddie) and *Helicoverpa armigera* (Hubner) may cause heavy yield loss and put forth major challenge for pest management in maize, a staple food crop of USA (Diffenbaugh *et al.,* 2008). Range extension in migratory species like *Helicoverpa armigera* (Hubner), a major pest of cotton, pulses and vegetables in North India is predicted with global climate warming (Sharma *et al.,* 2005; 2010). Subsequently, these ongoing shifts in insect-pest distribution and range due to changing climate may alter regional structure, diversity and functioning of ecosystems (IPCC, 2007).

### iii) Changes in Insect Phenology

In addition to the shift in space of species distributions, climate change has led to an ecological shift in time, with changes in species' phenology (timing of life history stages insects). It is one of the easiest impacts of climate change to monitor and is by far the most documented in this regard for a wide range of organisms (Root *et al.,* 2003). With increased temperatures, it is expected that insects will pass through their larval stages faster and become adults earlier. Therefore, expected responses in insects could include an advance in the timing of larval and adult emergence and an increase in the length of the flight period (Menéndez, 2007). Members of the order Lepidoptera are the best examples of such phenological changes. Changes in butterfly phenology have been reported in UK, where 26 of 35 species have advanced their first appearance (Roy and Sparks, 2000). Early adult emergence and an early arrival of migratory species have also been reported for aphids in the UK (Harrington *et al.,* 2007). Gordo and Sanz (2005) investigated climate impacts on four Mediterranean insect species *viz.* butterfly, bee, fly and beetle and indicated that all species exhibited changes in their first appearance date over the last 50 years, which was correlated with increases in spring temperature.

For insects, changes in phenology can be studied through long-term experiments with variable sowing dates for observing the appearance of pests on crops. Likewise, the timing of arrival of insect species can also be recorded through light traps, suction traps or pheromone traps. Analysis of long-term data on phenology would reveal changes in the timings of pest appearance under the climate change (Pathak *et al.*, 2012). Suction traps are being used to monitor aphids at The Rothamsted Insect Survey since 1964. Analysis of suction trap data has revealed that spring flights of the potato aphid [(*Myzus persicae* (Sulzer)] started two weeks earlier for every 1°C rise in combined mean temperature of January and February. Likewise, long-term data from several insect-recording schemes in Europe and North America have provided evidence for species becoming active, migrating or reproducing earlier in the year due to increases in temperatures that lead directly to increased growth rates or earlier emergence from winter inactivity (Roy and Sparks, 2000). Increasing temperatures have also allowed a number of species to remain active for a longer period during the year or to increase their annual number of generations.

Under the All India Coordinated Rice Improvement Programme (AICRIP) of ICAR, there is also widespread network of the Coordinating Centres all over India that collect light trap insect data round the year. Analysis of historical light trap data vis-à-vis current data can provide important information on the impacts of climate change on rice pests.

### iv) Increased Overwintering Survival

Being poikilotherms, insects have limited ability of homeostasis with external temperature changes. Hence they have developed a range of strategies such as behavioural avoidance through migration and physiological adaptations like diapause to support life under thermally stressful environments (Bale and Hayward, 2010). Diapause is a period of suspended developmental activities, the manifestation of which is governed by environmental factors like temperature, humidity and photoperiod. As an adaptive trait, diapause plays vital role in seasonal regulation of insect life cycles because of which the insects have better advantage to survive great deal of environmental adversities. There are two main types of insect diapause; aestivation and hibernation to sustain life under high and low temperature extremes respectively (Chapman, 1998). The studies have shown that, global warming is occurring notably in winter than in summer and is greatest at high latitudes (IPCC, 2007, IMD, 2010). Looking at the past 100 years climate profile of India, warming was more pronounced during winter season and it was the minimum and not the maximum temperature where significant increase was observed (IMD, 2010). Thus, insects undergoing a winter diapause are likely to experience the most significant changes in their thermal environment (Bale and Hayward, 2010).

Accelerated metabolic rates at higher temperatures shorten the duration of insect diapause due to faster depletion of stored nutrient resources (Hahn and Denlinger, 2007). Warming in winter may cause delay in onset and early summer may lead to faster termination of diapause in insects, which can then resume their active growth and development. This gives an important implication that increase in temperature in the range of 1°C to 5°C would increase insect survival due to low winter mortality, increased population built-up, early infestations and resultant crop damage by insect-pests under global warming scenario (Harrington *et al.,* 2001; Sharma *et al.,* 2005; 2010).

### v) Increase in Number of Generations

Some crop pests are "stop and go" developers in relation to temperature. They develop more rapidly during periods of time with suitable temperatures. Increased temperatures will accelerate the development of these types of insects possibly resulting in more cycles of generations per year (Awmack *et al.,* 1997). It has been estimated that with a 2°C temperature increase, insects might experience one to five additional life cycles per season (Yamamura and Kiritani, 1998). Warming could decrease the occurrence of severe cold events, which could in turn expand the over-wintering area for insect pests (Patterson *et al.,* 1999). Thus, global increase in temperature within certain favourable range may accelerate the rates of development, reproduction and survival in tropical and subtropical insects. Consequently, insects will be capable of completing more number of generations per year (Yamamura and Kiritani 1998; Petzoldt and Seaman, 2010).

### vi) Introduction of invasive alien species

Climate change can as well promote arrival and establishment of the exotic species. Risk of introduction invasive alien species, increase with global climate change. Even though the causes of biological invasions are manifold and multifaceted, changes in abiotic and/or biotic components of the environment (climate change, biological control) are recognised as primary drivers of species invasion (Dukes and Mooney, 1999; IPCC, 2007). Globalization and liberalization of world agricultural trade coupled with the rapid transport and communication means nowadays, have substantially and plausibly increased the chances of exotic introductions. According to the Convention on Biological Diversity (CBD), invasive alien species are the greatest threat to loss of biodiversity in the world and impose high costs to agriculture, forestry and aquatic ecosystems by altering their regional structure, diversity and functioning (Mooney and Hobbs, 2000; Sutherst, 2000; Timoney, 2003).

It is expected that global warming may exacerbate ecological consequences like introduction of new pests by altering phenological events like flowering times especially in temperate plant species (Fitter and Fitter, 2002; Parmesan and Yohe, 2003; Willis *et al.*, 2008) as several tropical plants can withstand the phenological changes (Corlett and LaFrankie, 1998). Invasion of new insect-pests will be the major problem with changing climate favouring the introduction of insect susceptible cultivars or crops (Gregory *et al.*, 2009).

### vii) Pest population dynamics and outbreaks

The population dynamics is the aspect of population ecology dealing with factors affecting changes in population densities. The seasonal effects of weather and ongoing changes in climatic conditions will directly lead to modifications in dispersal and development of insect species. The changes in surrounding temperature regimes certainly involve alterations in development rates, voltinism and survival of insects and subsequently act upon size, density and genetic composition of populations, as well as on the extent of host plant exploitation (Bale *et al.*, 2002). It may result in upsetting ecological balance because of unpredictable changes in the population of insect-pests along with their existing and potential natural enemies (Rao *et al.*, 2006; IPCC, 2007).

Changes in climatic variables have led to increased frequency and intensity of outbreaks of insect-pests. Outbreak of sugarcane woolly aphid *Ceratovacuna lanigera* Zehntner in sugarcane belt of Karnataka and Maharashtra states during 2002-03 resulted in 30% yield losses (Joshi and Virakamath, 2004; Srikanth, 2007). These situations of increased and frequent pest damage to the crops have made another big hole in the pockets of already distressed farmers by increasing the cost of plant protection and reducing the margin of profit.

### viii) Crop-pest interactions

The capacity of an herbivore insect to complete its development depends on the adaptation to both, the environmental conditions and the host plant. The global change of climate have been found to exert both bottom-up and top-down effects on the tri-tropic interactions between crops, insects and natural enemies by means of certain physiological changes especially related to host-suitability and nutritional status (Hare, 1992; Roth and Lindroth, 1995; Coviella and Trumble, 1999; Gutierrez, 2008). This has been shown by gypsy moth attacking the Red maple (*Acer rubrum* L.) and Sugar maple (*A. saccharum*) which had reduced larval weight, increased feeding time and prolonged development (Williams *et al.,* 2000). The large outbreaks observed in the expansion areas on the new hosts may be explained either by the high susceptibility of the hosts or by the inability of natural enemies to locate the moth larvae on an unusual hosts or environment (Stastny *et al.,* 2006).

Temperature and photoperiod have been found to affect profoundly the critical events such as stem elongation, flowering and fruiting in the life cycle of plants (Cleland *et al.,* 2007). Global warming lead increased temperatures may accelerate the life cycles in some of the plant species (Parmesan and Yohe, 2003: Fitter and Fitter, 2002: Willis *et al.,* 2008) which may affect significantly, feeding and reproduction patterns in associated insect-pests like aphids, jassids, mealybugs, etc. Such increases can greatly exacerbate the negative ecological and economical consequences (Timoney, 2003, Millennium Ecosystem Assessment, 2005).

### ix) Increased incidence of insect vectored plant diseases

Climate change may lead to more incidence of insect transmitted plant diseases through range expansion and rapid multiplication of insect vectors (Petzoldt and Seaman, 2010; Sharma *et al.,* 2005; 2010). Increased temperatures, particularly in early season, have been reported to increase the incidence of viral diseases in potato due to early colonization of virus-bearing aphids, the major vectors for potato viruses in Northern Europe (Robert *et al.,* 2000).

## Increased Level of $CO_2$

One of the most studied aspects of climate change is the effect of increasing concentrations of $CO_2$ on plants. Plants consist primarily of carbon and elevated $CO_2$ levels allow them to grow more rapidly because they can assimilate carbon more quickly. Greenhouse growers have known this for decades and add $CO_2$ to encourage plant growth. Similarly, because $CO_2$ increases the photosynthetic rates of most crop plants, scientists initially thought that increasing $CO_2$ would be a solution for the world's food supply (LaMarche *et al.,* 1984). In addition to enhanced growth, many crop plants become more drought-tolerant due to $CO_2$ enrichment. This is because openings in the leaves (stomata) that let $CO_2$ in and also let water vapor out. If there is high $CO_2$ concentration in the vicinity of leaf then the stomata need not open as much. It was suggested that under conditions of elevated $CO_2$, plants will produce better yields even when conditions are harsh (LaMarche *et al.,* 1984). Unfortunately, such optimistic predictions have not proven accurate. One reason for this is that insects also eat more when plants are grown under elevated levels of $CO_2$ to compensate their low nutritional quality.

A rise in $CO_2$ in atmosphere generally increases the carbon to nitrogen ratio due to accumulation of non- structural carbohydrates of plant tissues thereby reducing the nutritional quality for protein limited insects diluting the nitrogen content by 15-25% in the tissues (Coviella *et al.,* 1999). The expected reactions from herbivores to the increase in carbon to nitrogen ratio are compensatory feeding, concentrations of defensive chemicals in plants and competition between pest species. Insects may accelerate their food intake to compensate for reduced leaf nitrogen content (Holton *et al.,* 2003), although this is not always the case (Knepp *et al.,* 2005). However, the response of plants to increased $CO_2$ varies among species. Increased carbon to nitrogen ratios in plant tissue may slow insect development and increase the length of life stages vulnerable to attack by parasitoids. O' Neill *et al.* (2008) found that the life span of Japanese beetle, *Popillia japonica,* a major pest of soybean (*Glycine max*), is prolonged by 8.25% when fed on foliages developed under elevated $CO_2$. Besides, females fed on such foliages laid approximately twice as many eggs as compared to females fed on foliages grown under normal ambient conditions. A higher level of sugars like glucose, sucrose, and fructose in soybean foliages grown under higher $CO_2$ is considered to be a preferential factor for Japanese beetle, *P. japonica* to feed on them. Though, Casteel *et al.* (2008) showed in soybean a low level of deterrent phytochemicals under high ambient $CO_2$ that allows the insects, particularly the Japanese beetle to feed voraciously on the plants. Chen *et al.* (2004) also reported a higher fecundity in aphids with higher carbohydrate levels in food plants grown under elevated $CO_2$ level. Phytophagous insects may also develop adaptations to overcome higher carbon to nitrogen ratios, for example the pine sawfly, *Neodiprion lecontei,* showed an increase in the efficiency of nitrogen utilization when reared on plants treated with high $CO_2$ concentration (Williams *et al.,* 1994). However, other insect species seem unable to compensate the lower nutritional quality of the plants by increasing the efficiency of nutrient utilization (Brooks and Whitekar, 1999). The experiments of Lindroth *et al.* (1993) on three species of saturnid moths showed that the performance of caterpillars is only marginally affected when the nitrogen content of the leaves is reduced by 23% and the carbon to nitrogen ratio increased by 13-28%.

Impact of $CO_2$ on insect population via host plants can be studied through open top chambers (OTCs) and free air carbon dioxide enrichment (FACE). The OTCs are essentially plastic enclosures placed around a sample of an ecosystem. Air is drawn into a box by a fan, enriched with $CO_2$, and blown through the chamber. Open-top chambers are relatively inexpensive to build because they consist simply of an aluminum frame covered by panels of polyvinyl chloride plastic film.

Temperature control is affected by air movement and differential between inside and outside air temperatures has been found less than 1°C. Relative humidity within the open tops is directly related to transpiration rate and is always higher than in the external air. OTCs have long been known to modify the environment by altering light intensity, relative humidity, wind speed and direction, and other environmental factors. This is considered a shortcoming for studies on insect and disease occurrence as chambers may interfere with the dispersal of organisms or alter the plant's susceptibility to a given pest. Therefore, to separate out the effect of higher temperature and humidity on insect development from $CO_2$ effect on it, control should also be in an OTC at ambient $CO_2$ concentration. Rao *et al.* (2009) have conducted feeding trials with two foliage feeding insect species, *A. janata* and *S. litura* using foliage of castor plants grown under three concentrations *viz.*, 700 ppm, 550 ppm, 350 ppm (ambient) of $CO_2$ inside open top chamber (OTC) and 350 ppm $CO_2$ in the open. Biochemical analysis of the foliage showed that plants grown under the elevated $CO_2$ levels had lower N-content, and higher C-content, C/N ratio and polyphenols. Compared to the larvae fed on the ambient $CO_2$ foliage, the larvae fed on 700 ppm and 550 ppm $CO_2$ foliage exhibited higher consumption. The 700 ppm and 550 ppm $CO_2$ foliage was more digestible with higher values of approximate digestibility. The relative consumption rate of larvae increased, whereas the efficiency parameters, viz. efficiency of conversion of ingested food (ECI), efficiency of conversion of digested food (ECD) and relative growth rate (RGR) decreased in the case of larvae grown on 700 ppm and 550 ppm $CO_2$ foliage. The consumption and weight gain of the larvae were negatively and significantly influenced by the leaf nitrogen, which was found to be the most important factor affecting consumption and growth of larvae.

On the other hand, in FACE, plants are grown in open without any enclosure. The FACE technology facilitates modification of the environment around growing plants to future concentrations of atmospheric $CO_2$ under natural conditions of temperature, precipitation, pollination, wind, humidity, and sunlight. Therefore, FACE field data represent plant responses to concentrations of atmospheric $CO_2$ in a natural setting. FACE facilities have been developed and deployed ranging from the scale of 1–2m for low-stature crops or ecosystems to those of 10–20m diameter for field crop evaluation to those for young-to medium-aged forest stands and finally to mature trees of basically any size. Hamilton *et al.* (2005) used the FACE technology to create an atmosphere with $CO_2$ and $O_2$ concentrations similar to those predicted for the middle of the 21st century. During the early season, soybean grown under the elevated $CO_2$ atmosphere had 57% more damage from the insects like Japanese beetle, potato leafhopper, western corn rootworm and Maxican bean beetle than those grown under the ambient conditions and even required an insecticide treatment to continue the experiment. Measured increases in the levels of simple sugars in the soybean leaves might have stimulated the additional insect feeding.

### Precipitation

Distribution and frequency of rainfall may also affect the incidence of pests directly as well as through changes in humidity levels. It is being predicted that under the climate change, frequency of rainfall would decline while its intensity would increase. This would lead to heavy showers and floods on one hand and drought spells on the other. Under such situations, incidence of small pests such as aphids, jassids, whiteflies, mites, etc. on crops may be reduced as these get washed away by the heavy rains (Pathak *et al.*, 2012). The deviation of rainfall during monsoon and November and its relationship with level of *Helicoverpa armigera* (Hub.) damage severity showed higher November rainfall favoured higher infestation. Average rainfall is predicted to decrease in several regions and the occurrence of summer droughts is likely to increase. Root herbivore species responded differently to the summer rainfall manipulations. The larvae of dominant root chewing species, *A. lineatus*, were more numerous under enhanced rainfall, in contrast, abundance of the *Coccoidea lecanopsis formicarum* was unaffected by the rainfall manipulations (Karuppaiah and Sujayanad, 2012). Crops such as groundnut, cotton, chillies and coriander before, after and along with tobacco lead to higher incidence of *S. litura*. Resistance to pesticides, favourable weather conditions such as cyclonic weather and heavy rainfall followed by dry spell also contributed to its outbreak (Chari *et al.*, 1993). Lever (1969) analysed the relationship between outbreaks of armyworm, *Mythimna separate* (Walker) and to a lesser extent *Spodoptera mauritia* (Boisd.) and rainfall from 1938 to 1965 and observed that all but three outbreaks occurred when rainfall exceeded the average 89 cm. The effect of rainfall on pests can be studied by simulating various rainfall intensities through sprinklers. Aphid population on wheat and other crops was adversely affected by rainfall and sprinkler irrigation (Daebeler and Hinz, 1977; Chander, 1998). Masters *et al.* (1998) have carried out novel manipulations of local climate to investigate how warmer winters with either wetter or drier summers would affect the homopteran insects a major component of the insect fauna of grasslands.

## Impact of Climate Change on Pollinators and Pollination

Insects play vital role in providing various ecosystem services. One of the very important is pollination as they are excellent pollinators for many of the economically important crops (Murugan, 2006; Sidhu and Mehta, 2008). Approximately 73 per cent of the world's cultivated crops are pollinated by bees, 19 per cent by flies, 6.5 per cent by bats, 5 per cent by wasps, 5 per cent by beetles, 4 per cent by birds, and 4 per cent by butterflies and moths (Abrol, 2009). The pollinators in turn benefit by obtaining floral resources such as nectar, pollen or both. This mutualism has evolved over centuries and been helping both natural terrestrial ecosystems as well as man-made agro-ecosystems. Thus the entomophilies pollination is a fundamental process essential for the production of about one-third of the world human food (Klein *et al.,* 2007).

Climate change, an emerging global phenomenon, with a potential to affect every component of agricultural ecosystems, is reported to impact insect pollinators at various levels, including their pollination efficiency. According to Millennium Ecosystem Assessment report 2005, pollination is one of the 15 major ecosystem services currently under threat from mounting pressures exerted by growing population, depleting natural resource base and global climate change (Costanza *et al.,* 1987; Sachs, 2008). Earlier studies have clearly shown that the population abundance, geographic range and pollination activities of important pollinator species like bees, moths and butterflies are declining considerably with changing climate (FAO, 2008). The climatic factors like temperature and water availability have been found to affect profoundly the critical events like flowering, pollination and fruiting in the life cycle of plants (Cleland *et al.,* 2007). Many pollinators have synchronised their life cycles with plant phenological events. Impending climate change is expected to disrupt the synchrony between plant-pollinator relationships by changing the phenological events in their life cycles and may thus affect the extent of pollination (Kudo *et al.,* 2004; Ricketts *et al.,* 2008). The quality and the quantity of pollination have multiple implications for food security, species diversity, ecosystem stability and resilience to climate change (FAO, 2008).

Although pollination is a critical issue it appears to be neglected and overlooked for other ecosystem services such as water and air quality, climate regulation and food availability. The pollination services and associated risks are not addressed properly in determining the actions needed for conserving pollinators. The high degree of uncertainty regarding the risks related to pollination services implies the need for well focused research to understand scientifically the pollination processes.

## Impact of Climate Change on the Pest Management Strategies

### a) Breakdown of host plant resistance

Host plant resistance is one of the most environmental friendly components for managing harmful insect-pests of crops wherein the plant can lessen the damage caused by insect-pests through various mechanisms like antixenosis, antibiosis and tolerance (Painter, 1968; Dhaliwal and Dilavari, 1993). Expression of the host plant resistance is greatly influenced by environmental factors like temperature, sunlight, soil moisture, air pollution, etc. Changes in these climatic factors may alter the interactions between insect pests and their host plants (Sharma *et al.*, 2010). Under stressful environment, plant becomes more susceptible to attack by insect-pests because of weakening of their own defensive system resulting in pest outbreaks and more crop damage (Rhoades 1985). With global temperature rise and increased water stress, tropical countries like India may face the problem of severe yield loss in sorghum due to breakdown of resistance against midge *Stenodiplosis sorghicola* (Coq.) and spotted stem borer, *Chilo partellus* Swinhoe (Sharma *et al.*, 2005). There will be an increased impact on insect pests which benefit from reduced host defences as a result of the stress caused by the lack of adaptation to suboptimal climatic conditions. Some plants can change their chemical composition in direct response to insect damage to make their tissues less suitable for growth and survival of insect pests (Sharma, 2002).

Generally, $CO_2$ impacts on insects are thought to be indirect. Impact on insect damage will result from changes in nutritional quality and secondary metabolites of the host plants. Increased levels of $CO_2$ will enhance plant growth, but may also increase the damage caused by some phytophagous insects. In the enriched $CO_2$ atmosphere expected in the 21st century, many species of herbivorous insects will confront less nutritious host plants that will induce both lengthened larval developmental times and greater mortality (Coviella and Trumble, 1999). The effects of climate change on the magnitude of herbivory and direction of response will not only be species-specific, but also specific to each insect–plant system. Bark beetles, wood borers, and sap sucking insects benefit from severe drought (Bjorkman and Larsson, 1999; Huberty and Denno, 2004; Koricheva *et al.*, 1998), while *Spodoptera exigua* (Hub.) exhibited a reduced ability to feed on drought-stressed tomato leaf tissue, which contained higher levels of defense compounds as a result of the abiotic stress (English-Loeb *et al.*, 1997). In atmospheres experimentally enriched with $CO_2$, the nutritional quality of leaves declined substantially due to a dilution of nitrogen by 10–30% (Coley and Markham, 1998). Increased $CO_2$ may also cause a slight decrease in nitrogen-based defences (e.g. alkaloids) and a slight increase in carbon-based defences (e.g. tannins). Lower foliar nitrogen due to $CO_2$ causes an increase in food consumption by herbivores. Increase in amounts of simple sugars and down-regulation of gene expression for a protease-specific deterrent to coleopteran herbivores may have resulted in greater insect feeding (Zavala *et al.*, 2008). Elevated $CO_2$ decreases the induction of jasmonic acid and ethylene related transcripts (*lox7, aos, hpl,* and *acc1*) in soybean plants causing decreased accumulation of defenses (polyphenol oxidase, protease inhibitors, etc.) over time compared to plants grown under ambient conditions, suggesting that $CO_2$ exposure might have resulted in increased insect damage (Casteel, 2010). Problems with new insect pests will occur if climatic changes favour the introduction of nonresistant crops or cultivars into new areas. The introduction of new crops and cultivars could be one of the methods to take advantage of climate change (Parry, 1990; Parry and Carter, 1989).

### b) Transgenic crops

In recent advancement of integrated pest management, insect resistant transgenics expressing the *Bacillus thuringiensis* (Berliner) (*Bt*) insecticidal protein (delta-endotoxin) were developed (Kranti *et al.* 2005). However, these transgenic plants showed a reduction in the level of toxin protein during periods of high temperature, elevated $CO_2$ levels, or drought, leading to decreased resistance to insect pests (Chen *et al.*, 2005; Chen *et al.*, 2005a; Dong and Li, 2007). Cotton bollworm, *Heliothis virescens* (F.), destroyed *Bt* cottons due to high temperatures in Texas, USA (Kaiser, 1996). Similarly, *H. armigera* and *H. punctigera* damaged *Bt*-cotton in the second half of the growing season in Australia because of reduced production of *Bt* toxins in the transgenic crops (Hilder and Boulter, 1999). Cry1Ac levels decrease with plant age, resulting in greater susceptibility of the crop to bollworms during the later stages of crop growth (Adamczyk *et al.*, 2001; Greenplate *et al.*, 2000; Kranthi *et al.*, 2005; Sachs *et al.*, 1998). Possible causes for the failure of insect control may be due to inadequate production of the toxin protein, the effect of environment on transgene expression, locally resistant insect populations, and development of resistance due to inadequate management (Sharma and Ortiz, 2000). It is therefore important to understand the effects of climate change on the efficacy of transgenic plants for pest management.

### c) Natural enemies

Relationships between insect pests and their natural enemies will change as a result of global warming, resulting in both increases and decreases in the status of individual pest species. Changes in temperature will also alter the timing of diurnal activity patterns of different groups of insects (Young, 1982), and changes in interspecific interactions could also alter the effectiveness of natural enemies for pest management (Hill and Dymock, 1989). Quantifying the effect of climate change on the activity and effectiveness of natural enemies for pest management will be a major concern in future pest management programs. The majority of insects are benign to agro-ecosystems, and there is considerable evidence to suggest that this is due to population control through interspecific interactions among insect pests and their natural enemies–pathogens, parasites, and predators (Price, 1987). Oriental armyworm, *M. separata* populations increased during extended periods of drought (which is detrimental to the natural enemies), followed by heavy rainfall because of the adverse effects of drought on the activity and abundance of the natural enemies of this pest (Sharma *et al.*, 2002). Aphid abundance increases with an increase in $CO_2$ and temperature, however, the parasitism rates remain unchanged in elevated $CO_2$. Temperatures up to 25°C will enhance the control of aphids by coccinellids (Freier and Triltsch, 1996). Temperature not only affects the rate of insect development, but also has a profound effect on fecundity and sex ratio of parasitoids (Dhillon and Sharma, 2008 and 2009). The interactions between insect pests and their natural enemies need to be studied carefully to devise appropriate methods for using natural enemies in pest management.

## d) Biopesticides and Synthetic Insecticides

There will be an increased variability in insect damage as a result of climate change. Higher temperatures will make dry seasons drier, and conversely, may increase the amount and intensity of rainfall, making wet seasons wetter than at present. Current sensitivities on environmental pollution, human health hazards, and, pest resurgence are a consequence of improper use of synthetic insecticides. Natural plant products, entomopathogenic viruses, fungi, bacteria, nematodes, and synthetic pesticides are highly sensitive to the environment. Temperature is a major factor affecting insecticide toxicity either positive or negative and, thus, efficacy. The response relationship between temperature and efficacy has been found to vary depending on the mode of action of an insecticide, target species, method of application, and quantity of insecticide ingested or contacted (Johnson, 1990). Increased temperature will increase the activity of some of the insecticides. Diflubenzuron (an insect growth regulator (IGR)) caused rapid mortality at higher temperatures and was more efficient at 35°C (Amarasekare and Edelson, 2004). This was probably because this IGR is only effective when the insect moults

(Ware, 2000), and the insect growth rate and moulting rate increase at higher temperatures (Lactin and Johnson, 1995). However, the biological activity of the entomopathogenic fungus, *Beauveria bassiana* (Balsamo), is reduced at temperatures >25°C (Amarasekare and Edelson, 2004; Inglis *et al.*, 1999). Increase in temperature and UV radiation, and a decrease in relative humidity, may render many of the pest control tactics to be less effective, and such an effect will be more pronounced on natural plant products and the biopesticides. Entomopathogens used as biocontrol agents suffer from instability after exposure to solar radiation, especially in the ultraviolet (UV) portion of the spectrum. Several studies have reported a significant decrease in biological activity of entomopathogens, viz. NPV, GV, *Beauveria* and *Bt* (up to 90%) within a few days (Broome *et al.*, 1974; Ignoffo *et al.*, 1977; Jones and McKinley, 1986). Another effect of increased temperature and UV radiation may be to slow down the activity even without the loss of activity due to UV radiation; as a result, more time may be required to achieve insect mortality (Moscardi, 1999; Szewczyk *et al.*, 2006). Larvae continue to feed and damage crops until shortly before death. Chen and McCarl (2001) estimated that pest treatment costs under the 2090 projections of climate exhibit increases of 3–10% for corn, soybeans, cotton and potatoes and mixed results for wheat, and show a $200 million per year projected loss to society due to climate change related pesticide treatment cost effects in the USA. Therefore, there is a need to develop appropriate strategies for pest management that will be effective under situations of global warming in the future. Farmers will need a set of pest control strategies that can produce sustainable yields under climatic change.

## Implications for Food Security

The greatest challenge in the coming century is to double the present levels of food production to meet the needs of ever increasing population by sustainable use of shrinking natural resource base (Deka *et al.*, 2008). The aggravating pest problems under changing climate regimes are expected to intensify the yield losses; threatening the food security of the countries with high dependency on agriculture. The climate change is likely to affect the extent of entomophilies pollination by disrupting the synchrony between plant-pollinator life cycles (Kudo *et al.*, 2004), with an estimated risk of reduction in world food production by one-third (Klein *et al.*, 2007). This has major implication for food and nutritional security (FAO, 2008). This may have direct bearing on the livelihood of the rural poor as their survival is directly linked to outcomes from food production systems. The increased food prices resulting from declining food production may also impact negatively the urban population (IPCC, 2007; Chahal *et al.*, 2008).

## Adaptation of agriculture to changing pest scenario due to climate

No doubt, understanding and dealing with the problem of abiotic stresses and crop insect pest interactions under the influence of changing climate is difficult task. Some of the strategies that we would feel useful in tackling the issue are pointed out below.

### *Sensitization of Stakeholders about Climate Change and its Impacts*

Considering the impacts of future climate change on sustainability and productivity of agriculture, especially in the developing countries like India, there is an urgent need to sensitize the farmers, extension workers and other stakeholders involved in supply chain management about the climate change associated changes in incidence of pests and diseases of major crops in their regions and the different adaptation strategies to cope with the situation. This can be achieved through organization of awareness campaigns, training and capacity-building programmes, development of learning material and support guides for different risk scenarios of pest, etc.

### *Farmers' Participatory Research for Enhancing Adaptive Capacity*

The decision making ability and adaptive capacity of farmers can be enhanced through the integration of a farmers' participatory and multidisciplinary research approach involving research and developmental organizations and farmers as equal partners. This will help to improve the channels of communication between researchers and farmers for dissemination of knowledge and information regarding the current advances in the provision of weather and climate information, weather based agro-advisory services for facilitating operational decisions at farm level. A decision support system (DSS) involving mechanisms for collection and dissemination of information on insect-pest data under diverse environmental conditions for improved assessments well in advance needs to be developed. In view of changing pest scenario due to climate, we recommend that our future research programmes should focus on the search for more general forms of resistance against various classes of insects or diseases under abiotically stressful environments.

### *Promotion of Resource Conservation Technologies (RCTs)*

Shrinking resource base due to anthropogenic developmental activities is a major challenge ahead for humanity. Conservation of natural resources can be promoted by giving incentives to the farmers those who are adopting environmental conserving pest controlling activities such as organic farming, bio-control, integrated pest management, habitat conservation for important insect pollinators, etc. Strategies for adaptation and coping could benefit from combining scientific and indigenous technical knowledge (ITK), especially in developing countries where technology is least developed. ITK is helpful to adapt the adverse effects of changing climate. *e.g.* application of natural mulches helps in suppression of harmful pests and diseases besides moderating soil temperatures and conservation of soil moisture. Further more study towards integrating indigenous adaptation measures in global adaptation strategies and scientific research is required.

## Climate Change : Challenge Ahead

In addition to the strategies discussed above, we need to decide the future line of research for combating the pest problems under climate change regimes.

### Breeding Climate-Resilient Varieties

In order to minimize the impacts of climate and other environmental changes, it will be crucial to breed new varieties for improved resistance to abiotic and biotic stresses. Considering late onset and/ or shorter duration of winter, there is chance of delaying and shortening the growing seasons for certain Rabi/ cold season crops. Hence we should concentrate on breeding varieties suitable for late planting and those can sustain adverse climatic conditions and pest and disease incidences.

### Rescheduling of Crop Calendars

Global temperature increase and altered rainfall patterns may result in shrinking of crop growing seasons with intense problems of early insect infestations. As such certain effective cultural practices like crop rotation and planting dates will be less or no effective in controlling crop pests with changed climate. Hence there is need to change the crop calendars according to the changing crop environment. The growers of the crops have to change insect management strategies in accordance with the projected changes in pest incidence and extent of crop losses in view of the changing climate.

### GIS Based Risk Mapping of Crop Pests

Geographic Information System (GIS) is an enabling technology for entomologists, which help in relating insect-pest outbreaks to biographic and physiographic features of the landscape, hence can best be utilized in area wide pest management programmes. How climatic changes will affect development, incidence, and population dynamics of insect-pests can be studied through GIS by predicting and mapping trends of potential changes in geographical distribution (Sharma *et al.*, 2010) and delineation of agro-ecological hotspots and future areas of pest risk (Yadav *et al.*, 2010).

## Screening of Pesticides with Novel Mode of Actions

It has been reported by some researchers that the application of neonicotinoid insecticides for controlling sucking pests induces salicylic acid associated plant defense responses which enhance plant vigour and abiotic stress tolerance, independent of their insecticidal action (Gonias *et al.*, 2003; Thielert, 2006, Horii *et al.*, 2007; Chiriboga *et al.*, 2009; Ford *et al.*, 2010). This gives an insight into investigating role of insecticides in enhancing stress tolerance in plants. Such more compounds needs to be identified for use in future crop pest management.

## Conclusion

Climate change now a day is globally acknowledged fact. It has serious impacts on diversity, distribution, incidence, reproduction, growth, development, voltisim and phenology of insect pests. Climate changes also affect the activity of plant defense and resistance, bio-pesticides, synthetic chemicals, invasive insect species, expression of *Bt* toxins in transgenic crops. Considering such declining production efficiency due to depleting natural resource base, serious consequences of climate change on diversity and abundance of insect-pests and the extent of crop losses, food security for 21st century is the major challenge for human kind in years to come. Being a tropical country, India is more challenged with impacts of looming climate change. In India, pest damage varies in different agro-climatic regions across the country mainly due to differential impacts of abiotic factors such as temperature, humidity and rainfall. This entails the intensification of yield losses due to potential changes in crop diversity and increased incidence of insect-pests due to changing climate. It will have serious environmental and socio-economic impacts on rural farmers whose livelihoods depend directly on the agriculture and other climate sensitive sectors.

Dealing with the climate change is really tedious task owing to its complexity, uncertainty, unpredictability and differential impacts over time and place. Understanding abiotic stress responses in crop plants, insect-pests and their natural enemies is an important and challenge ahead in agricultural research. Impacts of climate change on crop production mediated through changes in populations of serious insect-pests need to be given careful attention for planning and devising adaptation and mitigation strategies for future pest management programmes. Therefore, there is a need to have a concerted look at the likely effects of climate change on crop protection, and devise appropriate measures to mitigate the effects of climate change on food security.

## References

Abrol, D.P. (2009). Plant-pollinatorinteractions inthe context of climate change - an endangered mutualism. Journal of Palynology, 45:1-25.

Adamczyk, Jr. J.J.; Adams, L.C. and Hardee, D.D. (2001). Field efficacy and seasonal expression profiles for terminal leaves of single and double *Bacillus thuringiensis* toxin cotton genotypes. Journal of Economic Entomology, 94: 1589-1593.

Aggarwal, P.K. (2008). Climate change and Indian Agriculture: impacts, adaptation and mitigation, Indian J. Agric. Sci., 78: 911-919.

Alfred, J.R.B. (1998). Faunal Diversity in India: An Overview. In: *Faunal Diversity in India* (Eds.: Alfred, J.R.B. et al.). ENVIS Centre, Zoological Survey of India, Calcutta, p. 1-495.

Amarasekare, K.G., Edelson, J.V. (2004). Effect of temperature on efficacy of insecticides to differential grasshopper (Orthoptera: Acrididae). J. Econ. Entomol., 97: 1595–1602.

Anand, T. and Pereira, G.N. (2008). Butterflies-The Flying Jewels of the Western Ghats. http://www.daijiworld.com/chan/exclusive_arch.asp?ex_id=1003.

Awmack, C.S.; Woodcock, C.M. and Harrington R. (1997). Climate change may increase vulnerability of aphids to natural enemies. Ecol. Entomol., 22: 366-368.

Bale, J.; Masters, G. J.; Hodkins, I. D.; Awmack, C.; Bezemer, T. M.; Brown, V. K.; Buterfield, J.; Buse, A.; Coulson, J. C.; Farrar, J.; Good, J. E. G.; Harrigton, R.; Hartley, S.; Jones, T. H.; Lindroth, R. L.; Press, M. C.; Symrnioudis, I.; Watt, A. D. and Whittaker, J. B. (2002). Herbivory in global climate change research: direct effects of rising temperature on insect herbivores. Global Change Biol. 8: 1-16.

Bale, J. S. and Hayward, S. A. L. (2010). Insect overwintering in a changing climate. The J. Exp. Biol. 213: 980-994.

Battisti, A., Stastny, M., Buffo, E. and Larsson, S. (2006). A rapid altitudinal range expansion in the pine processionary moth produced by the 2003 climatic anomaly. Global Change Biology, 12: 662-671.

Bjorkman, C. and Larsson, S. (1999). Insects on drought-stressed trees: four feeding guilds in one experiment. In: Lieutier, F.; Mattson, W. J. and Wagner, M. R. (Eds.), Physiology and Genetics of Tree-Phytophage Interactions. International Symposium, Gujan, France, pp. 323–335.

Bokhorst, S., Huiskes, A., Convey, P., van Bodegom, P.M., and Aerts, R. (2008). Climate change effects on soil arthropod communities from the Falkland Islands and the Maritime Antarctic. Soil Biology and Biochemistry 40: 1547-1556.

Brooks, G.L. and Whittaker, J. B. (1999). Responses of three generations of a xylem-feeding insect, *Neophilaenus lineatus* (Homoptera), to elevated $CO_2$. Global Change Biol., 5: 395-401.

Broome, J.R., Sikorowski, P.P. and Nee, W.W. (1974). Effect of sunlight on the activity of nuclear polyhedrosis virus from *Malacosoma disstria*. J. Econ. Entomol., 67: 135–136.

Cammell, M.E. and Knight, J.D. (1991). Effects of climate change on the population dynamics of Crop pests. Adv. Ecol. Res., 22: 117- 62.

Cannon, R.J.C. (1998). The implications of predicted climate change for insect pests in the UK, with emphasis on non-indigenous species. Global Change Biol., 4: 785–796.

Carroll, A.L., Taylor, S.W., Regniere, J. and Safranyik, D.L. (2004). Effects of climate change on range expansion by the mountain pine beetle in British Columbia. In: Mountain Pine Beetle Symposium: Challenges and Solutions (Eds.: Shore, T.L., Brooks, J.E. and Stone, J.E.), Natural Resources Canada, Canadian Forest Service, Pacific Forestry Centre, Victoria, British Columbia. pp. 223-232.

Casteel, C.L. (2010). Impacts of climate change on herbivore induced plant signaling and defenses. Ph.D. Dissertation, University of Illinois, Urbana-Champaign, IL. USA.

Casteel, C.L., O'Neill, B.F., Zavala, J.A., Bilgin, D.D., Berenbaum, M.R. and Delucia, E.H. (2008).Transcriptional profiling reveals elevated $CO_2$ and elevated O3 alter resistance of soybean (*Glycine max*) to Japanese beetles (*Popillia japonica*). Plant Cell and Environment, 31: 419–434.

Chahal, S.K., Bains, G.S. and Dhaliwal, L.K. (2008). Climate change: Mitigation and Adaptation. Proceedings of International Conference on Climate Change, Biodiversity and Food Security in the South Asian Region, 3-4 November, 2008, Punjab State Council for Science and Technology, Chandigarh and nited Nations Educational, Scientific and Cultural Organization, New Delhi, pp.12.

Chand, R. and Raju, S.S. (2009). Instability in Indian Agriculture during different phases of technology and policy, Indian J. Agric. Econ. 64: 187-207.

Chander, S. (1998). Infestation of root and foliage/earhead aphids on wheat in relation to predators. Indian J. Agric. Sci., 68: 754-755.

Chapman, R.F. (1998). The Insects-Structure and Function, Cambridge University Press, 4th Edition. pp. 788.

Chari, M.S., Raghupathirao, G., Ramaprasad, G. and Sreedhar, U. (1993). Pest epidemics ITobacco: A recent ternd. Lead papers of the National Seminar on changing scenario in pest and pest management in india. Pp. 96-102. Plant Protection Association of India, Hyderabad.

Chen, C. and McCarl, B. (2001). Pesticide usage as influenced by climate: a statistical investigation. Clim. Change 50: 475–487.

Chen, D.H., Ye, G.Y., Yang, C.Q., Chen, Y. and Wu, Y.K. (2005). The effect of high temperature on the insecticidal properties of *Bt* Cotton. Environ. Exp. Bot., 53: 333–342.

Chen, F.J., Wu, G. and Ge, F. (2004). Impacts of elevated $CO_2$ on the population abundance and reproductive activity of aphid, *Sitobion acenae* Fabricius feeding on spring wheat. J. Appl. Entomol. 128: 723-730.

Chen, F.J., Wu, G., Ge, F., Parajulee, M.N., Shrestha, R.B., (2005a). Effects of elevated $CO_2$ and transgenic *Bt* cotton on plant chemistry, performance, and feeding of an insect herbivore, the cotton bollworm. Entomol. Exp. Appl. 115: 341–350.

Chiriboga, A., Herms, D.A. and Royalty, R.N., Effects of imidacloprid (Merit®2F) on physiology of woody plants and performance of two spotted spider mite, fall webworm, and imported willow leaf beetle.Bayer Environmental Science.

https://kb.osu.edu/dspace/bitstream/handle/1811/36552/ACHIRIBOGA%20OARDC%20Poster%20Apr09.pdf, 2009.

Cleland, E.E., Chuine, I., Menzel, A., Mooney, H.A., Schwartz, M.D. (2007). Shifting plant phenology in response to global change. *TREE*, 22: 357-365.

Coley, P.D. and Markham, A. (1998). Possible effects of climate change on plant/herbivore interactions in moist tropical forests. Climatic Change 39: 455-472.

Corlett, T.T. and LaFrankie, Jr. J.V. (1998). Potential impacts of climate change on tropical Asian forests through and influences on phenology. Climate Change 39: 439-454.

Costanza, R., d'Arge, R., de Groot, R., Farber, S., Grasso, M., Hannon, B., Limburg, K., Naeem, S., O'Neill, R.V., Paruelo, J., Raskin, R.G., Sutton, P. and van den Belt, M. (1987). The Value of the World's Ecosystem Services and Natural Capital, Nature, 387: 253-260.

Coviella, C.E., Trumble, J.T. (1999). Effects of elevated atmospheric carbon dioxide on insect-plant interactions Conserv. Biol. 13: 700–712.

Daebeler, F. and Hinz, B. (1977). Compensation of aphid injury to sugarbeet by means of supplementary rain. Arch Phytopathol. Pflanzenschutz, 13(3): 199-205.

De, U.S. and Mukhopadhyay, R.K. (1998). Severe heat wave over the Indian subcontinent in 1998 in perspective of global climate, Curr. Sci., 75: 1308-1315.

Deka, S., Byjesh, K., Kumar, U. and Choudhary, R. (2008). Climate change and impacts on crop pests- a critique. ISPRS Archives XXXVIII-8/W3 Workshop Proceedings: Impact of climate change on Agriculture, pp. 147-149.

Dewar, R.C., Watt, A.D. (1992). Predicted changes in the synchrony of larval emergence and budburst under climatic warming. Oecologia 89: 557–559.

Dhaliwal, G.S. and Dilawari, V.K. (1993). Advances in Host Resistance to Insects. Kalyani Publishers, New Delhi.

Dhawan, A.K., Singh, K., Saini, S., Mohindru, B., kaur, A., Singh, G. and Singh, S. (2007). Incidence and damage potential of mealybug, *Phenacoccus solenopsis* Tinsley, on cotton in Punjab. Indian J. Ecol., 34: 110-116.

Dhillon, M.K., Sharma, H.C. (2008). Temperature and *Helicoverpa armigera* food influence survival and development of the ichneumonid parasitoid, *Campoletis chlorideae*. Indian J. Plant Prot. 36: 240–244.

Dhillon, M.K., Sharma, H.C. (2009). Temperature influences the performance and effectiveness of field and laboratory strains of the ichneumonid parasitoid, *Campoletis chlorideae*. BioControl 54: 743–750.

Diffenbaugh, N.S,, Krupke, C.H., White, M.A. and Alexander, C.E. (2008). Global warming presents new challenges for maize pest management. Env.Res. Letter, 3: 1-9.

Dong, H.Z. and Li, W.J. (2007). Variability of endotoxin expression in *Bt* transgenic cotton. J. Agron. Crop Sci. 193: 21–29.

Dukes, J.S. and Mooney, H.A. (1999). Does global change increase the success of biological invaders? TREE, 14: 135-139.

Elphinstone, J. and Toth, I.K. (2008). *Erwinia chrysanthemi* (*Dikeya spp.*) - The facts. Potato Council, Oxford, U.K.

English-Loeb, G., Stout, M.J., Duffey, S.S. (1997). Drought stress in tomatoes: changes in plant chemistry and potential nonlinear consequences for insect herbivores. Oikos 79: 456–468.

FAO, Food and Agriculture Organization of the United Nations (2008). Rapid Assessment of Pollinators' Status. FAO, Rome, Italy.

Fitter, A.H. and Fitter, R.S. (2002). Rapid changes in flowering time in British plants. Science 296: 1689-1691.

Fleming, R.A., Volney, W.J. (1995). Effects of climate change on insect defoliator population processes in Canada's boreal forests: some plausible scenarios. Water Air Soil Pollut. 82: 445–454.

Ford, K.A., Casida, J.E. and Chandranb, D. (2010). Neonicotinoid insecticides induce salicylate associated plant defense responses. PNAS. 107: 17527-17532.

Freier, B. and Triltsch, H. (1996). Climate chamber experiments and computer simulations on the influence of increasing temperature on wheat-aphid-predator interactions. Aspects of Applied Biology 45: 293-298.

Gonias, E.D., Oosterhuis, D.M., Bibi, A.C. and Brown, R.S. (2003). Yield, growth and physiology of TrimaxTM Treated Cotton. Summaries of Arkansas Cotton Research, pp. 139-144.

Gordo, O. and Sanz, J.J. (2005). Phenology and climate change: a long-term study in a Mediterranean locality. Oecologia. 146: 484–495.

Greenplate, J.T., Penn, S.R., Shappley, Z., Oppenhuizen, M., Mann, J., Reich, B. *et al.* (2000). Bollgard II efficacy: quantification of total lepidopteran activity in a 2-gene product. In: Dugger, P., Richter, D. (Eds.), Proceedings, Beltwide Cotton Conference National Cotton Council of America, Memphis, TN, USA, pp. 1041–1043.

Gregory, P.J., Johnson, S.N., Newton, A.C. and Ingram, J.S.I. (2009). Integrating pests and pathogens into the climate change/ food security debate, J. Exp. Bot., 60: 2827-2838.

Gutierrez, A.P., Ponti, L., d'Oultremont, T., Ellis, C.K. (2008). Climate change effects on poikilotherm tritrophic interactions. Clim. Change 87: S167–S192.

Hahn, D.A. and Denlinger, D.L. (2007). Meeting the energetic demands of insect diapause: nutrient storage and utilization. J. Insect Physiol, 53: 760-773.

Hamilton, J.C., Dermody, O., Aldea, M., Zangerl, A.R., Rogers, A., Berenbaum, M.R., and Delucia, E. (2005). Anthopogenic changes in tropospheric composition increase susceptibility of soybean to insect herbivory. Enviro. Entomology, 34 (2): 479-4.8

Hampson, G.F. (1908). The moths of India- Supplementary papers to the Volume in the Fauna of British India. Ser.III, Pt.X, J.Bombay Nat. Hist.Soc., 18: 257-271.

Hare, J.D. (1992). Effects of plant variation on herbivore-natural enemy interactions. In *Plant Resistance to Herbivores and Pathogens: Ecology, Evolution and Genetics.*(eds. Fritz, R.S., Simms, E.L.), University of Chicago Press, Chicago, Illinois, USA, pp. 278-298.

Harrington, R., Fleming, R. and Woiwod, I. (2001). Climate change impacts on insect management and conservation in temperate regions: can they be predicted? Agric. For. Entomol., 3: 233–240.

Harrington, R., Clark, S.J., Weltham, S.J., Virrier, P.J., Denhol, C.H., Hullé, M., Maurice, D., Roun-sevell, M.D. & Cocu, N. (2007). Environmental change and the phenology of European aphids. Global Change Biology, 13: 1556-1565.

Hilder, V.A. and Boulter, D. (1999). Genetic engineering of crop plants for insect resistance - a critical review. Crop Protection, 18: 177-191.

Hill, M.G. and Dymock, J.J. (1989). Impact of Climate Change: Agricultural/ Horticultural Systems. DSIR Entomology Division Submission to the New Zealand Climate Change Program. Department of Scientific and Industrial Research, Auckland, New Zealand, pp. 16.

Holton, M.K., Lindroth, R.L., Nordheim, E.V. (2003). Foliar quality influences tree-herbivore-parasitoid interactions: effects of elevated CO2, O3, and plant genotype. Oecologia, 137: 233–244.

Horii, A., McCue, P. and Shetty, K. (2007). Enhancement of seed vigour following insecticide and phenolic elicitor treatment, Biores. Tech., 98: 623-632.

Huberty, A.F. and Denno, R.F. (2004). Plant water stress and its consequences for herbivorous insects: a new synthesis. Ecology 85: 1383–1398.

IARI News, Brown plant hopper outbreak in rice, 24: 1-2 (2008).

Ignoffo, E.M., Hostetter, D.L., Sikorowski, P.P., Sutter, G., Brooks, W.M. (1977). Inactivation of entomopathogenic viruses, a bacterium, fungus, and protozoan by an ultraviolet light source. Environ. Entomol. 6: 411–415.

IMD. 2010. Annual Climate Summary (2010). India Meteorological Department, Pune. Government of India, Ministry of Earth Sciences, pp. 27.

Inglis, G.D., Duke, G.M., Kawchuck, L.M., Goettel, M.S. (1999). Influence of oscillating temperature on the competitive infection and colonization of the migratory grasshopper by *Beauveria bassiana* and *Metarhizium flavoviride*. Biol. Control, 14: 111–120.

IPCC (2001). Climate Change (2001). Impacts, Adaptation, and Vulnerability. Cambridge University Press, Cambridge, UK, p. 1032.

IPCC (2007). Climate Change- Impacts, Adaptation and Vulnerability. In: (Eds.: Parry, M.L., Canziani, O.F., Palutikof, J.P., van der Linden, P.J., Hanson, C.E.) Cambridge University Press, Cambridge, UK, pp. 976.

IRRI News (2009). Pest Outbreaks in India. Rice today, 6. www.ricenews.irri.org

Johnson, D.L. (1990). Influence of temperature on toxicity of two pyrethroids to grasshoppers (Orthoptera: Acrididae). J. Econ. Entomol. 83: 366–373.

Jones, K.A. and McKinley, D.J. (1986). UV inactivation of *Spodoptera littoralis* nuclear polyhedrosis virus in Egypt: assessment and protection. In: Samson, R.A. Vlak, J.M., Peters, D. (Eds.), Fundamental and Applied Aspects of Invertebrate Pathology. Proceedings, IV International Colloqium on Invertebrate Pathology, Wageningen, The Netherlands, p. 155.

Joshi, S. and Viraktamath, C.A. (2004). The sugarcane woolly aphid, *Ceratovacuna lanigera* Zehntner (Hemiptera: Aphididae): its biology, pest status and control. *Curr. Sci.,* 87: 307-316.

Kaiser, J. (1996). Pests overwhelm *Bt* cotton crop. Nature 273: 423.

Kannan, R. and James, D.A. (2009). Effects of climate change on global diversity: a review of key literature, *Tropical Ecol.,* 50: 31-39.

Karuppaiah, V. and Sujayanad, G.K. (2012). Impact of Climate Change on Population Dynamics of Insect Pests. World Journal of Agricultural Sciences 8 (3): 240-246.

Klein, A.M., Vaissiere, B., Cane, J.H., Steffan-Dewenter, I., Cunningham, S.A., Kremen, C. and Tscharntke, T. (2007). Importance of pollinators in changing landscapes for world crops. Proceedings of the Royal Society B, 274: 303-313.

Knepp, R.G., Hamilton, J.G., Mohan, J.E., Zangerl, A.R., Berenbaum, M.R. and DeLucia, E.H. (2005). Elevated $CO_2$ reduces leaf damage by insect herbivores in a forest community. New Phytologist. 167: 207-218.

Koricheva, J., Larsson, S. and Haukioja, E. (1998). Insect performance on experimentally stressed woody plants: a meta-analysis. Annu. Rev. Entomol. 43: 195–216.

Kranti, K.R., Naidu, S., Dhawad, C.S., Tatwawadi, A., Mate, K., Patil, E., Bharose, A.A., Behere, G.T., Wadaskar, R.M. and Kranti, S. (2005). Temporal and intraplant variation in Cry1Ac expression of Bt cotton and its influence on the development of cotton bollworm, *Helicovera armigera* (Hubner), (Noctuidae, Lepidoptera), Curr. Sci., 89: 291-298.

Kremen, C., Colwell, R.K., Erwin, T.L., Murphy, D.D., Noss, R.F. and Sanjayan, M.A. (1993). Terrestrial arthropod assemblages: Their use in conservation planning, Conserv. Biol., 7: 796-808.

Kuchlein, J.H. and Ellis, W.N. (1997). Climate-induced changes in the microlepidoptera fauna of the Netherlands and the implications for nature conservation. J. Insect conserv. 1: 73-80.

Kudo, G., Nishikawa, Y., Kasagi, T. and Kosuge, S. (2004). Does seed production of spring ephemerals decrease when spring comes early? Ecol. Res., 19: 255-259.

Lactin, D.J., Johnson, D.L. (1995). Temperature dependent feeding rates of *Melanoplus sanguinipes* nymphs (Orthoptera: Acrididae) in laboratory trials. Environ. Entomol. 24: 1291–1296.

Lal, M. (2003). Global climate change: India's monsoon and its variability. J. Env. Studies Policy, 6: 1-34.

LaMarche, V.C., Grabyll, D.A., Fritts, H.C. and Rose, M.R. (1984). Increasing atmospheric carbon dioxide: Tree ring evidence for growth enhancement in natural vegetation. Sci. 22: 5-21.

Lever, R.J.W. (1969). Do armyworm follow the rain? Wild Crops, 21: 351-352.

Lindroth R.L., Kinney K.K. and Platz, C.L. (1993). Responses of deciduous trees to elevated atmospheric $CO_2$: productivity, phytochemistry and insect performance. Ecol. 74: 763-777.

Logan, J.A., Regniere, J. and Powell, J.A. (2003). Assessing the impacts of global warming on forest pest dynamics. Frontiers in Ecol. Env., 1: 130–137.

Masters, G.J., Brown,V.K., Clarke, I.P., Whittaker, J.B. and Hollier, J.A. (1998). Direct and indirect effects of climate change on insect herbivores: *Auchenorrhyncha* (Homoptera). Ecol. Entomol., 23(1): 45-52.

Menéndez, R. (2007). How are insects responding to global warming? Tijdschrift voor Entomologie. 150: 355–365.

Merrill, R., Gutie´rrez, D., Lewis, O., Gutie´rrez. J., Diez, S. and Wilson, R. (2008). Combined effects of climate and biotic interactions on the elevational range of a phytophagous insect. J. Animal Ecol., 77: 145–155.

Millennium Ecosystem Assessment. (2005). Ecosystems and Human Well-being: Synthesis, Island Press, Washington, DC.

Mooney, H.A. and Hobbs, R.J. (2000). (eds.), Invasive Species in a Changing World, Island Press Washington, DC.

Moscardi, F. (1999). Assessment of the application of baculoviruses for control of Lepidoptera. Annu. Rev. Entomol. 44: 257–289.

Murugan, K. (2006). Bio-diversity of insects, Curr.Sci., 91: 1602-1603.

NATCOM (2004). India's initial national communication to the United Nations framework- convention on climate change. Ministry of Environment and Forests, pp. 268.

Newton, A.C., Begg, G., and Swanston, J.S. (2009). Deployment of diversity for enhanced crop function. Annals of Applied Biology 154: 309-322.

O'Neill, Zangerl, A.R., DeLucia, E.H. and Berenbaum, M.R. (2008). Longevity and Fecundity of Japanese Beetle (*Popillia japonica*) on Foliage Grown Under Elevated Carbon Dioxide.

Painter, R.H. (1968). Insect resistance in crop plants. University Press of Kansas, Lawrence.

Parmesan, C. (2007). Influences of species, latitudes and methodologies on estimates of phonological response to global warming. Global Change Biol., 13: 1860–1872.

Parmesan, C., Yohe, G. (2003). A globally coherent fingerprint of climate change impacts across natural systems. Nature 421: 37–42.

Parry, M.L. (1990). Climate Change and World Agriculture. Earthscan, London, UK, 157 pp.

Parry, M.L., Carter, T.R. (1989). An assessment of the effects of climatic change on agriculture. Climate Change, 15: 95–116.

Pathak, H. Aggarwal, P.K. and Singh, S.D. (2012). Climate Change Impact, Adaptation and Mitigation in griculture: Methodology for Assessment and Application Division of Environmental Sciences, Indian Agricultural Research Institute New Delhi, pp 111-130.

Patterson, D.T., Westbrook, J.K., Joyce, R.J.V., Lingren, P.D. and Rogasik, J. (1999). Weeds, insects and diseases. Clim. Change 43: 711–727.

Petzoldt, C. and Seamann, A. (2010). Climate Change Effects on Insects and Pathogens. Climate change and Agriculture: Promoting Practical and Profitable responses 6-16; Accessed online at http://www.climatean dfarming.org/pdfs/ FactSheets/III.2Insects.Pathogens. pdf.

Porter, J.H., Parry, M.L., Carter, T.R. (1991). The potential effects of climate change on agricultural insect pests. Agric. Forest Meteorol. 57: 221–240.

Price, P.W. (1987). The role of natural enemies in insect populations. In: Barbosa, P., Schultz, J.C. (Eds.), Insect Outbreaks. Academic Press, San Diego, CA, USA, pp. 287–312.

Rao, G.G.S.N., Rao, A.V.M.S., Rao, V.U.M. (2009). Trends in rainfall and temperature in rainfed India in previous century. *In:* Global climate change and Indian Agriculture case studies from ICAR network project, (Ed.: PK Aggarwal), ICAR Publication, New Delhi. pp.71-73.

Rhoades, D.F. (1985). Offensive-defensive interactions between herbivores and plants: their relevance in herbivore population dynamics and ecological theory. American Nat., 125: 205-238.

Ricketts, T.H., Regetz, J., Steffan-Dewenter, I., Cunningham, S.A., Kremen, C., Bogdanski, A., Gemmil-Herren, B., Greenleaf, S.S., Klein, A.M., Mayfield, M.M., Morandin, L.A., Ochieng, A. and Viana, B.F. (2008). Landscape effects on crop pollination services: are there general patterns? Ecology Letters, 11, 499-515.

Robert, Y., Woodford, J.A.T. and Ducray-Bourdin, D.G. (2000). Some epidemiological approaches to the control of aphid-borne virus diseases in seed potato crops in northern Europe. Virus Res., 71: 33–47.

Root, T.L., Price, J.T., Hall, K.R., Schneider, S.H., Rosenzweig, C. and Pounds, J.A. (2003). Fingerprints of global warming on wild animals and plants. Nature. 421: 57-60.

Roth, S.K. and Lindroth, R.L. (1995). Elevated atmospheric $CO_2$ effects on phytochemistry, insect performance and insect–parasitoid interactions. Global Change Biol., 1: 173–182.

Roy, D.B., Sparks, T.H. (2000). Phenology of British butterflies and climate change. Global Change Biol. 6: 407–416.

Sachs, E.S., Benedict, J.H., Stelly, D.M., Taylor, J.F., Altman, D.W., Berberich, S.A., *et al.* (1998). Expression and segregation of genes encoding *Cry1A* insecticidal proteins in cotton. Crop Sci. 38: 1–11.

Sachs, J. (2008). Common Wealth: Economics for a Crowded Planet. Allen Lane, London.

Samra, J.S. and Singh, G. (2004). Heat Wave of March 2004: Impact on Agriculture, Indian Council of Agricultural Research, New Delhi, p. 32.

Samways, M. (2005). Insect Diversity Conservation. Cambridge University Press, Cambridge, pp. 342.

Sharma, H.C. (2010). Effect of climate change on IPM in grain legumes. In: Fifth International Food Legumes Research Conference (IFLRC V), and the Seventh European Conference on Grain Legumes (AEP VII), 26–30 April 2010, Anatalaya, Turkey.

Sharma, H.C. (2002). Host plant resistance to insects: principles and practices. In: Sarath Babu, B. Varaprasad, K.S. Anitha, K. Prasada Rao, R.D.V.J. Chandurkar, P.S. (Eds.), Resources Management in Plant Protection, Vol. 1. Plant Protection Association of India, Rajendranagar, Hyderabad, Andhra Pradesh, India, pp. 37–63.

Sharma, H.C., Dhillon, M.K., Kibuka, J., Mukuru, S.Z. (2005). Plant defense responses to sorghum spotted stem borer, *Chilo partellus* under irrigated and drought conditions. Int. Sorghum Millets News, 46: 49–52.

Sharma, H.C., Ortiz, R. (2000). Transgenics, pest management, and the environment. Curr. Sci. 79: 421–437.

Sharma, H.C., Srivastava, C.P., Durairaj, C., Gowda, C.L.L. (2010). Pest management in grain legumes and climate change. In: Yadav, S.S., McNeil, D.L., Redden, R., Patil, S.A. (Eds.), Climate Change and Management of Cool Season Grain Legume Crops. Springer Science + Business Media, Dordrecht, The Netherlands, pp. 115–140.

Sidhu, A.K. and Mehta, H.S. (2008). Role of butterflies in the natural ecosystem with special reference to high altitude (Pangi Valley, Himachal Pradesh). *In:* Proceedings of International Conference on Climate Change, Biodiversity and Food Security in the South Asian Region, 3 - 4 November, 2008. Punjab State Council for Science and Technology, Chandigarh and United Nations Educational, Scientific and Cultural Organization, New Delhi, pp.36.

Srikanth, J. (2007). World and Indian scenario of sugarcane woolly aphid. In Woolly Aphid Management in Sugarcane (Eds. Mukunthan, N., Srikanth, J., Singaravelu, B., Rajula Shanthy, T., Thiagarajan, R. and Puthira Prathap, D.), Extension Publication, Sugarcane Breeding Institute, Coimbatore, 154: 1-12.

Sutherst, R. (2000). Climate change and invasive species: a conceptual framework. *In:* Mooney, H.A., Hobbs, R.J. (Eds.), Invasive Species in a Changing World. Island Press, Washington, DC, pp. 211–240.

Szewczyk, B., Hoyos-Carvajal, L., Paluszek, M., Skrzecz, I. and Lobo de Souza, M. (2006). Baculoviruses re-emerging biopesticides. Biotechnol. Adv. 24: 143–160.

Thomas, C.D., Cameron, A., Green, R.E., Bakkenes, M., Beaumont, L.J., Collingham, Y.C., Erasmus, B.F.N., Fereira de Sigueira, M., Grainger, A., Hannah, Hughes L., Huntley B., Jaarsveld van A.S., Midgley, G.F., Miles, L., Ortega-Huerta M.A., Townsend, P.A., Phillips, O.L. and Williams, S.E. (2004). Extinction risk from climate change. Nature. 427: 145–148.

Timoney, K.P. (2003). The changing disturbance regime of the boreal forest of the Canadian Prairie Provinces. Forest Chronicle, 79: 502-516.

Volney, W.J.A. and Fleming, R.A. (2000). Climate change and impacts of boreal forest insects. Agric. Ecosyst.Env., 82: 283-294.

Walther, G.R.; Post, E.; Convey, P.; Menzel, A.; Parmesan, C. and Beebee, T. J. C. (2002). Ecological responses to recent climate change. Nature, 416: 389-395.

Ware, G.W. (2000). The Pesticide Book, Fifth ed. Thompson Publication, Fresno, CA.

Whittaker, J.B., Tribe, N.P. (1996). An altitudinal transect as an indicator of responses of a spittlebug (Auchenorrhyncha: Cercopidae) to climate change. Eur. J. Entomol. 93: 319–324.

Whittaker, J.B. and Tribe, N.P. (1998). Predicting numbers of an insect (*Neophilaenus lineatus*: Homoptera) in a changing climate. J. Anim. Ecol. 67: 987–991.

Williams, R.S., Lincoln D.E. and Thomas R.B. (1994). Loblolly pine grown under elevated $CO_2$ affects early instar pine sawfly performance. Oecologia. 98: 64-71.

Williams, R., Norby, R.J. and Lioncln, D.E. (2000). Effects of elevated $CO_2$ and temperature grown red and sugar maple on gypsy moth performance, Global Change Biol. 6: 685-695.

Willis, C.G., Ruhfel, B., Primack, R.B., Miller Rushing, A.J. and Davis, C.C. (2008). Phylogenetic patterns of species loss in Thoreau's woods are driven by climate change. Proceedings, National Academy of Sciences, USA, 105, 17029-17033.

Woiwod, I. (1997). Detecting the effects of climate change on Lepidoptera. J. Insect Conser., 1: 149- 158.

Xannepuccia, A.D., Ciccihino, A., Escalante, A., Navaro, A., and Isaach, J.P. (2009). Differential responses of marsh arthropods to rainfall-induced habitat loss. Zoological Studies, 48: 173-183.

Yadav, D.S., Subhash, C. and Selvraj, K. (2010). Agroecological zoning of brown plant hopper, {*Nilaparvata lugens* (Stal.)} incidence on rice {*Oryza sativa* (L.)}. J. Scientific Ind. Res., 69: 818-822.

Yamamura, K. and Kiritani, K. (1998). A simple method to estimate the potential increase in the number of generations under global warming in temperate zones. Appl. Entomol. *Zool.,* 33: 289-298.

Young, A.M. (1982). Population Biology of Tropical Insects. Plenum Press, New York, USA, 511 pp.

Zavala, J.A., Cast, C.L., DeLucia, E.H. and Berenbaum, M.R. (2008). Anthropogenic increase in carbon dioxide compromises plant defense against invasive insects. Proc. Natl. Acad. Sci. U.S.A. 105: 5129–5133.

Zvereva, E.L. and Kozlov, M.V. (2010). Responses of terrestrial arthropods to air pollution: a metaanalysis. Environmental Science Pollution Research International, 17: 297-311.

❑❑❑

# Section 4

## Sustainable Agriculture through Alternate Farming and Policy Interventions

# 18

# Sustainable Agriculture with Agroforestry Adoption to Climate Change

**Abhishek Raj and M.K. Jhariya**

## Introduction

Sustainable agriculture is the production of food, fiber, or other plant or animal products using farming techniques (including agroforestry) that protect the environment, public health, human communities, and animal welfare. This form of agriculture enables us to produce healthy food without compromising future generations ability to do the same. Sustainable agriculture describes farming systems that are capable of maintaining their productivity and usefulness to society indefinitely. Such systems must be resource-conserving, socially supportive, commercially competitive and environmentally sound. It is a system of agriculture that will maintain its productivity over the long run. Sustainable farming could be organic farming, biodynamic, perma-culture, agro-ecological systems and low input. The goal of sustainable agriculture is to minimize adverse impacts to the immediate and off-farm environments while providing a sustained level of production and profit. Sustainable agriculture is influenced by environmental climate, soil types and the various crops and types of farm practices employed (Olorunfemi, 2013).

Agroforestry, the incorporation of trees into farming systems has enormous potential to mitigate the effects of drought, prevent desertification and restore degraded soils. Agroforestry can also help to boost food production (for humans as well as animals) and provide alternative sources of nutrition or income when crop yields are low. With climate change expected to lead to unpredictable seasons in the future, placing even greater pressure on agricultural systems, food production and food prices, agroforestry is a viable option to help buffer farmers against the impacts" (WAC, 2012). Therefore, agro-forestry has many potential, such as enhance the overall (biomass) productivity, soil fertility improvement, soil conservation, nutrient cycling, micro-climate improvement, carbon sequestration, bio drainage, bio energy and biofuel etc. (Bargali *et al.* 2004, 2009; Fanish and Priya, 2013). Moreover, tree component of agroforestry produce tangible products in sustainable way. The incorporation of babul trees in agroforestry and babul based traditional agroforestry systems helps farmers to strengthen their socioeconomic conditions as well as to help conserve environment and biodiversity too (Raj, 2015a). The central India forms one of the major ecosystems of the Indian subcontinent (Raj and Toppo, 2014; Toppo *et al.*, 2014) and having store house of gum producing states (Raj *et al.*, 2015; Das, 2014; Raj, 2015b).

## Concept and prospect of sustainable agriculture and agroforestry

The world's rapidly growing population leads to nine billion mouths to feed by 2050 and facing climate change are necessitates conventional forms of agriculture that are unsustainable. These problems can be addressed by maintaining sustainability in agriculture, which integrates three main goals- i) environmental health, ii) economic profitability, iii) social and economic equity. Thus, sustainable agriculture and land management practices are the better solution changing weather pattern and whole climate variability. Farming practices such as agroforestry plays best management practices for making sustainability in agricultural and forestry production. Agroforestry contributes a vital role in Indian economy and has potential to satisfy three objectives, *viz.* to protect and ameliorate the environment, enhance sustainable production of economic goods on a long-term basis and improve socioeconomic condition of rural people (Jhariya *et al.*, 2015; Raj *et al.*, 2014a). The poor, particularly the rural poor, depend on nature for many elements of their livelihoods, including food, fuel, shelter and medicines (Jhariya and Raj, 2014). In the present scenario of climate change, agro-forestry practices, emerging as a viable option for combating negative impacts of climate change (Singh *et al.*, 2013; Singh and Jhariya, 2016). Plantation of multipurpose trees (MPTs) species gives multiple benefits such as increasing productivity, ecosystem stability and biological diversity to degraded land. Further due to importance of neem in social forestry, agroforestry, reforestation and rehabilitation of the wasteland and degraded industrial lands it helps to combat desertification, deforestation and soil erosion and to reduce excessive global temperature (Jhariya *et al.*, 2013). As per Raj *et al.* (2014b) the soil biological attributes are also responsible for determination and maintenance of physical properties of soil. Similarly, Beekeeping and apiculture also play an important role in the sustainable agriculture. Both emerge as key source of non-land based income to marginal and small farmers, helps to generating off-farm employment for sustainable livelihoods (Painkra *et al.*, 2016).

## World and Indian scenario

Agriculture itself causes climate change. Agriculture as currently practiced with monocultures and agrochemicals in a globalized production system, is a major contributor to climate change, causing emissions of greenhouse gases (GHGs) through changes in land use and soil losses/ degradation, through agricultural technologies and from livestocks (Rathore *et al.* 2014). An agricultural sector contributed by the emission of greenhouse gases (GHGs) to climate change, originated from fertilizers; ruminant digestion, rice cultivation and fuel use (Abhilash *et al.,* 2015). $CO_2$ emission from agriculture field 5.1-6.1 Gt eq/yr (10-12% of total global anthropogenic emissions of GHGs). $CH_4$ contributes 3.3 Gt $CO_2$-eq/yr and $N_2O$ 2.8 Gt $CO_2$- eq/yr, agriculture accounts for about 60% of $N_2O$ and about 50% of $CH_4$. Despite large annual exchanges of $CO_2$ between the atmosphere and agricultural lands, the net flux is estimated to be approximately balanced with $CO_2$ emissions around 0.04 Gt $CO_2$/ yr only (Vermeulen *et al.,* 2013). By using integrated farming practices, we can utilize the livestocks organic manures instead of agrochemical. This not only sustains farm productivity but also maintain ecosystem and environment.

Nair *et al.* (2009) has reported that area currently under agroforestry worldwide is 1,023 million ha. According to Dhyani *et al.* (2013) in India the current area under forestry is estimated at 25.32 Mha, or 8.2% of total geographical area of the country. This includes 20.0 Mha in cultivated lands (7.0 Mha in irrigated and 13.0 Mha in rainfed areas) and 5.32 Mha in other areas such as shifting cultivation (2.28 Mha), home gardens and rehabilitation of problem soils (2.93 Mha). Nair and Nair (2003) estimated the extent of alley-cropping, silvopastoral, windbreaks and riparian buffers in the USA as 235.2 mha. Kumar (2006) estimated the area of homegarden in South and Southeast Asian homegarden as 8.0 Mha.

## Climate change mitigation and adaptation strategy

Climate change is the most important global environmental challenge which is faced by all living organism including humans and disturb natural ecosystems, agriculture and health. This change in climate and weather patterns results unsustainable agricultural production. In this situation, agroforestry plays a viable option to mitigate climate change and reduce global warming by absorbing green house gases ($CO_2$) through the process of carbon-sequestration. Therefore, agroforestry, a form of Climate Smart Agriculture, is a promising adaptation option for smallholder farmers throughout the developing world. The diverse adaptive benefits of agroforestry have been captured in case examples and scientific studies in developing countries in Asia, Africa, and Central and South America (Colin, 2013).

**Table 1.** The carbon absorption capacity of different agroforestry models.

| Agroforestry model | Carbon storage capacity | Region | Source |
|---|---|---|---|
| Agrisilviculture system (aged 11 years) | 26.0 tC/ha | Semiarid region | NRCAF (2005) |
| Block plantation (aged 6 years) | 24.1–31.1 tC/ha | Central India | Swamy *et al.* (2003) |
| *Populus deltoides* 'G-48' + wheat | 18.53 tC/ha | | |
| Silvopasture | 31.71 tC/ha | Himachal Pradesh | Verma *et al.* (2008) |
| Agrisilviculture | 13.37 tC/ha | | |
| Agri-horticulture | 12.28 tC/ha | | |
| Silvopastoralism (aged 5 years) | 6.55 Mg $ha^{-1}$ $y^{-1}$ | Kerala, India | Kumar *et al.* (1998a) |
| Indonesian homegardens (aged 13.4 years) | 8.00 Mg $ha^{-1}$ $y^{-1}$ | Sumatra | Roshetko *et al.* (2002) |

In 2009, the Food and Agriculture Organization of the United Nations coined a new term, "climate-smart agriculture" (CSA). CSA is not a specific tool or technology, but an approach to improving the productivity and sustainability of agriculture under climate change, for the purposes of advancing food security and strengthening the resilience of rural livelihoods (FAO, 2013b). Similarly, the FAO defines climate-smart agriculture as "agriculture that sustainably increases productivity, resilience (adaptation), reduces/removes GHGs (mitigation), and enhances achievement of national food security and development goals" (Stoorvogel *et al.*, 2015). Therefore, climate smart agriculture is defined by three objectives: firstly, increasing agricultural productivity to support increased incomes, food security and development; secondly, increasing adaptive capacity at multiple levels (from farm to nation); and thirdly, decreasing greenhouse gas emissions and increasing carbon sinks (Campbell *et al.*, 2014).

## Conclusion

The ever growing population in world needs more diversified food for their sustenance. Sustainable agriculture meet all the concerned issue related to food scarcity, climate change and economics problems etc.

Agroforestry and climate-smart agriculture emerge as a robust technology to check and address these problems. Therefore, sustainable agriculture is an integrated system of plant and animal production practices, addressing food security problem by making feeds to people, mitigate adverse effects of climate change by enhancing environmental quality, sustain economic viability and enhance quality of life.

## References


Abhilash, P.C., Tripathi, V., Dubey, R.K. and Edrisial, S.A. (2015). Coping with changes: adaptation of trees in a changing environment. Trends Plant Sci., 20(3): 137-138.

Bargali, S.S., Bargali, K., Singh, L., Ghosh, L. and Lakhera, M.L. (2009). *Acacia nilotica* based traditional agroforestry system: effect on paddy crop and management. Current Science, 96(4): 581-587.

Bargali, S.S., Singh, S.P. and Pandya, K.S. (2004). Effects of *Acacia nilotica* on gram crop in a traditional agroforestry system of Chhattisgarh plains. International Journal of Ecology and Environmental Sciences, 30(4): 363-368.

Campbell, B.M., Thornton, P., Zougmore, R., Asten, P.V. and Lipper, L. (2014). Sustainable intensification: What is its role in climate smart agriculture? Current Opinion in Environmental Sustainability, 8: 39–43.

Colin, M. (2013). "Agroforestry and Smallholder Farmers: Climate Change Adaptation through Sustainable Land Use". Capstone Collection. Paper 2612.

Das, I., Katiyar, P. and Raj, A. (2014). Effects of temperature and relative humidity on ethephon induced gum exudation in *Acacia nilotica.* Asian Journal of Multidisciplinary Studies, 2(10): 114-116.

Dhyani, S.K., Handa, A.K. and Uma (2013). Area under agroforestry in India: An assessment for present status and future perspective. Indian Journal of Agroforestry, 15(1):1-11.

Fanish, S.A. and Priya, R.S. (2013). Review on Benefits of Agro Forestry System. Int. J. of Education and Research, 1(1):1-12.

FAO (2013b). Climate smart agriculture sourcebook. Retrieved from: http://www.fao.org/climatechange/climatesmart/en/

Jhariya, M.K. and Raj, A. (2014). Human welfare from biodiversity. Agrobios Newsletter, 12(9): 89–91.

Jhariya, M.K., Bargali, S.S. and Raj, A. (2015). Possibilities and Perspectives of Agroforestry in Chhattisgarh. In: Precious Forests - Precious Earth, PP. 237-257. Dr. Miodrag Zlatic (Ed.), ISBN:978-953-51-2175-6, InTech, DOI: 10.5772/60841.

Jhariya, M.K., Raj, A., Sahu, K.P. and Paikra, P.R. (2013). Neem – a tree for solving global problem. Indian Journal of Applied Research, 3(10): 66–68.

Kumar, B.M. (2006). Carbon sequestration potential of Tropical homegardens. In: Tropical Homegardens: A time tested example of sustainable agroforestry. Advance in Agroforestry 3. Edited by Kumar, B.M. and Nair, P.K.R. (eds.), Springer, Dordrecht, the Netherlands, pp. 185-204.

Kumar, B.M., George, S.J., Jamaludheen, V. and Suresh, T.K. (1998a). Comparison of biomass production, tree allometry and nutrient use efficiency of multipurpose trees grown in wood lot and silvopastoral experiments in Kerala, India. For. Ecol. Manage., 112: 145–163.

Nair, P.K.R., Kumar, B.M. and Nair, V.D. (2009). Agroforestry as a strategy for carbon sequestration. J. Plant Nutr. Soil Sci., 172:10–23.

Nair, P.K.R. and Nair, V.D. (2003). Carbon storage in North American Agroforestry systems. In: The potential of U.S. forest soils to sequester carbon and mitigate the green house effect, Boca Raton, FL. Edited by Kimble, J., Heath, L.S., Birdsey, R.A. and Lal, R. (eds), CRC Press LLC. PP. 333-346.

NRCAF (2005). Annual Report. National Research Centre for Agroforestry, Jhansi.

Olorunfemi, I.E. (2013). The Role of Climate, Soil and Crop on Sustainable Agriculture in Nigerian Ecological Zones: A Brief Overview. Scientia Agriculturae, 4(1): 26-36.

Painkra, G.P., Bhagat, P.K., Jhariya, M.K. and Yadav, D.K. (2016). Beekeeping for Poverty Alleviation and Livelihood Security in Chhattisgarh, India. Pp. 429-453. *In:* Innovative Technology For Sustainable Agriculture Development, Edited by Sarju Narain and Sudhir Kumar Rawat (Ed.). ISBN: 978-81-7622-375-1. Biotech Books, New Delhi, India.

Raj, A. and Toppo, P. (2014). Assessment of floral diversity in Dhamtari district of Chhattisgarh. Journal of Plant Development Sciences, 6(4): 631-635.

Raj, A. (2015a). Gum exudation in *Acacia nilotica*: effects of temperature and relative humidity. In Proceedings of the National Expo on Assemblage of Innovative ideas/ work of post graduate agricultural research scholars, Agricultural College and Research Institute, Madurai (Tamil Nadu), pp 151.

Raj, A. (2015b). Evaluation of Gummosis Potential Using Various Concentration of Ethephon. M.Sc. Thesis, I.G.K.V., Raipur (C.G.), pp 89.

Raj, A., Haokip, V. and Chandrawanshi, S. 2015. *Acacia nilotica:* a multipurpose tree and source of Indian gum Arabic. South Indian Journal of Biological Sciences, 1(2): 66-69.

Raj, A., Jhariya, M.K. and Pithoura, F. (2014a). Need of Agroforestry and Impact on ecosystem. Journal of Plant Development Sciences, 6(4): 577-581.

Raj, A., Jhariya, M.K. and Toppo, P. (2014b). Cow dung for ecofriendly and sustainable productive farming. International Journal of Scientific Research, 3(10): 42-43.

Rathore, A.K., Jhariya, M.K., Jain, R. and Kumar, S. (2014). Agriculture: cause, victim as well as mitigator of climate change. Eco. Env. & Cons., 20(3): 995-1000.

Roshetko, M., Delaney, M., Hairiah, K. and Purnomosidhi, P. (2002). Carbon stocks in Indonesian homegarden systems: Can smallholder systems be targeted for increased carbon storage? Am. J. Alt. Agr., 17: 125–137.

Singh, N.R., Jhariya, M.K. and Raj, A. (2013). Tree crop interaction in agroforestry system. Readers Shelf, 10(3): 15–16.

Singh, N.R. and Jhariya, M.K. (2016). Agroforestry and Agrihorticulture for Higher Income and Resource Conservation. Pp. 125-145. In: Innovative Technology For Sustainable Agriculture Development, Edited by Sarju Narain and Sudhir Kumar Rawat (Ed.). ISBN: 978-81-7622-375-1. Biotech Books, New Delhi, India.

Stoorvogel, J.J., Hendriks, C. and Claessens, L. (2015). When is agriculture climate-smart? A call for proper soil management. Conference paper: 7$^{th}$ Congress of the European Society for Soil Conservation "Agro ecological assessment and functional-environmental optimization of soils and terrestrial ecosystems" Moscow, Russian Federation.

Swamy, S.L., Puri, S. and Singh, A.K. (2003). Growth, biomass, carbon storage and nutrient distribution in Gmelina arborea Roxb. stands on red lateritic soils in central India. Bioresource Technology, 90: 109–126.

Toppo, P., Raj, A. and Harshlata (2014). Biodiversity of woody perennial flora in Badal Khole sanctuary of Jashpur district in Chhattisgarh. Journal of Environment and Bio-sciences, 28(2): 217-221.

Verma, K.S., Kumar, S. and Bhardwaj, D.R. (2008). Soil organic carbon stocks and carbon sequestration potential of agroforestry systems in H.P. Himalaya Region of India. Journal of Tree Sciences, 27(1): 14–27.

Vermeulen, S.J., Challinor, A.J., Thornton, P.K., Campbell, B.M., Eriyagama, N., Vervoort, J.M., Kinyangi, J., Jarvis, A., Laderach, P., Ramirez-Villegas, J., Nicklin, K.J., Hawkins, E. and Smith, D.R. (2013). Addressing uncertainty in adaptation planning for agriculture. PNAS, 110(21): 8357-8362.

World Agroforestry Centre (2012). Surviving drought through agroforestry. Retrieved from: http://worldagroforestry.org/newsroom/highlights/surviving—drought—through—agroforestry.

❑❑❑

# 19

# Agroforestry Interventions for Rehabilitation of Sustainability and Climate Change Mitigation in Degraded *Jhum* Lands of North Eastern Hill Region

**Nirmal and R.A. Alone**

## Introduction

North Eastern Himalaya region of India comprises of the states of Arunachal Pradesh, Manipur, Meghalaya, Mizoram, Nagaland, Sikkim and Tripura (Kukreti, 2000) lies between 21.5° N to 29.5° N latitude and 85.5° E to 97.5° E longitude. The total area of the region is 1,83,753 $km^2$ thousand hectares of which about 9,729 ha (52.95%) is estimated to be under various type of forest including fallow land under regeneration in shifting cultivation with 220 different ethnic groups (Moral, 1997) and a total population of 1,01,39,990 (Table 1.) Northeast India is mountainous and a unique ethnic region with typical land tenure system with climate ranging from tropical to alpine type and altitudinal variation from 15 m to 5000 m amsl. Sustainable realizations of equitable livelihood in the region call upon conservation of bio-resources, environmental externalities, management of fragility and social sensitivities. Traditionally several niche based agroforestry systems of harmonizing woody biomass growth, crop production, livestock rearing and micro-enterprising have been evolved and indigenous technical knowledge generated to enhance income, employment and subsistence. The high rainfall agro-ecologies are endowed with unique social capital of tribe based leadership or headship for managing open access, common properties and forest resources, sharing of goods and services. Some of their traditional systems like bun cultivation were evolved to manage soil acidity, fertility, pests and diseases by soil sterilization. The region has a distinction for some of the rarest species of yak, mithun, pigs, flowers, mushrooms etc. in symphony with natural flora and ethnic culture making it one of the 12 megadiversities in the world. Soil and water conserving economic systems based on perennials like bamboos, mandarins, areca nut, cardamom, medicinal plants and flowers befitted rainfall, topographical and geological conditions. Traditional shifting cultivation with 20-30 years of cycle managed by tribal heads was quite equitable with social justice, restorative, regenerative, organic and sustainable. However, increasing population, migration or settlement from neighbouring regions, consumerism, better communications and changing life styles are multiplying pressure or strains on the natural resources. Free trading, improvements in marketing, transport communication, education and banking infrastructure demands anticipatory research and development processes. Integrated management of multiple uses of natural, bio-resources, livestock and human resources are fundamentally vital issues.

**Table 1.** Population Status of North East Hill Region

| States | Population (as per Census 2011) | Area(Sq Km) | Percentage All India | | Person (per Sq Km) |
|---|---|---|---|---|---|
| | | | Population | Area | |
| Arunachal Pradesh | 1382611 | 83743 | 0.11% | 2.54% | 16.51 |
| Manipur | 2721756 | 22327 | 0.22% | 0.67% | 122.17 |
| Meghalaya | 2964007 | 22429 | 0.24% | 0.68% | 132.15 |
| Mizoram | 1091014 | 21081 | 0.09% | 0.64% | 51.75 |
| Nagaland | 1980602 | 16579 | 0.16% | 0.50% | 119.46 |
| Sikkim | 6,07,688 | 7096 | 0.05% | 0.21% | 85.63 |
| Tripura | 36,17,032 | 10492 | 0.29% | 0.31% | 344.93 |
| Total NEH | 10139990 | 183753 | 1.16% | 5.55% | 124.66 |
| All India | 121,00,00,000 | 32,87,263 | | | 374.17 |

Based on: Census (2011)

Harmonizing or synchronizing symphony of forests on the ridges, planting of fruit and other trees on slopes, crops in the table lands, fish culture in valley ponds, pigs, or poultry or fruit trees and mushrooms on pond embankments are important. Different components of agroforestry system provide year round employment, flow of income, goods, services and communities remain occupied. It is also based on multiple uses of inputs, cycling and recycling of water, nutrients and energy fluxes. Reuse and re-capturing of nutrients and recycling of water, nutrients and other inputs ensure highest utilization efficiency, minimal environmental loading and pollution of resources. However, sharing of income, goods, services, responsibilities and accountability across the ridges to valley in non-conflicting manner is key to success.

## Agriculture scenario in NEH region

The NEH region is a classic case of economic under-development. The region is characterized by high dependence on agriculture, low levels of modern input use, traditional farming techniques, lack of farm mechanization, subsistence farming, low level of productivity, poor infrastructure, etc. The agricultural practices of North East India are of two types - (i) Shifting cultivation, (*Jhum*), and (ii) Settled or plains agriculture (Seitinthang, 2014). About three-fourths of her population, depends on agriculture and other allied activities. The region's agriculture is characterized by low crop intensity (117%). Agriculture provides livelihood to 70% of the region's population. In Mizoram, around 51% of the population lives in rural areas and is dependent on agriculture. The figure in Sikkim is high at 89% (Goswami, 2006). However, the pattern of agricultural growth has remained uneven across regions. The states continue to be net importers of food grains even for their own consumption.

The region has a forest cover of more than 67 per cent against the national average of 22.8 per cent. The North-East accounts for about 8 per cent of the total geographical area of the country but has only 3.4 per cent of the total land put under agricultural purposes (CMIE, 2007). Further, the region contributes only 2.8 per cent to the total food grain production of the nation (RBI, 2005).

Poor soil condition, short sunshine hours, excessive humidity and frequent floods are the natural constraints in these states. The undulating nature of the terrain surrounding the foothills and the high density of rainfall require proper water management and the prevention of soil erosion. The system of shifting cultivation (*jhum*) is widely practised in hilly regions, accounting for roughly 30 per cent of the total area under settled agriculture.

## Shifting cultivation in NEH region

*Jhum* kheti also known as slash and burn or shifting cultivation, is the most primitive and popular agroforestry system practiced in North East hill region. *Jhum* in the region is a complex system with wide variation that depends upon the distinct ecological characteristics in the area and cultural diversity among various tribal clans. However, the basic cropping practice is quite similar. Usually all the essential crops such as paddy, maize, tapioca, colocasia, millets, sweet potato, ginger etc. are grown on the same piece of land as mixed crop. *Jhum* in its most traditional form is not a very unsustainable land use practice particularly when the *Jhum* cycle is more than 20 years. The soils get enough time to rejuvenate and restore their health and productive capacity. However, with increase in population pressure on land resources, the *Jhum* cycle is getting reduced very fast and has reached at 4 years at present. This makes the system unstable and leads to severe land degradation as a result of soil erosion and associated factors such as reduction in soil organic matter, nutrients etc. There is decline of forest cover due to shifting cultivation in the NEH region although the degree varies from one state to the other. The net decrease in forest cover during 2011 and 2013 was 625 km$^2$ (FSI, 2011; FSI, 2013). Total area under *Jhum* also varies among the different hill states. Basic statistical data published by North Eastern Council Secretariat (2015) revealed that maximum area under shifting cultivation is in Manipur followed by Arunachal Pradesh (Table 2). In terms of percentage of the total geographical area, Mizoram (51.90%) and Tripura (32.43%) are the most severely affected by *jhum* cultivation. For instance in Arunachal Pradesh different tribes like Adi, Galo, Monpa, Tangsa, Apatani, Memba, Khamba, Miniyoung, Pasi, Bokar, Bori, Panngi, Aka, Ramo, Mismi, Mizi, Nocte, Wancho, etc. are involved in shifting cultivation over an area of 3600 km$^2$. The farmers maintain high species diversity in *jhum* agro-ecosystem (Majumdar and Arunachalam, 2004; Ramakrishnan, 1992; Satapathy and Bujarbaruah, 2005). The whole socio economic structure, organization and institutions of various population groups of the region are woven around shifting cultivation (Borthakur, 1992). It is argued that in shifting cultivation there is no deliberate choice of planting woody perennials (Lundgren and Nair, 1983), hence this type of land use has become ecologically unsound due to decrease in shifting cultivation cycle as a consequence of increase in population pressure. Sood *et al.* (2002b) and Rahman *et al.* (2011) based on their systematic field observations and literature review have delineated the following causes of shifting cultivation:

- High land : man ratio
- Physical, socio-cultural isolation of the state from the rest of the world
- Community ownership of the land acting as disincentive for permanent cultivation
- Integration of *jhuming* with social and cultural aspects of the tribes
- Lack of knowledge of consequences of *jhuming*

- Higher returns to labour in shifting cultivation
- Lack of know-how, equipments, credit and other facilities for settled cultivation
- Higher agricultural production of agricultural crops in initial 2-3 years
- Lack of infrastructural facilities like input sale centres, lack of markets, etc.

**Table 2.** Area under shifting cultivation in different states of North East

| State | Geographical area ($km^2$) | Annual area under Shifting cultivation ($km^2$) | Fallow period (in year) | Minimum area under shifting cultivation one time or other ($km^2$) | No. of families practicing shifting cultivation (thousand) | Area under jhum (% of geographical area) |
|---|---|---|---|---|---|---|
| Arunachal Pradesh | 83743 | 700 | 3-10 | 2100 | 54 | 6.25 |
| Manipur | 22327 | 900 | 4-7 | 3600 | 70 | 18.10 |
| Meghalaya | 22429 | 530 | 5-7 | 2650 | 52.3 | 14.42 |
| Mizoram | 21087 | 630 | 3-4 | 1890 | 50 | 51.90 |
| Nagaland | 16579 | 190 | 5-8 | 1913 | 116 | 12.22 |
| Tripura | 10492 | 223 | 5-9 | 1119 | 93 | 32.43 |
| Sikkim | 7096 | - | - | - | - | - |

*Source.* NEC Basic statistics for North East India 2015

Roy *et al.* (2002), Rao and Bhattacharyya (2005)

## Potentials of Agroforestry

Agroforestry as a sustainable production system for food and trees products on farmland and rural development has gained importance throughout the world (Rethy *et al.,* 2003). Mountainous forests provide productive and protective benefits not only to the mountain inhabitants but also to a large number of lowland people. The brunt of the losses of deforestation in mountains is not only limited to the region but also has consequences for other regions. For instance, the annual cost of deforestation in the Himalayas was estimated to be $1 billion in the Ganges valley in 1980 (Myeres, 1999). The greatest value of mountains is as source of world's major rivers (Price, 1998). Half of the world's population is dependent on mountains for water for various uses. The Himalayas, one of the most important mountainous ranges of the world, are ecologically fragile and subsistence agriculture constitutes the main source of livelihood (Tucker, 1983). The Himalayan forests with a rich biodiversity are presently decreasing in area due to large-scale exploitation of fuel wood, fodder and timber (Khosla and Khurana, 1996). Indian Eastern Himalayas, is one of the biodiversity hot spots of the world. Deforestation and unsustainable use of natural resources in the region, causes floods in the adjacent plains of Assam and other adjoining countries and have regional as well as global consequences in terms of floods, droughts, loss of biodiversity and other environmental hazards like sea level rise through melting of permanent glaciers. Moreover tree and tree products play an important role in livelihoods of the local people. Thus it is important to conserve forests and increase the tree cover in this region for the benefit of the people of the state, country and international community as a whole. Agroforestry is one of the approaches which have potential for sustainable use of the land in the region. The potential uses of scientific agroforestry for North East include:

i) Maximization of overall production of the crops and degraded lands,
ii) Reduction of loss of soil as well as nutrients through reduction in runoff,
iii) Addition of organic manure to soils,
iv) Utilization of the wastelands, rehabilitation of abandoned and current shifting cultivation area,
v) Improvement of physical conditions of soils, water holding capacity
vi) Release and recycling of elements by affecting biochemical nutrients cycling,
vii) Moderating effect on extreme conditions particularly the floods in North East and adjoining areas.
viii) Lowering effect on the water table in areas where water table is high,
ix) Provision of employment opportunities to rural tribal people,
x) Increase of farm income of the tribals of the region, and
xi) Reduction of the pressure of existing forests, increase in effective tree cover and conservation of plant biodiversity; (Singh, 1993; Sood *et al.,* 2000).

At present the efforts are to conserve the forests, provide ecological security and meet the subsistence requirement of the tribals for food, fuelwood, fodder and necessary plant products and agroforestry is now emerging as an appropriate alternative for the NEH region, various agroforestry systems can be designed and adopted depending on needs of the farmers. However, marshy and swampy areas holds potential for agrisilvifishery system (Sood *et al.,* 2000).

## Agroforestry systems and practices

The systematic work on documentation of traditional agroforestry systems and development of scientifically designed agroforestry systems in NEH region is almost non-existent. Trend to understand and develop scientifically tested agroforestry system is now gaining importance and studies in this direction have already been started (Singh, 1998; Singh *et al.*, 1994; Singh *et al.*, 1997, Singh *et al.*, 2000). Based on review of literature the agroforestry systems existing in the state has been delineated below:

### i) Agrisilviculture

Agriculture is the major economic activity in NEH region. It plays an important role in the overall socio - economic development in the state. Shifting cultivation (*Jhum*) a primitive sequential form of agroforestry system is commonly practiced to more or less extent by the people in all agroclimatic zones of the North Eastern states. However its extent in tropical zone, where wet rice cultivation is very common, is very less compared to other agroclimatic zones of the region.

Now with spread of education, more job opportunities, increased development, increased opportunities for marketing tree products, out-migration for education and jobs and consequent decreasing hands in the family to work on various labour intensive *jhum* operations like felling of trees, a sense of sedentary cultivation is getting inculcated among the shifting cultivators. Now many existing plants are retained around the *jhum* fields, banana and tapioca are being planted on the boundary of *jhum* fields and inside many agricultural crops are grown (Sundriyal *et al.*, 2002). This type of practice has now become common in tropical and subtropical zones of the region where population density is high and demand for value tree products is high. Further planting of *Terminalia myriocarpa* and *Morus laevigata* along the periphery of *jhum* areas is also becoming popular in tropical and subtropical zones.

Besides this simultaneous form of agroforestry systems are developing in NEH region and these types of systems are still in their infancy. These types of systems are mainly limited to tropical zone and some of subtropical zone where few scattered trees, bamboos and banana are grown mainly around the periphery of crop land. Raising of *Piper nigrum* with *Erythrina indica* forms one of the common agrisilviculture system in the region.

### ii) Agrisilvipastoral system

In this type of system the farmers cultivate agricultural crops along with grasses/ rearing of livestock and the forest trees to meet their food, fodder and other tree-based needs. For example *Thysanolaena maxima* grass is raised along the margin of terrace of cropland mainly in Subansiri district and adjoining areas of Assam and Arunachal Pradesh. Agrisilvipastoral systems have been developed by the tribals of the state based on long trial and error approach over centuries. This type of the systems is common than sequential agroforestry (shifting cultivation). It protects soil from erosion as well as provides fodder. *Saccharum spontaneum* is naturally occuring grass in the region. Homesteads with rearing of livestock are one of agrisilvipastoral type of agroforestry systems which exists in the region (Sood *et al.*, 2000). These homesteads are used for growing vegetables, spices, ornamental plants, medicinal plants, fruit trees and other forest trees. Homesteads are prevalent in almost all agroclimatic zones of the region. The crop diversity in such homesteads is very high. Bamboos are also one of the important non timber forest product cultivated in and around the homestead gardens.

#### iii) Hortisilviculture system

This is specialized form of agrisilviculture system which occurs in the region. The most common form of this system is cultivation of tea (*Camellia sinensis*) interspersed mainly with trees of *Dipterocarpus* sp, *Shorea robusta, Tectona grandis, Albizzia lebbeck, Macarnage denticulate, Schima wallichii* and *Acacia* spp. Besides grasses of various kinds are grown around the periphery of tea plantations. This type of system is of frequent occurrence up to an elevation of about 1500 m above mean sea level. These fast growing trees not only provide fuel, timber, and shade to tea but also some of them fix nitrogen and enrich the soil.

Orange and banana alongwith some other forestry tree species are planted for shade along with pineapple and its occurrence is common upto lower limits of temperate zones (Sing *et al.*, 1994). Such type of hortisilviculture system is successfully practiced in Arunachal Pradesh, Nagaland, Mizoram, Meghalaya, Manipur and Tripura mainly in the tropical, subtropical zones. Coffee is also one of the important crops raised under the shade tree of *Dipterocarpus macrocarpus.* Forestry tree species are grown on periphery of fruit trees to a limited extent almost in all agroclimatic zones of the state. Sikkim is well known for its Large cardamom growing under all kinds of shade trees.

#### iv) Silvipastoral practices

It is also one of the common agroforestry systems in Arunachal Pradesh. Bamboo groves, cane, pine groves and tree groves surrounded by the pasture land with their natural vegetation is found in almost all the agroclimatic zones of the state. This system is mainly dominated in Apatani Plateau of Upper Subansiri district and adjoining temperate and subtropical parts of Arunachal Pradesh. The most common bamboo species grown in this system are: *Dendrocalamus hamiltonii, Dendrocalamus strictus, Melocanna bambusoides, Arundinaria grifithii*and *Arundinaria callosa*. The common cane species under silvipastures include *Calamus erectus, Calamus flagellum* and *Calamus floribundus*. The common grasses in silvipastoral system include *Cynodondactylon, Imperate cylindrica, Thysanolaena maxima, Chrysopogon* spp, *Paspalum* spp and *Setaria* spp. The Apatani people of the area freely graze the Mithun (*Bos frontalis*) and other livestock in the pasture (Sood *et al.*, 2000).

### v) Agrihorticultural practices

In this type of system agricultural crops along with horticultural woody perennials have concurrently grown in various parts of the region. Arecanut with Agricultural crops and on cropland also forms agrihorticultural system which is common in lower altitudes of the region. In middle altitudes, vegetables are cultivated in home garden along with the fruit trees. The common vegetables and cereals cultivated are cucumber, millet, pumpkin, maize, tomato, lai sag, ginger, soybean, onion and french bean. They also grow maize along with tobacco, millets and fruit trees mainly plum, apple and pear (Patwary *et al.*, 2000).

### vi) Other Agroforestry Practices

Besides above other practices are also prevalent in some parts of the region. Agrisilvifishery system is also most prevalent in Apatani Valley of Upper Subansiri district of Arunachal Pradesh (Rethy *et al.*, 2000). The farmers reared fish in paddy fields surrounded by scattered trees. The cocoon of muga or eri is most delicious food among the tribal peoples of Arunachal Pradesh. To meet this type of food requirement tribals rear silkworms under agrisilvisericulture system (Dhyani *et al.*, 1996).

## Scientific exploration of Agroforestry

A few scientifically designed agroforestry systems/ traditional agroforestry systems exist in various parts of North Eastern India. Notwithstanding this scientifically designed site specific agroforestry systems are very few in the region. Further scientific studies on existing traditional agroforestry systems in NEH region are now gaining importance.

A number of new agroforestry systems have been developed by ICAR research complex for North East Hill Region, Shillong and other research agencies (Singh, 1999). These include sericulture based agroforesty, silvipasture on waste land, agrihorticuture systems and bamboo based agroforesty, however, their potential for different agroclimatic conditions of the region is yet to be explored.

Besides ICAR, Research complex for North East Hill Region, other scientific institutions have been working on existing traditional agroforesty systems in the region like North East Regional Institute of Science and Technology (NERIST), Arunachal Pradesh (Rethy *et al.*, 2000; Sood *et al.*, *2000a*), North East Hill University (NEHU), Shillong (Dhyani, 1997; Tripathi *et al.*, 1996), North East Hill University, Mizoram Campus, Aizawal. These institutions have been studying the categorization of exiting agroforestry systems, identification of their systems units, biological and economic productivity of different existing agroforestry systems, species preferences in different agroclimatic zones, examining constraints and potentials of different agroforestry systems identified and developing strategies to improve the adoption and productivity of these systems. However, tree-crop interactions were not the component of these studies. Singh *et al.* (1997), and Singh (1998) designed and evaluated some scientific agroforestry interventions for slopy degraded lands in tropical and agroecological zones of the the region they tested agisilvicultural, silvipastoral, agrihorticultural, agrisilvihorticultural and pure silvicultural models with different forestry species, horticultural species and crop combinations to fulfill the objectives of soil conservation and economic profitability. The growth and survival of all tree species is higher in all agroforestry models compared to pure silviculture. This enhances biomass and economic yield under these agroforestry models. The main crops include maize, millets, upland paddy, cabbage, cauliflower, pea, radish, ginger and turmeric. Trees included *Anthocephalus cadamba, Melia azaderach, Leucaena leucocephala, Dalbergia sissoo, Terminalia bellerica, Gmelina arborea, Morus* spp. Horticultural component include *Emblica* spp, Kinnow, Peach, Pear, Guava, Pine apple, jamun, Papaya. The pastoral species grown were *Scyto santhes, Medicago sativa,* Guinea grass (Singh *et al.*, 1997). This study did not include social acceptability of agroforestry systems by the tribal people of the region. Sood *et al.*, (2002) designed *Anthecephalus caamba* in the system with respect to nitrogen application.

Keeping in view the above, it can be concluded that comprehensive studies to delineate existing agroforestry systems, their constraints and potentials, species preference are yet lacking in the region.

## Important species suitable for Agroforestry systems

There have been few comprehensive studies to document existing and potential agroforestry species (herbs, shrubs and trees) for different climatic zones of different agroclimatic zones of the region (Sood *et al.*, 2000). A broad list is given below (Table 3).

In the lower limits of alpine zone a few woody perennials like *Taxus baccata, Populus ciliata, Betula* spp, *Rhododendron* spp, *Berberis asiatica, Berberis wallichiana, Euryaacuminata, Vaccinium venosum*are grown in silivipastoral form. The alpine zone above 4000 m remains covered with snow for major part of the year and is considered as tree line (Kukreti, 2000).

**Table 3.** Important Species Suitable for Agroforestry in NEH region

| **Tropical Zone up to 900m Elevation** | | |
|---|---|---|
| **Herbs** | **Shrubs** | **Trees** |
| *Acorus calamus* | *Phyllostachys bambusoides* | *Erythrina stricta* |
| *Agave Species* | *Phylocanthus* spp. | *Litsea monopetala* |
| *Amaranthus* spp. | *Calamus erectus* | *Bambusa polymorpha* |
| *Ananus cosmosus* | *Calamus tenuis* | *Bambusa tulda* |
| *Costus speciousus* | *Callicarpa arborea* | *Bambusa vulgaris* |
| *Cymbopogon citratus* | *Musa* spp. | *Emblica officinalis* |
| *Dioscorea floribundus* | *Polymorphum* spp. | *Macranga denticulate* |
| *Diplazium esculentum* | *Maesa chisia* | *Castanopsis indica* |
| *Musa* spp. | *Dendrocalamus hamiltonii* | *Cephalostachyum pergracile* |
| *Phrynium pubinerve* | *Gloriosa superba* | *Dendrocalamus giganteus* |
| *Piper longum* | *Murraya koenighii* | *Elaeocarpus floribundus* |
| *Vetiveria zizanoides* | *Manihot esculenta* | *Dillenia indica* |
| | *Pseudostachyum* spp. | *Moringa oleifera* |
| | *Sarcochlamys pulcherima* | *Parkia roxburghii* |
| | *Solanum torrum* | *Sapindus mukorossi* |
| | *Zanthoxylum alatum* | *Spondias pinnata* |
| | | *Spondia axillaries* |
| | | *Bambusa arundinacea* |
| | | *Bambusa balcoa* |
| **Subtropical and temperate zone 900 – 2000 m** | | |
| **Herbs** | **Trees** | **Trees** |
| *Aconitum ferox* | *Albizia mollis* | *Populus gamblei* |
| *Coptis teeta* | *Alnus nepalensis* | *Pyrus* spp. |
| *Dichranopteris linearis* | *Erythrina indica* | *Salix daptoniana* |
| *Fagopyrum cymosum* | *Exbucklandia populnea* | *Dendrocalamus hamiltonii* |
| *Panax pseudoginseng* | *Juglans regia* | *Dendrocalamus membranaceous* |
| *Podophyllum hexandrum* | *Lindera* spp. | *Dendrocalamus strictus* |
| *Pteradium acquilinum* | *Malus* spp. | *Melocanna bambusoides* |
| **Shrubs** | *Myrica esculenta* | *Oxytenanthera albociliata* |
| *Artemesia* spp. | *Pinus* spp. | *Arundinaria maling* |
| *Zanthoxylum* spp. | *Populus ciliata* | |

Based on: Choudhary and Haridasan, 1993; Patwary *et al.,* 2000; Singh, 1993

## Future research and needs of agroforestry

Agroforestry as a sustainable land use strategy of rural development has received peripheral attention in NEH region until recently. Now this strategy of farming is attaining attention of the policy makers, planners, academicians and researchers in the region (Sundriyal *et al.,* 2002). It has been now realized that there is a vast scope to introduce new agroforestry system and to improve the old agroforestry system in the region which will help in sustainable land use, sustain and improve the livelihoods of the tribal farmers and control and improve the shifting cultivation practices (Singh *et al.,* 1999; Singh *et al.,* 2000). However, agroforesry systems are very site specific and there is need to study their constraints and potentials case-to-case basis. Notwithstanding, Sood *et al.,* (2000) has listed the common constraints of agroforestry land use in North Eastern Region:

- Lack of proper documentation of agroforesry systems, systems units, species preference in existing agroforestry systems in different agroclimatic zones of the region.
- Lack of knowledge of constraints and potentials of existing agrorestry systems with respect to their biological and economic productivity.
- Farming practices that are based on the experience of the tribals and lack scientific inputs.
- Lack of sufficient technical know-how of agroforestry amongst the planners and executors of agroforestry programmes.
- Community ownership of land acting as disincentives for settled agriculture and agroforestry.
- Lack of systematic wide-scale research endeavors in agroforestry in the state.
- Lack of extension facilities for agroforestry.
- Lack of systematic endeavor to design scientifically viable agroforestry systems for different zones of the region.

To improve the existing agroforestry systems with respect to their biological and economic productivity and sustainability the following are broad areas which merit immediate attention (built on Sood *et al.,* 2000):

i) Survey, Diagnosis and Design

- Survey and evaluation of existing land use problems and diagnosing agroforestry alternatives to solve the land use problem.
- Diagnosis of land area suitable for agroforestry.
- Design of site specific agroforestry system having high productivity, suitability and adaptability to cultural aspects of tribal.

ii) Identificaton and evaluation of plant species for agroforestry

- Preparation of inventory of existing agroforestry plant species (agriculture, horticulture, forestry and grass species) for different climatic zone of NEH region.
- Evaluation of potential tree species for their suitability for different agro climatic condition.
- Breeding and genetic improvement of the plant species.

iii) Agroforestry management

- Studies of trees crop interactions to find suitable tree-crop combinations for different climatic zones of the region.
- Identification of nitrogen fixing trees for different agroforestry systems in different agroclimatic zones.
- Evaluation of effect of nitrogen fixing tree species on yield of crops.
- Nutrient dynamics, water regimes and soil fertility status in various tree-crop combination under various agroforestry systems.
- Effect of adoption of different agroforestry systems on plant biodiversity.
- Nodulation ecology and microbial population studies on different potential agroforestry systems.

iv) Socio-economic, cultural aspects and extension

- Knowledge of socio-economic status of the tribals, their attitude and perceptions towards trees and tree cultivation.
- Socio economic and cultural constraints for practicing agroforestry.
- Traditional belief, ethos, and taboos of different tribal communities associated with cultivation of trees and their conservation should be evaluated.

- Availability of various resources with the tribals for the practicing of agroforestry.
- Cost-benefit estimation of various agroforestry systems and their environmental impacts.
- Education and training of forestry planners and managers in biophysical, socio-economic and extension aspects of agroforestry.

In NEH region, three areas have become important for considering the domain of agroforestry (Pant and Pandey, 2004; Singh, 1999)

- Development of alternative to shifting cultivation.
- Improvement in traditional agroforestry systems.
- Rehabilitation of degraded hill slopes.

According to Pathak and Singh (2000), the major components of the priority agroforestry systems are agrisilviculture (*jhum* land), and agrihorticulture. Sundriyal (2001) suggested contour-hedgegrow agroforestry with multipurpose fast growing trees for hilly regions of North East. Sood (2003) has argued that agroforestry uptake should not be looked upon in isolation of livelihood strategies of the farmers. Trees play an important role in the livelihoods of the farmers: secure provision of food and essential subsistence goods, provide cash for purchase of outside goods and services, savings (resource accumulated to meet future planned needs and emergencies), provide social security (secure access to subsistence goods and productive resources), and reduce the risks to livelihood. Tree management is an integral part of rural livelihood strategies. The issues facing growing of trees on farmland cannot be seen in isolation of farmers overall livelihood strategies (Malla, 2000). The role of the trees in the livelihoods of the farmers is dependent on the biophysical status of the farms owned; tree resource use, socio-economic, psychological and attitudinal aspects of the farmers (Sood, 2003; Sood and Mitchell, 2004). Thus to design biophysically suitable and socially acceptable agroforestry systems all the biophysical, socio-economic, perceptional and attitudinal aspects of the farmers should be taken into account. In NEH region these aspects need careful attention due to frequent variation in biophysical as well as socio-economic aspects in different areas and between different tribes. Further agroforestry has long gestation period and thus requires long period of experimentation and study. Traditional agroforestry systems are products of centuries of trial and error by the farmers that have adapted them to long ecological, socio-cultural and economic factors (Christanty and Iskandar, 1985). Thus knowledge of existing agroforestry can act as useful resource to reduce the long period of experimentation and study to design new agroforestry systems, particularly in North East hill region of India where traditional agroforestry exists in various forms.

**Table 4.** Comparison of *Jhum* cultivation and agroforestry

| Factor | *Jhum* cultivation | Agroforestry |
|---|---|---|
| **Ecological factor** | | |
| Fragility | High | Low |
| Homeostasis | Internal | External control |
| Biodiversity | High | Low (restricted) |
| Leaf area | Less | More** |
| Carbon sequestration | Low | High** |
| Regeneration | Natural | Artificial/managed |
| Dispersal | From surrounding natural vegetation | Man made |
| Potential nutrient loss | High | Low |
| Pathogen/ disease attack | Less risk | More risk |
| Connectance (biotic interdependence) | Low | High |
| Energy source | Low | High |
| Nutrient source | High* | High* |
| Irrigation | Rainfed | Moderate |
| Soil fertility management | Poorly managed | Well-managed |
| Tillage | No tillage | Minimum tillage |
| Carrying capacity of land | Low | High |
| Ecological status | Complex | Complex |
| **Economic factor** | | |
| Labour | Intensive | Systematic |
| Inorganic fertilizer | Not used | Used sometimes |
| Monetary input-output | Low | High |
| Production/yield | Low | High |
| Land required | Large (extensive cultivation) | Relatively less (Semi-intensive cultivation) |
| **Socio-cultural factor** | | |
| Approach to cultivation | Slashing and burning followed by cropping | Trees grown with crops |
| Cropping pattern | One rotation | More than one rotation |
| Harvesting pattern | Multiple harvest | Multiple harvest |

| | | |
|---|---|---|
| Integration of animal component | Poorly integrated | Well-integrated |
| Cultural value | Traditional value | Intervention |
| Local adaptability | More | Less |
| Sustainability | Diversity conserved | Production sustained |

* In jhum cultivation, the initial slashing and burning becomes a huge one-time source of soil nutrients, while in agroforestry systems the litter and fine roots of the tree component continually add plant nutrients into the soil. The latter could be further improved by selecting biological $N_2$-fixing tree species to promote natural nutrient cycling.** Due to presence of woody perennials.

Based on: Arunachalam *et al.* (2002)

## Conclusion

Agroforestry as sustainable land use and alternative to *jhum* cultivation is now gaining importance in NEH region. Different types of traditional agroforestry systems and practices are prevalent in the region.

Documentation of different existing agroforestry systems, system units and their biophysical and socio-economic evaluation would be an important strategy to improve the existing and designing new agroforestry systems. This would also reduce the long experimentation period required to develop the new systems. There is also a need to study the role of the trees in the livelihoods of different tribes of North East hill region. Understanding of biophysical and socio-economic aspects of the farms and tribes would help in designing biophysically suitable and socially acceptable agroforestry systems for NEH region. Training of forestry planners and executors coupled with right extension approach would boost the wider acceptability of agroforestry for sutainable agriculture in North East Himalayan region.

## References

Arunachalam, A., Khan, M.L. and Arunachalam, K. (2002) Balancing traditional jhum cultivation with mordern agroforestry in eastern Himalaya-A biodiversity hotspot. Curr. Sci. 83(2): 117-118.

Borthakur, D.N. (1992). Agriculture in Northeastern region with special reference to hill agriculture. Beecee Prakashan, Guwahati

Census (2011). Primary census abstracts, Registrar General of India, Ministry of Home Affairs, Government of India, Available at: http://www.censusindia.gov.in, Accessed on 20 July 2015.

Centre for Monitoring Indian Economy (CMIE). (2007). Economic Intelligence Service: Agriculture. Mumbai: CMIE.

Choudhury, S.S. and Haridasan, K. (1993). Water conservation via-a-vis forestry in Arunachal Pradesh. Arunachal Pradesh Forest News. 11(1): 5-17.

Christanty, L. and Iskandar, J. (1985). Development of decision making and management in traditional agroforestry: Examples in West Java. In: Rao, Y.S., Vergara, N.T. and Lovelace, G.W. (Eds) *Community forestry: socio-economic aspects.* Regional Office of Asia and Pacific (RAPA), FAO, Bangkok. pp. 199-214.

Das, A.P. and Pradhan, S.K. (1990). Shifting cultivation in North East India.

Dhyani, S.K.(1997). Analysis of the effect of agroforestry systems on productivity and soil characteristics. *Ph.D. Thesis,* NEHU, Shillong.

Dhyani, S.K., Chauhan, D.S., Kumar, D., Kushwaha, V. and Lepcha, S.T. (1996). Sericulture based agroforestry system for hilly area of north-east India. Agroforestry Systems. 34:247-258.

FSI. (2011). State of Forest Report 2011.

FSI. (2013). State of Forest Report 2013.

Goswami, P.J. (2006). Changing Agricultural Scenario in North East India in the Era of Globalisation. In Deb Bimal J. and Ray B. Datta. (Eds) *Changing Agricultural Scenario in North East India.* Concept Publishing Company. New Delhi, India.

Khosla, P.K. and Khurana, D.K. (1996). Agroforestry for sustainable development in the Himalayas: Potential and scope. In: GOI (Ed) *Farming systems in the Himalayas.* Department of Agriculture and Cooperation, Ministry of Agriculture, Government of India, New Delhi. pp. 85-110.

Kukreti, M.C. (2000). Vegetation and its status in Himalayan region. Indian Forester. 126: 175-179.

Lundgren, B.O. and Nair, P.K.R. (1983). Agroforestry for soil conservation. *In*: Proc, 2nd International Conference on soil erosion and soil conservation, Hawaii.

Majumder, M. and Arunachalam, A. (2004). Agricultural crop diversity and management of jhum cultivation systems in Northeast India. *In:* Pant, RM. And Pandey, A. (Eds) Proceedings of seminar on Environment, Agriculture and sustainable life support systems: Perspectives of NER, Organised by North Eastern Regional Institute of Science and Technology, Nirjuli on November 6:.45-62.

Malla, Y.B. (2000). Farmer's tree management strategies in a changing rural economy and factors influencing decisions on tree growing in Nepal. International Tree Crop Journal 10: 247-266.

Moral, Dipankar (1997). North-East India as a Linguistic Area. Mon-Khmer Studies. 27: 43–53.

Myers, N. (1999). The next green revolution: its environmental underpinnings. Current Science. (http://www.iisc.ernet.in/~currsci/feb25/articles16.htm)

North Eastern Council Secretariat. (2015). Basic Statistics of North Eastern Region, Directorate of Printing & Stationary, Government of Meghalaya, Shillong-793001

Pant, R.M. and Pandey, A. (2004). *Proceedings of seminar on Environment, Agriculture and sustainable life support systems: Perspectives of NER*, Organised by North Eastern Regional Institute of Science and Technology, Nirjuli on November 6, 2004.

Pathak, P.S. and Singh, K.A. (2000). Sustainable farming systems in the hilly regions of India with special reference to North Eastern Region, *In*: Kohli, R.K., Singh, H.P., Dhir, K.K., Batish, D.R. and Khurana, D.K. (Eds). Man and Forests. ISTS, Solan and Punjab University, Chandigarh. pp. 313-346.

Patwary, B. C., Kumar, S.R., Mani, P. and Panigrahy, N. (2000). (Eds) Proceedings of Brain Storming Session on Water Resources Problems of North Eastern Region Organixed by North Eastern Regional Centre, NIH, Roorkee.

Price, M.F. (1998). Mountains globally important ecosystems. *Unasylva* 49(195). (http://www.fao.org/docrep/w9300e/w9300e03.htm).

Rahman, S.A., Rahman, M.F. and Sundarland, T. (2011). Causes and consequences of shifting cultivation and its alternative in hill tracts of eastern Bangladesh. Agroforest. Syst. 84(2): 141-155.

Ramakrishnan, P.S. (1992). Shifting agriculture and sustainable development. Man and Biosphere Series: V.10. The Parthenon Publishing Group, Paris. 424 pp.

Rao Nageswar, P.P. and Bhattacharyya, K.C. (2005). Application of geographic informationsystems and remote sensing in assessment and management of natural resources. *In*.Agroforestry in North-Eastern India: Opportunities and Challenges. (*Eds*) B.P. Bhatt and K.M. Bujarbaruah. ICAR Research Complex for NEH Region, Umiam, Meghalaya. pp. 15-31.

Reserve Bank of India (RBI). (2005). Handbook of Statistics on the Indian Economy, 2004– 05, Mumbai: RBI.

Rethy, P., Sood, K.K., Singh, B. Chakrabarty, A. and Kago, D. (2000). Unique Agri Piscicultural practice of Apatani tribe in Zero valley of Arunachal Pradesh. Journal of Ecobiology, 12(3): 185-188.

Rethy, P., Dabral, P.P., Singh, B. and Sood, K.K. (2003) (Eds) Forest Conservation and Management. International Book Distributors, Dehradun, Uttarakhand, India.

Roy, P.S., Singh, S., Aggrawal, S., Gupta, S., Joshi, P.K., Aung, S., Singh, I.J. and Shukla,Y. (2002). TRESS II-Tropical forest assessment in India and north Myanmar. In: Forestcover assessment in Asia: Proceedings of the international workshop on Asia forest coverassessment and conservation issues. Feb 12-14, Dehradun, India, pp. 61-100.

Saikia, S. (2014). Shifting cultivation and environmental problems in Northeast India. Journal of Economics and Bussiness Springs. 1.1: 52-59

Satapathy, K.K. and Bujarbaruah, K.M. (2005). Slash and burn agriculture: Its practice, effects and problem of development. *In* Agroforestry in North-Eastern India: Opportunities and Challenges. (*Eds*) B.P.Bhatt and K.M. Bujarbaruah. ICAR Research Complex for NEH Region, Umiam, Meghalaya. pp. 1-14.

Seitinthang, L. (2014). Cropping pattern of Northeast India: An appraisal. American Research Thoughts 1(1): 488-498.

Singh, B., (1998). Final technical report of project financed under Himalayan Eco-development research programmes (IERP), by GB Pant Institute of Himalayan Environment and Development, Almora.

Singh, B., Rethy, P. and Arunachalam A.S. (1994). Prospects of agroforestry at Arunachal Pradesh. Indian Journal of Hill Farming 7(2): 119-123.

Singh, B., Rethy, P. and Singh, A.B (1997). Agroforestry intervention for sustainable productivity in Arunachal Pradesh. Journal of Nature Conservation 9(1): 97-106

Singh, B., Rethy, P. and Sood, K.K. (2000). Role of agroforestry systems in the reclamation of wasteland in Arunachal Pradesh. *In*: Arunachalam, A. and Khan, M.L. (Eds) *Sustainable Magement of forests*. International Book Distributors, Dehra Dun, Uttranchal, India. pp. 387-393.

Singh, B., Solanki, G.S., Sood, K.K and Rethy, P. (1999). Agro-horticultural practices and suitable management of degraded lands in North East India. Paper Presented in National Seminar on Natural Disaster Management, NERIST, Nirjuli, October 13, 1999.

Singh, K.A. (1999). Resource management and productivity enhancement through agroforestry in the Eastern hilly agroecosystem in India. Indian Journal of Agroforestry 1(1): 63-72.

Singh, N. B. (1993). Present and future of agroforestry in Arunachal Pradesh. J. Eco. Tax. Bot. 17: 519-533.

Sonowal, D.K, Bhatt, B.K, Satapathy, K.K, and Baishya, B. (2006). Agroforestry in North East India: Opportunities and Challenges

Sood, K.K. (2003). Factors affecting tree growing in traditional agroforestry systems of Western Himalaya, India. Ph.D. Thesis (Unpublished). University of Aberdeen, Aberdeen, U.K.

Sood, K.K., Singh, B. and Rethy, P. (2002a). Growth performance of *Anthocephalus cadamba* plantation under nitrogen application under silvopastoral system. Journal of Nature Conservation 14(1):67-74.

Sood, K.K., Singh, B. and Rethy, P. (2002b). Shifting cultivation in Arunachal Pradesh - Causes and control. *In*: Sundriyal, R.C., Singh, T. and Sinha, G.N. (Eds.). Arunachal Pradesh: Environmental Planning and Sustainable Development. HIMAVIKAS Occasional Publication, G.B. Pant Institute of Himalayan Environment and Development, Almora., pp.161-165.

Sood, K.K., Singh, B. and Rethy, P., (2000). Agroforestry approaches for sustainable development in Arunachal Pradesh in next millennium. In: Kohli, R.K., Singh, H.P., Dhir, K.K., Batish, D.R. and Khurana, D.K. (Eds). Man and Forests. ISTS, Solan and Punjab University, Chandigarh, pp. 461-470.

Sood, K.K. and Mitchell, C.P. (2004). Do socio-physchological factors matter in agroforestry planning: Lessons from smallholder traditional agroforestry systems? Small-scale Forest Economics, Management and Policy, 3(2):239-255.

Sundriyal, S.C. (2001). Hill agro-ecosystem and sustainable development in Arunachal Pradesh. Arunachal Forest News 19:186-206.

Sundriyal, R.C., Singh, T. and Sinha, G.N. (2002). (Eds) Arunachal Pradesh: Environmental Planning and Sustainable Development. HIMAVIKAS Occasional Publication, G.B. Pant Institute of Himalayan Environment and Development, Almora.

Tripathi, R.S., Tiwari, B.K. and Pandey, H.N. (1996). State of Environment of Meghalaya. North Eastern Hill University, Shillong.

Tucker, R.P. (1983). The British colonial system and forests of Western Himalayas 1815-1914. In: Tucker, R.P. and Richards, J.F. (Eds) Global deforestation and the nineteenth century world economy. Duke Press Policy Studies, Durham, N.C. pp. 146-166.

❑❑❑

# 20

# Non-conventional Feed Resources for Sustainable Livestock Farming in Changing Climate Scenario in Arunachal Pradesh

**R.A. Alone, Doni Jini, Nirmal, M. Kanwat, Anup Chandra Jitendra Kumar, M.S. Baruah and R. Bhagawati**

## Introduction

In livestock production system, it is traditional to feed cereals, cakes and meals in the ration. With increase in human population and urbanization the area under fodder or cereal production cannot be increased. There is a huge gap between demand and supply of all kinds of feeds and fodders. Also, rise in global temperature is threatening food and fodder security as a whole. Climate change affects natural resources (such as water sources, land and pastures), biodiversity and livestock health. This has a direct effect on livestock production and livestock systems and as such it is not desirable for the livelihood of the people especially those residing in hilly region. The situation is acute in developing countries where annual feed deficits and increasing animal populations are major problems.

In India, there is already a shortage of 25,159 and 117 million tonnes of concentrates, green forages and crop residues, constituting respectively a shortage of 32,20 and 25 per cent of the requirement (Ravi Kiran *et al.*, 2012). In North eastern region poor quality and shortage of proper livestock feed is a great impediment in enhancing livestock productivity. Only a few numbers of feed mills are in operation in the region, which are unable to produce quality animal feed and meet the demand for feed.

In West Siang district of Arunachal Pradesh livestock husbandry is an integral component of prevailing farming systems. This region accounts 6.9 per cent cattle population, 9.3 per cent mithun, 6.91 per cent goat,

11.6 per cent pigs and 8.4 per cent poultry population of the state. But due to remoteness, inaccessibility and unaffordable cost of concentrate feed, the rural hill farmers of this region has evolved low input production system which mainly dependent on local vegetation's, crop residues. Hence, emphasis need to be given to non-conventional feed resources for filling the gap for need and supply.

Non-conventional feed resources (NCFR) generally refer to all those feeds that have not been traditionally used for feeding livestock and are not commercially used in the production of livestock feeds. As per (FAO, 1985) some of the distinctive features for non - conventional feed are as follows:

- They are the end products of production processes and consumption that have not been used, recycled or salvaged.
- They are mostly of organic origin and can be obtained either in a solid, slurry or liquid form.
- The economic value of these non-conventional feed resources is usually less than the cost of their collection and transformation for use and consequently, they are discharged as wastes.
- Feed crops which generate valuable NCFR are usually excellent sources of fermentable nutrient molecules such as cassava and sweet potato and this is an advantage to livestock especially ruminants due to their ability to utilize inorganic nitrogen and non-protein nitrogenous sources.
- The majority of feeds of crop origin are bulky poor-quality cellulosic roughages with high crude fiber and low nitrogenous content that are suitable for feeding mostly ruminants.
- Some of these feeds contain anti-nutritional components which have deleterious effects on the animals and not enough is known about the nature of the activity of these components and ways of alleviating their effects.
- Non-conventional feed resources have considerable potential as feed materials and some of their value can be increased if there are economically viable technological means for converting them into some useable products.
- Substantial information is required on chemical composition, nutritive value, the presence of anti-nutritional components and value in feeding systems.

## Generation of NCFR

Most non-conventional resources are usually regarded as waste. When it is recycled or reprocessed and converted into valuable products for human and livestock they become new feed materials of importance. In addition, they can be used to supplement the existing limited feed resources.

## Limitation of NCFR

They are numerous factors may account for the limited use of several crop residues, AIBPs and are listed below :

- Low nutritive value.
- Short period of utilization (seasonal availability).
- High moisture content (e.g. citrus pulp, tomato pulp, olive cake, etc.).
- High cost of handling and transportation from the production site (tomato pulp, olive cake, grape marc, etc.) to the farm.
- Farmers are not aware of the nutritive value of some feed sources and the way for their efficient integration in livestock feeding.
- Competition with alternative users (fuel, compost, etc.).
- Presence of anti-nutritional factors (phenolic compounds mainly tannins, saponins, etc.).
- Dehydration may cause loss of protein value (Maillard reactions).
- Lipid peroxidation (rancidity of high fat products, e.g. olive cake).
- Mould growth (aflatoxins) may cause toxicity.

Some of the commonly available non-conventional feed resources available in the region are as follows

## Maize

The leftover of maize consist of (a) the stalk, (b) the husks, skins and trimmings and (c) the cobs. The grain is recovered from the cob by de-husking the ear. The stalk which is usually referred to as stover (Devendra,1985) is obtained during harvesting of maize. This portion along with the husk, skin and trimmings are usually fed to ruminants. The harder portions are used for making silage or compost.

## Rice

Rice is one of the major components in the region. The rice when sent for milling large amounts of byproducts like rice straw and rice husks (15-17%), broken rice (4-5%) and rice bran (6-10%) are produced.

## Rice straw

It is used for feeding of ruminants (Doyle, Devendra and Pearce 1986). Rice straws contain only 3 to 5 percent crude protein. Animals supplemented with only straw diet will usually not gain any weight and very often will actually lose weight. Hence, enrichment of straw with nitrogen/protein and energy is necessary. For good growth on straw diets, a level of 8 to 10 percent protein is needed for young stock.

## Rice husk

As a roughage source only small levels, of about 5% rice husk in the total diets, can be included to give the best results. Support for this finding is found in the results of practical feeding trials of Tillman *et al.,* (1969) who demonstrated that 3% raw husk gave the best dressing percentage in bullocks, and in digestibility studies with steers where a 5% level of husk gave a higher digestible energy value than the 20% level (White, Reynolds and Hembry, 1971).

## Rice bran

Rice bran results from the physical abrasion and separation of the hull from rice grain, which is used for human consumption. Rice bran is produced when the hull and fragments of the hull are blended with some of the germ. When rice bran is supplemented at 0.4 percent of body weight, it has approximately the same energy value as corn grain fed at the same level (Foster *et al,* 1993). Fresh rice bran contains sufficient oil which is easily available and palatable source of energy in the pig diets. But with higher addition in diet, the digestibility of energy and protein goes on decreasing with the increasing level of fat and fibre. However, the metabolizability of the energy and protein was better in pigs fed Rice bran in the diet.

## Broken rice

It comprises mainly of germs, chipped and broken kernels. Because of its low fibre content and high energy value, its valuable energy gets, especially for substituting other energy feeds in diets for pigs and poultry. Several studies have shown that broken rice can easily substitute maize without any loss in performance. Broken rice has been used up to levels of 30, 40, and 50%, compared to a maize diet (50% inclusion) in growing pigs and there was no significant between the two groups of experimental animals (Mellish, *et al.*,1973). Khajarern, *et al.*, (1979) in Thailand used 50.7% level inclusion of broken rice in broiler diets and did not observe any loss in the performance of the birds.

## Rice brew waste

This is the material that remains after grains have been fermented during the making process. These materials can be fed as dried brewer's grains. Brewers' dried grain (BDG) when produced is sun dried to reduce the moisture content up to a level of 8-10 %. In swine less than 5per cent should be included in the rations as the low energy level, low lysine and tryptophan values may limits its use in pig ration. In poultry ration with enzyme feed supplementation at a certain level should be provided. BDG with Kemzyme-HF @ 0.75 g/kg diet was incorporated in broiler ration at a level of 5 % for economic production (Swain *et al.*, 2005) .In Arunachal, popular form of pig feeding is distillery waste from rice mixed with other feeds, such as rice bran and broken rice mainly for feeding boar for fattening purpose. The following mixing ratio is commonly used in combination with distillery waste i.e. Rice bran (2 kg)+ broken rice (1 kg)+ distillers' residues till gravy (Jini *et al.*, 2009).

## Damaged Soybean grain

When properly roasted it can be used with minimum number of feed ingredients for pigs and poultry. It contains around 43-48% proteins. Simple roasting is sufficient to detoxify its anti-nitritional effect. The roasted grains can be fed in poultry ration upto 20 per cent.

## Rice bean grain

The protein content varies from 14 to25 per cent, with high amounts of vitamins, calcium and iron. When properly treated, it can be added in the diet up to 30-40% on DM basis without any adverse effect on growth performance.

## Ragi

Cost of feed per kg meat production was lowest in chicks fed diet with 50 % maize replaced by ragi. It may be concluded that ragi could replace up to 50 % maize in the diet of laying Gramapriya chicks. Results suggested that unground bajra and ragi could replace maize completely in the diet of laying hens without affecting the egg production, egg weight, feed efficiency and other quality parameters in addition to production of stronger shell (Swain *et.al,* 2009).

## Sugarcane

Sugarcane and sugarcane residues may be considered as an alimentary alternative for feeding ruminants. Cane tops are palatable and cattle can be maintained entirely on cane tops with little supplement of protein either as concentrates mixture or leguminous feeds. Sugarcane leaves have high crude fiber content (40-42% DM) and the leaves are also rich in soluble carbohydrates. They are potential feed resource for ruminant in the dry season (Pate, 1981).The residue after juice extraction from the sugarcane plant referred as bagasse constitutes approximately 15-20% of sugarcane tops with a moisture content of about 50% renders it suitable feed for ruminants especially dairy cows (Randal, *et al.,* 1967).

## Banana waste

Banana wastes are by-products of banana harvesting and packing for human consumption. It include small-sized, damaged bananas, banana peels, leaves, young stalks and pseudo stems, which can be fed to livestock. Fresh plantain and banana fruits may be ensiled with molasses, grass, legumes, rice bran etc to feed ruminants. Whole, fresh banana leaves, stalks and pseudo stems are chopped and directly fed to pigs either fresh or sun-dried. Its inclusion in ration depress the utilization of the nitrogen. Hence, it is necessary to provide both energy and nitrogen supplementation during the preparation of ration. Banana foliage and wheat straw (75:25) ensiled with molasses and urea (Baloch *et al.,* 1988) could replace 50 percent of green maize in the rations of lactating cows/ buffaloes without altering milk production. Dried banana pseudostems have been fed to goats and sheep at levels of 20-50% in diets with no adverse effects, but daily weight gains were depressed (Poyyamozhi and Kadirvel, 1986). Banana peels are widely used by small, marginal and landless farmers as complementary feeds for ruminants in the tropics. Their nutritive value is similar to that of cassava or citrus peels. In grass-fed Zebus, the addition of 15-30% banana peels in the diet increased weight gain significantly without causing health problems or affecting palatability (Hernan Botero, Enrique Toro and Rios, 2000). In goats, dry ripe plantain peels can replace up to 100 percent maize without adversely affecting growth performance, and were found to be an economical source of carbohydrates (Aregheore, 1998).

## Pineapple wastes

Pineapple by-products consist basically of the residual pulp, peels, stem and leaves. During production of pineapple, rough handling of fruits and exposure to adverse environmental conditions during transportation and storage can cause up to 55% of product waste (Nunes *et al.*,2009).Dried and ensiled pineapple waste can be used as supplemental roughage and could replace 50% roughage in the total mixed ration for dairy cattle (Sruamisri,2007). On feeding twenty four cross bred local goats for 80 days was found that dehydrated pineapple by-products would increase the digestibility with increase in weight of the animals (Costa *et al.*, 2007).

## Citrus waste

The residue left after extraction of the juice is called citrus pulp (50-70 percent of the fruit by weight). It contains 60-65 percent peel, 30-35 percent internal tissues and up to 10 percent seeds (Crawshaw, 2004). Dried citrus pulp can replace 20 percent concentrate in dairy cattle (Assis *et al.*, 2004) and up to 30 per cent in lactating ewes (Fegeros *et al.*, 1995) without adversely affecting DM intake, rumen metabolites, digestibility, milk yield or milk protein and fat contents. Citrus pulp silage can be included at 5-10 percent in the diet of growing and finishing pigs. Weight gain and feed intake tend to increase improved carcass quality by increasing meat content and carcass length, and reducing back fat (Cerisuelo *et al.*, 2010). In poultry with limits up to 5-10 percent did not affect feed intake, egg production and egg weight in laying hens (Yang and Chung, 1985).

## Tapioca

It can be used for feeding of pigs, cattle goat and poultry. The roots are suitable for feeding of young animals and fattening pigs. Properly dried whole tuberous roots can replace maize in non-ruminant rations if the HCN does not exceed 100 ppm. Tapioca root meal can complete substitute for maize in diets for dairy cows producing approximately 12 kg of milk per day. Cassava could totally replace oats in the concentrate feed of cattle (Mathur *et al.*, 1969). Maize can be substituted by cassava root meal upto 15 % in broiler rations (Kinabo, 1977). In crossbred pigs, Das *et al.* (2012) reported that, tapioca root meal can be included up to a level of 50 percent in the diet of pigs in Nagaland.

## Taro

Taro leaves has great potential as animal feed in particular for pigs. But due to their oxalate content the leaves should be soaked, washed or cooked and dried or ensiled before feeding them to livestock. An trial of inclusion of boiled taro cocoyam in the diets of weaned pigs carried out by Agwunobi *et al.* (2002) revealed that there was no significant difference in feed intake, weight gain and feed efficiency upto 50% replacement of maize. Chittavong Malavanh *et al.* (2007) who successfully fed ensiled taro leaves to pigs during reproduction and lactation. *Colocasia esculenta* meal effectively replace 25% maize in the diets of broiler finishers for raw sundried and 50% for boiled sundried (Mohammed *et al.*, 2009).

## Sweet Potato

Sweet potato is cut into slices or mixed with vines and household waste. Sweet potato is an ideal livestock feed, as the roots can provide a source of energy and the leaves can provide protein .Study carried out in Nigeria recommended 27 and 30 % levels of sweet potato in the starter and finisher diets (Agwunobi, 1999). Boiled sweet potato tubers could be fed to the level of 40% of total dry matter intake to the weaned piglet for better growth rate and nutrient utilization whereas up to 60% of total dry matter intake to grower pig along with good quality protein supplement for better growth performance. The sweet potato vines can serve as a nutritive and palatable feed for cattle. The unmarketable and poorly developed tubers can also be utilized in animal feed.

## Browse Foliage

The Arunachal Pradesh is one of the biodiversity 'hot spot' in the world. The forests in mid-hills are very rich in flora and fauna diversity. The locally available fodder trees, shrubs and herbs have high potential values as sources of feed for domestic livestock and wildlife. These can be successfully incorporated into production system to provide additional feed resources for use in mixed diets of livestock. Numerous tree and shrub species have been investigated.

*Gmelina arborea*: This is a deciduous tree of medium size grows up to 40 meters in height and has diameter of 60-100cm (Jenson, 1995). It is fast growing tree, specimens shows about 4 rings per inch of radius. The leaves are simple and are more or less heart shaped and are usually 10-25cm long and 5-18cm in width. *Gmelina arborea* is commonly found in forests of mid-hills of Arunachal Pradesh. *Gmelina* is found in subtropical and tropical climates except in very hot and temperate climate. It is grown up to 1000 m altitude. It can only tolerate drought for short period. It is moderately frost hardy. Severe frost may kill the shoots, but they recover and sprout again from the base. The tree grows well in climates with mean annual temperatures of 21-28$^{0}$C (Jenson, 1995) and grows best in deep well drained base rich soils with a pH between 5.0 and 8.0. *Gmelina* is suitable to grow as a shade tree. It can easily be planted along boundaries of the field crops. *Gmelina* can be grown among agricultural crops. In this system of cropping, *Gmelina* is benefited due to watering and manuring done to agricultural crops. Rotations of 5-8 years reported to be commonly followed. The recent studies by Amata and Lebari (2011) on the nutrient profile of the fresh leaves reveal appreciable levels of crude protein (14% DM), crude fiber (6.7% DM), ash content (1.3% DM) and ether extract (12.7% DM). Metabolizable energy sources for livestock values were found to be appreciable (1368 Kcal $Kg^{-1}$) an indication of its suitability as an energy sources for livestock diets. The study revealed appreciable levels of essential amino acids and this indicates that the leaves could be a good source of protein. The recent studies by Amata and Iwelu (2012) show the potentials of the fresh fruit pulps of *Gmelina arborea* as non-conventional feeding material.

*Moringa oleifera*: Fugile (2001) has described *Moringa* as a miracle tree and has been shown to have numerous uses, amongst which are the coagulant properties of the seeds. Studies by Foidi *et al.* (2001) have shown that it can be used as a protein meal in livestock diet as it contains approximately 60% crude protein. These studies have also shown that the leaves of the plant are edible and highly nutritious and provide approximately 25% crude protein. Total digestibility of these leaves is high. The leaves are free of anti-nutritive factors such as tannins, phenols and saponins. These studies have also revealed that the iron content of the leaves is up to 582 mg/kg DM, The β-carotene is up to 400mg/kg and vitamin C content is as high as 9.2 g/kg.

*Trema orientalis*: It is an evergreen shrub or tree up to 18 m in height. A short basally swollen bole, heavy branching and rounded to spreading crown. It has an extensive root system that enables it to survive long periods of drought. Leaves simple, alternate, stipulate, along drooping branches, to 14 cm long, papery, rough to the touch and dull above, short grey hairs below, the edge finely toothed all round, blade unequal sided. *T. orientalis* is found in the lowland humid tropics. The studies on the nutrient profile of the leaves by Heli *et al.* (2008) show that it contains DM (38.30%), CP (10.60%), EE (1.39%), CF (25.6%), NFE (56.20%) and TA (6.2%) revealing its potential as non-conventional feed source.

*Terminalia catappa*: The tree grows to 35 m (115 ft) tall, with an upright, symmetrical crown and horizontal branches. *Terminalia catappa* has corky, light fruit that are dispersed by water. The seed within the fruit is edible when fully ripe, taste almost like almond. This tree belongs to the family Combretaceae. Some of its common names include tropical almond, Indian almond, Jangli badam, Java almond. It has nutty fruits that taste like commercially grown almonds. The leaves contain several flavonoides (Such as kaempferol or quercetin), several tannins (Such as punicalin, punicalagin ortercatin), saponines and phytosterols. Due to this chemical richness, the leaves (and the bark) are used in different medicines for various purposes. *Terminalia catappa* requires bright sunlight, moist and well drained soil. It is salt and draught tolerant and can be planted in frost free areas. The studies on the nutrient profile of the leaves by Amata and Lebari (2010) and that of the seeds by Amata and Nwagu (2012) reveal the potentials of this plant as a non-conventional feed source for livestock feeding in the state.

*Bauhinia purpurea*: It is a small to medium-sized deciduous fast-growing shrub or tree with a round, symmetrical, moderate dense crown to 10 m tall, young branches becoming glabrous or nearly so (glabrescent). It belongs to the family Leguminosae. Leaves are simple, alternate, base rounded to shallow-cordate, up to 12 cm x 12 cm, deeply 2-lobed at apex up to 1/3-1/2, ca. 7-12 cm long, and equally wide, margin entire and the surfaces smooth and glabrous. It occurs at lower elevations especially frequent along the valleys in its native habitat. It demands plenty of light and requires good drainage. Severe frost kills the leaves of seedlings and saplings, but they recover during summer. The species is frost-hardy but least drought-hardy compared to other species of Bauhinia. The studies on the nutrient profile of the leaves by Heli *et al.* (2008) indicate that it contains DM (40.4%), CP (14.9%), EE (4.20%), CF (17.4%), NFE (58.3%) and TA (5.2%).

*Ficus hirta*: It is an evergreen shrub or a small tree with a much-branched crown; it can grow up to 10 metres tall. The plant is sometimes harvested from the wild for local use as a food and medicine. It is found in forests and forest margins at low elevations. In secondary jungle, waysides, waste-places, and by the edge of the forest, from the lowland ascending to 2,000 metres. The studies on the nutrient profile of the leaves by Heli *et al.* (2008) found that it contains DM (29.7%), CP (19.9%), EE (2.3%), CF (29.8%), NFE (40.20%) and TA (7.8%) revealing its potential as non-conventional feed source.

Bamboo: Bamboo is a member of the grass family 'Poaceae'. Two important characters which make majority of bamboo distinct from other grasses are: (i) Woody perennial habit and (ii) peculiar flowering and seeding behavior. It is a fast growing, wide spread, renewable, versatile, low-or-no cost, environment enhancing resource with potential to improve livelihood security in the years to come, in both rural and urban areas. Arunachal Pradesh occupies an important position among the bamboo bearing states of India. Bamboo forms a major constituent of the forest vegetation of Arunachal Pradesh. Tropical, subtropical and temperate species are found well distributed in the State. In Arunachal Pradesh, which has about 46 bamboo species, the bamboo flora is seen up to an elevation of 2000 m or even more. The studies on the nutrient profile of the leaves of various bamboo species by Singh K.A. (1999) found that the CP content in the bamboo leaves of various plant species varied from 9 to 19 percent and the CF content varied in the range of 18 to 34 percent. The leaves are also found rich in mineral content. This makes the bamboo as a potential non-conventional feed resource for sustainable livestock farming.

## Conclusion

In Arunachal Pradesh, a distinctive gap exists between the requirements and supplies of nutrients for livestock. It is desirable that adequate feed resources should be built up as one of the mitigation strategy in response to changing climate scenario. The state has considerable amounts of crop residues such as straws and plantation wastes. In addition, majority of the geographical area of the state is under forest vegetation. However, the lack of knowledge about nutritional values and difficulty in handling and using for extended periods may account for their limited use. Therefore, it is essential to increase feeds by improving the nutritive value of crop residues, encouraging agro and social forestry and utilizing other non-conventional feed resources in various permutation and combinations. Crop residues, plantation wastes and browse foliage are increasingly becoming important as feeds in the near future as human and livestock populations are expanding. Special attention should be given to efficient integration of multipurpose fodder shrubs and trees as fodder bank in feeding calendars of livestock under harsh climates. The active involvement of farmers could provide a platform for discussion of the suitability of the different technologies discussed in this paper, thereby enabling scientists to develop, modify or refine their technologies as per the needs of the farming community in changing climate scenario.

## References

Abdulrashid, M. and Agwunobi L.N. (2009). Taro cocoyam (Colocasia esculenta) meal as feed ingredient in poultry. Pakistan Journal of Nutrition, 8(5): 668-673.

Agwunobi, L.N. (1999).Performance of broiler chickens fed sweet potato meal (*Ipomea batatas* L.) diets. Tropical Animal Health and Production, 31: 383-389.

Agwunobi, L.N., Awukam P.O., Cora, O.O. and Isika, M.A. (2012). Studies on the use of Colocasia esculenta (*Taro cocoyam*) in the diets of weaned pigs. Tropical Animal Health and Production, 34: 241-247.

Amata, I.A. and T. Lebari. (2011). Comparative evaluation of the nutrient profile of four selected browse plants in the tropics, recommende1d for use as non-conventional livestock feeding materials. African Journal of Biotechnology, 10(64): 14230-14233.

Amata, I.A. and E.E. Iwelu. (2012). Changes in the proximate composition and anti- nutritional content of the fruits of Gmelina arborea tree during growth and development. International journal of Innovations in Bio-Sciences. 2(3): 126-129.

Amata, I.A. and K.M. Nwagu. (2012). Comparative evaluation of the nutrient profile of the seeds of four selected tropical plants and maize. International Journal of Applied Biology and Pharmaceutical Technology, 4(1): 200-204.

Anindo, D., Toé, F., Tembely, S., Mukasa-Mugerwa, E., Lahlou-Kassi, A. and Sovani, S. (1998). Effect of molasses-urea-block (MUB) on dry matter intake, growth, reproductive performance and control of gastrointestinal nematode infection of grazing Menzram lambs. Small Ruminant Research, 27: 63-71.

Assefa , A and G, Animut.(2013). Supplementation of Raw, Malted and Heat Treated Grass Pea (*Lathyrus sativus*) Grain on Body Weight Gain and Carcass Characteristics of Farta Sheep. International Journal of Soil and Crop Sciences, 1: 1-6.

Assis, A.J., Campos, J.M.S., Filho, S., Queiroz, A.C., Lana, R., Euclydes, R.F., Neto, J.M., Magalhaes, A.L.R. and Mendonça, S. (2004). Citrus pulp in diets for milking cows. Intake of nutrients, milk production and composition. Revista Brasileira de Zootecnia, 33: 242-250.

Baloch, G.M., Soomro, F.M., Isani, G.B.and Carpenter, J.R. (1988). Utilization of banana plant silage as a source of roughage for dairy cows. Journal of Dairy Science, 71(1): 132.

Ben Salem, H., Nefzaoui, A., Messaoudi, A. and Ben Arif, T. (1995).A traditional technique as an alternative to plastic sheet for covering urea-treated straw. Digestibility and growth trials. Annal. Zootech, 45(1): 119.

Bhar, R., Pathak, N.N., and S., Paul. (2001). Performance of crossbred (Landrace × Local Indian) grower pig on maize replaced wheat bran and deoiled rice bran based diets. Indian Journal of Animal Science, 71: 165-168.

Cerisuelo, A., Castello, L., Moset, V., Martinez, M., Hernandez, P., Piquera, O., Gomez, E., Gasa, J. and Lainez, M. (2010). The inclusion of ensiled citrus pulp in diets for Growing pigs: effects on voluntary intake, growth performance, gut microbiology and meatquality. Livestock Science, 134: 180–182.

Chittavong, M., Preston, T.R. and Ogle, B. (2008). Ensiling leaves of Taro (Colocasia esculenta (L.) Shott) with sugar cane molasses. *Livestock Research for Rural Development.* Volume 20, supplement.http://www.cipav.org.co/lrrd/lrrd20/supplement/mala1.htm

Costa, R.G., Correia, M.X.C., Da Silva, J.H.V., De Medeiros, A.N. and DeCarvalho, F.F.R. (2007). Effect of different levels of dehydrated pineapple by-products on intake, digestibility and performance of growing goats. Small Ruminant Research, 71: 138-143.

Crawshaw, R. (2004).Co-product Feeds: Animal Feeds from the Food and Drinks Industries. Nottingham University Press.

Das, K.C., Tzudir, I. and Savino, I .N. (2012). Effect of replacement of maize with tapioca (*Manihotesculenta*) root meal on the performance of growing crossbred pigs. Livestock research for rural development,24: 197.http://www.cipav/lrrd./lrrd24/11/tzud24197.htm

Doyle, P.T., Devendra, C. and Pearce, G.R. (1986). Rice Straw as a Feed for Ruminants. International Development Program of Australian Universities and Colleges, x + 117 pp.

FAO, (1985). FAO/ APHCA Publication: Non-conventional feed resources in Asia and the pacific (2nd ed). FAO regional office for Asia and the Pacific, Bangkok, 1985.

Foidl, N., Makkar, H.P.S. and Becker, K. (2001). Potential of *Moringa olifera* for agricultural and industry uses. In: The Miracle Tree. The Multiple Attributes of moringa, Fugile, L.G. (ed). CTA, Wageningen, The Netherlands, pp. 45-76.

Forster, L.A. Jr., Goetsch, A.L., Galloway, D.L. and Johnson, Z.B. (1993). Feed intake, digestibility and live weight gain by cattle consuming forage supplemented with rice bran and corn. Journal of Animal Science, 71: 3105.

Fugile, L.G. (ed). (2001). The Miracle Tree. The multiple attributes of Moringa, CTA, Wageningen, The Netherlands.

Heli, T., Thakuria, K., Sharma, D.N., Milli, B.C., & Baruah, K.K. (2008). Natural Fodders Commonly Relished by Mithun in Arunachal Pradesh. Indian Journal of Animal Nutrition, 25(3): 201-205.

Jensen, M. (1995). Trees commonly cultivated in South East Asia: An illustrated field guide: FAO Regional Office for Asia and Pacific (RAP), Bangkok, Thailand.

Jini, D.,Lalmuanpuii, Deb,R and Meena , H.R. (2010). An overview on the status and strategy for improving pig farming in Arunachal Pradesh. The Asian Journal of Animal Science, 4(2): 253-258.

Mathur, M.L., Sampath, S. R and Gosh, S.N. (1969). Studies on tapioca: effect of 50 and 100 % replacement of oats by tapioca in the concentrate mixture of dairy cows. Indian Journal of Dairy Science, 22:193.

Magdeleine, M., Boval, C ., Philibert, M.L., Borde, A. and Archimède, H. (2010). Effect of banana foliage (Musa paradisiaca) on nutrition, parasite infection and growth of lambs. Livestock Science, 131: 234-239.

Mellish, K.S., Devendra, C. and Mahendranathan, T. (1973). The utilization of rice by growing male pigs in Malaysia. MARDI Reserach Bulletin, 1: 41-47.

Nunes M.C.N., Emond J.P., Rauth M., Dea S. and Chau K.V. (2009). Environmental conditions encountered during typical consumer retail display affect fruit and vegetable quality and waste. Postharvest Biology and Technology. 51: 232-241.

Pate F,M., Beardsley, D.W and Jayes, B.W. (1971). Chopped sugarcane tops as feedstuff for cattle and horses. Everglades Station Mimeo Report EES 71-5 Pate FM (1981). Fresh chopped sugar cane in growing steer diets. Journal of Animal Science, 53: 881-888.

Pereira, E.S., Regadas, F., Freitas, J.G.L., Neiva, E.R J. and Candido, M.J.D. (2009). Energetic value from by-product of the Brazil agroindustria. Archivos de Zootecnia, 58: 455-458.

Poyyamozhi, V.S. and Kadirvel, R. (1986). The value of banana stalk as a feed for goats. Animal Feed Science and Technology, 15: 95-100.

Preston,T,R and Leng, R.A (1976). Sugar cane for cattle production: present constrains, perspectives and priorities. Tropical Animal Production.

Randal, P.F., Soldevila, M. and Salas, B. (1967). A complete ration composed of concentrates and sugarcane bagasse vs a conventional ration of Pangola grass and supplemented concentrates for milk production. Journal of Agric. Univ. Puerto Rico, 51: 167-176.

Scerra, V., Caparra, P., Foti, F., Lanza, M. and Priolo, A. (2000). Citus pulp and wheat silage as an ingredient in lambs diets: Effects on growth and carcass and meat quality. Small Ruminant Research, 2019: 1-6.

Singh, K.A. (1999). Nutrient contents in tree fodders and bamboo leaves of eastern Himalaya. Indian Journal of Animal Nutrition, 16(2): 178-182.

Sruamisri S. (2007). Agricultural wastes as dairy feed in Chiang Mai. Animal Science Journal, 78: 335-341.

Swain, B.K., Chakurkar, E.B. and Barbuddhe. (2009). Effect of replacement of maize with bajra (Pennisetumtyphoides) or ragi (Eleusinecoracana) on the performance of laying hens. Proceeding of animal nutrition association world Conference, 14-17 February, pp 243.

Swain, B.K., Sundaram, R.N.S. and Barbuddhe, S.B. (2005). Effect of Feeding Brewery Dried Grain (BDG) with Enzyme Supplement on the Performance of Broilers. Indian Journal of Poultry Science, 40: 26-31.

Tillman, A.D., Fur, R.D., Hansen, K.R., Sherrod, L.B. and World, J.D. Jr. (1969). Utilization ofrice hulls in finishing cattle rations. Journal of Animal. Science, 29: 792-796.

Yang, S.J. and Chung, C.C. (1985). Feeding value of dried citrus by-products fed to layer. Korean Journal of Animal Science, 27: 239-245.

❑❑❑

# 21

# Role of Extension in Managing Climate Change

Dileep Kumar

## Introduction

Natural phenomena such as solar radiation, volcanic activity and ocean circulation influence climate and contribute to its variability. Evidence suggests that climate is now also changing as a result of human activities, such as the emission of greenhouse gases and changing land uses. The fourth assessment report of the Intergovernmental Panel on Climate Change (IPCC, 2007) concluded that the global average mean temperature and the frequency of hot extremes, heat waves, and heavy precipitation will very likely increase in response to increased concentrations of greenhouse gases in the atmosphere.

Society will have to make decisions in the coming years about how to adapt to a changing climate. Climate variability and climate change create risks to all sectors of the economy. Climate is already a prime factor in 9 out of 10 disasters, many of which cost billions of dollars and thousands of lives. Effective preparation for the possible effects of climate change requires the engagement of resource managers, planners, public works officials, local managers, community development specialists, businesses, residents, and property owners. The challenge is to provide these diverse stakeholders with trusted, useful, science-based information so that they in turn can make informed decisions.

The Extension programs contribute to and advance the decision-making processes and outcomes of stakeholders through the provision of high-quality science based information. Extension programs are stakeholder-driven and address the issues that are most frequently brought to the attention of Extension agents, such as the causes and potential impacts of climate change, extreme event frequency, sustainability, carbon resource management, public policy, community development, water resource management, and public health. By identifying needs, vulnerabilities, strategies, and information, Extension facilitates and ensures the implementation of science-based solutions to the climate challenge dissemination and promotion of emission reduction strategies to help mitigate climate change would be a new activity under the existing climate Extension program. Management of forestry and agricultural activities is regarded as an important option for greenhouse gases (GHG) mitigation. Activities in these sectors can reduce and avoid the release into the atmosphere of the three most important GHGs: carbon dioxide ($CO_2$), methane ($CH_4$), and nitrous oxide ($N_2O$). The main goal would be to shift agriculture from a net source to a net sink for greenhouse gases

Although the GHG sources and sinks in the forestry and agriculture sectors of world are minor portions of the total emissions profile, they represent significant potential for offsetting and reducing the projected increases in emissions over future decades. Activities with potential for GHG mitigation include afforestation, improved forest management and protection, soil carbon sequestration, agricultural $CH_4$ and $N_2O$ mitigation, and biofuels offsets.

Soil carbon sequestration has additional appeal because practices that enhance soil carbon also improve soil quality and fertility. Soils store carbon for long periods of time as stable organic matter, which reaches an equilibrium level in natural systems that is determined by tillage and other management practices, climate, soil texture, and vegetation. When native soils are disturbed by agricultural tillage, fallow, or residue burning, large amounts of $CO_2$ are released. However, a significant portion of the carbon captured by plants through photosynthesis can be sequestered by soils managed with direct seeding and other techniques that minimize soil disturbance. Irrigation can enhance carbon sequestration over native soil levels by overcoming the moisture limitation to increased plant biomass production. Examples of management practices with the potential to increase soil organic carbon include: Agriculture alters the terrestrial nitrogen cycle as well. Through nitrogen fertilization and improper water management, nitrogen is more prone to being lost to ground or surface water and to the atmosphere. $N_2O$, a common emission from agricultural soils, is a potent greenhouse gas (296 times more than $CO_2$). Atmospheric concentrations of $N_2O$ have increased by 15% during the past two centuries but reductions can be achieved through improved nitrogen management.

Methane emissions from livestock include enteric emissions from ruminant activity in cows and manure management emissions. Methane traps heat about 21 times more effectively than does $CO_2$. Other than reducing the number of cows, there is little opportunity to reduce enteric emissions.

On the other hand, emissions from manure management are affected by management options. Most modern dairies utilize a lagoon system for animal waste treatment, a practice that leads to large emissions of methane and nitrous oxide. Closed-system anaerobic digestion of the manure has the potential to eliminate most methane emissions from lagoons while conserving more nutrients and also producing a renewable energy source.

Livestock production in world includes both confined animal operations and pastured animals. The increase in production and concentration of intensive livestock operations along with increased urbanization of rural regions have resulted in greater awareness and concern for the proper storage, treatment, and utilization of livestock manure. Pastured animals offer limited opportunity for managing livestock manure to lessen greenhouse gas emissions. The principal opportunities for altering manure management, therefore, occur in dairy and poultry operations with confined livestock. Greenhouse emissions from broilers account for more than twice that of dairy cows, while the layer population produces around one-sixth of the emissions of dairy operation

There are several ways that extension systems can help farmer's deal with climate change. These include adaptation and contingency measures for what cannot be prevented. Extension can help farmers prepare for greater climate variability and uncertainty, create contingency measures to deal with exponentially increasing risk, and alleviate the consequences of climate change by providing advice on how to deal with droughts, floods, and so forth. Extension can also help with mitigation of climate change. This assistance may include providing links to new markets (especially carbon), information about new regulatory structures, and new government priorities and policies. Discussed below are ways in which extension can help with adaptation and mitigation:

*Technologies and management information*: Extension traditionally has played a role in providing information and promoting new technologies or new ways of managing crops and farms. Extension also links farmers to researchers and other actors in the innovation system. Farmers, extension agents, and researchers must work together on farmers' fields to prioritize, test, and promote new crop varieties and management techniques. While extension must now go beyond such methods, there is still a need for simple technology transfer in order to increase resilience to climate change and mitigate GHG emissions. Today's farmers will need to be able to quickly respond to climate change and adeptly manage risk. This will be especially challenging for extension in terms of knowledge and information systems. Farmers need to have access to this kind of information- be it climatic information, forecasts, adaptive technology innovations, or markets—through extension and information systems.

Extension agents can introduce locally appropriate technologies and management techniques that enable farmers to adapt to climate change by, for example, developing and disseminating local cultivars of drought-resistant crop varieties with information about the crops' advantages and disadvantages. Additionally, extension staff can share with farmers their knowledge of cropping and management systems that are resilient to changing climate conditions such as agro forestry, intercropping, sequential cropping, and no-till agriculture. Some of these practices have the following advantages:

*Resource management*: Tree planting can also help to improve soil, prevent soil erosion, and increase biodiversity. It is important to provide farmers with information about how the various options will potentially increase income and yields, protect household food security, improve soils, enhance sustainability, and generally help to alleviate the effects of climate change. At the same time, extension staff can play an important role in transferring indigenous technical Knowledge to help farmers worldwide. A core challenge for extension in the future is to shift from providing "packages" of technological and management advice to, instead, supporting farmers with the skills they need to choose the best option to deal with the climate uncertainty and variability and to make informed decisions about if and how to engage in new markets for carbon emissions. Some farmers will also need access to new technologies and management options in those areas where climate change renders their current farming systems enviable.

*Capacity development*: One of extension's major activities over time has been adult and non-formal education. This role continues today and is even more important in light of climate change. In addition, extension is also responsible for providing information using techniques ranging from flyers and radio messages to field demonstrations. Recent innovative extension activities include the adult education and experiential learning approaches utilized in farmer field schools, an extension and education approach already working with farmers on issues of climate change. Climate Field Schools (CFSs) have been established.

## The Important Role of Extension Systems

Climate change will initiate extreme events like sudden onset disasters and new vectors of human and livestock diseases. Evidences emerging that the biggest impacts will be in the form of small droughts, floods, and other events that cause severe hardship but do not attract the attention of the international community. The capacity of farmers to cope with such different forms of risk will become ever more crucial, and extension efforts must pay special attention to educating farmers about their options to enhance resilience and response capacity. There is a need for capacities to engage new sets of actors, including humanitarian agencies. Education must thus move beyond technical training to enhance farmers' abilities for planning, problem solving, critical thinking, and prioritizing, negotiating, building consensus and leadership skills, working with multiple stakeholders, and, finally, being proactive. Capacity development is important within extension as well. Extension agents have traditionally been trained only in technical expertise and often lack "soft" skills such as communication, development of farmer groups, systems thinking, knowledge management, and networking. To improve outcomes in rural development, farmers and extension agents need new skills that will require agricultural education and extension curriculums to include valuing and understanding the knowledge and experiences of rural people and co-learning (that is, farmers and extension agents learning together rather than extension agents training farmers in a one-way information transfer). There are many different ways to inform and educate farmers about adaptation options. Climate change adaptation funding should focus on extension systems and programs that incorporate a good understanding of what practices and skills are needed to best promote activities that help in the climate change effort and on increasing the capacity of extension agents and farmers, where needed.

### Facilitating, brokering, and implementing policies and programs

Another role of extension, which will be critical for climate change issues, is that of acting as an honest broker, bringing together different actors within the rural sector. Traditionally this has meant linking farmers to transport agents, markets, and inputs suppliers, with climate change, it will be increasingly important for the extension system to link farmers and other people in rural communities directly with voluntary and regulated carbon markets, private and public institutions that disseminate mitigation technologies, and funding programs for adaptation investments. Increased access to meteorological information will be imperative.

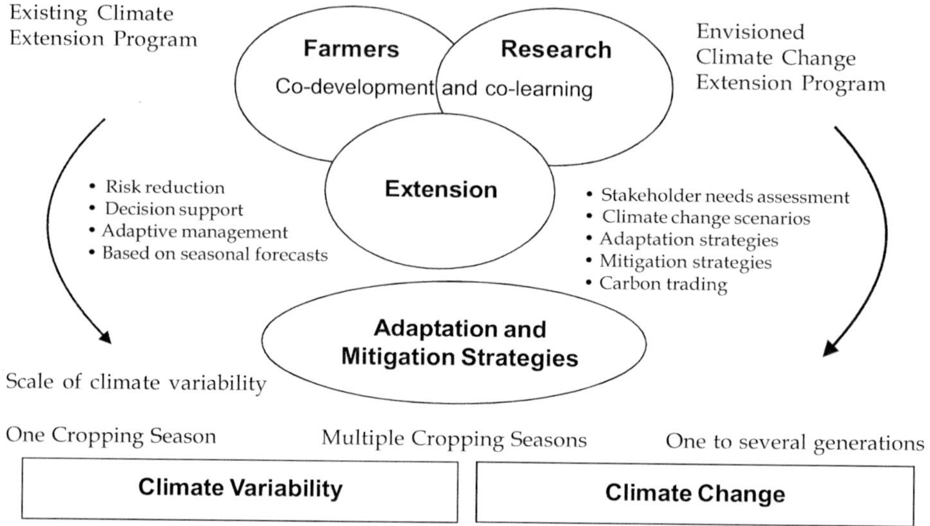


**Figure 1.** Framework of a Combined Climate Variability and Change Extension Program

## Extension also has an enormous challenge in bringing together

- Develop climate action plans for country
- Conduct community-wide greenhouse gas inventories of country
- Develop methodologies to incorporate climate information and forecasts in to Water Management Districts and their decision-making processes for water withdrawals and allocations
- Operate web-based climate information and decision support systems to reduce agricultural production and natural resource management risks related to climate variability and change
- Determines carbon footprint of agricultural and food systems
- Forecast forest fire risks
- Addresses public health issues related to climate variability and change
- Provides climate education professional development opportunities for country extension faculty
- Provides climate educational programs and curricula for youth
- Design energy-efficient, low-impact communities
- Changing planting or harvest dates are effective, low-cost options. The major risk could be shifting to a different market window with lower prices

- Changing varieties is another low-cost option, although some varieties can be more expensive or require investments in new planting equipment. Examples are the development of new peanut varieties resistant to Tomato Spot Wilt Virus (TSWV) disease, a major threat to peanut production in the southeast U.S., and the increased adoption of genetically modified cotton varieties resistant to certain types of herbicides and pests
- Increased use of irrigation, fertilizer, herbicide, and pesticide may be necessary to achieve maximum benefits from increased atmospheric $CO_2$. Climate change is also likely to increase weed and pest pressure in most cases
- Changing crop species or livestock produced could bring new profits, but is a risky and more expensive option because the necessary infrastructure or marketing mechanisms may not exist locally
- Investments in new irrigation or drainage systems or other capital items are likely to be essential if climate change increases climate variability
- Adoption of conservation and no-tillage practices
- Optimize crop rotations by using legumes, rotations crop-pasture, green manures
- Improved fertilization to stimulate biomass production and root growth
- Optimize manure management
- Promotion of land use shifts that enhance soil organic matter (e.g., forest, wetlands), mixed cropping systems that combine annual and perennial crops (e.g., agroforestry). Technologies and management information
- Capacity Development
- Facilitating and Implementing Policies & Programs
- Institutional linkages should be fostered for agricultural sustainability
- Since climate change could exacerbate rainfall variability, close collaboration between meteorological and Agricultural services will be necessary for a more effective use of climate forecasts. Extension services need to be strengthened and agents provided with the necessary equipments and logistics so that they can reach farmers more easily with agricultural technologies for adaptation in the face of changing climate

- Development of Special rural micro-credit Schemes : Because of the lack of adequate rural financial facilities, small holder farmers have often been by-passed by new Technologies. The agricultural bank that exists usually targets either big commercial farms or comes up with Collaterals as pre-conditions for accessing such loans. As such the loans should be free from interests and collaterals in order for the poor farmer to access and adapt to the impacts of the climate change
- Improved Extension and information delivery:Information delivery is critical in the process of enhancing the adaptive capacities of the rural areas to climate Change. Information on weather or new technologies could be transmitted to the farmers using rural radio and other
- Media and gatherings such as traditional ceremonies. The rapid development of mobile telephony is now opening up
- New opportunities should be exploited fully to reach the otherwise remote and unreachable areas
- Existing technology options should be made more available and accessible
- Climate change will almost surely make life even harder for the world's poorest and most vulnerable populations. Therefore, avoidance should be made in restricting their capacity to adapt by limiting their options. Technology options, in particular, should become more available
- Human Capital Development
- Agriculture needs to become professionalized with educational training incentives and development of human capital. in the direction of crop and livestock production. There is need for effective capacity to strengthen the most
- Vulnerable group in agricultural production with requisite knowledge and information necessary for climate change Adaptation
- Climate Change Adaptation Funding. Climate change adaptation funding should focus on extension systems and programs that incorporate a good understanding of what practices and skills are needed to best promote activities that help in the climate change effort

## Adaptation

- Water conservation techniques (precision farming);
- Increase (intensification) or reduce (intensification) land size cultivated;
- Move to different site;
- Increase irrigation;
- Soil conservation;
- Shading and shelter/mulching;
- Farming Operations;
- Planting of different crops;
- Planting of different varieties;
- Early planting;
- Early harvesting when dry soil is expected;
- Apply more or fewer inputs (fertilizers, agro-chemicals);
- Different planting dates;
- Treat seeds with fungicides before sowing;
- Mixed farming;
- Zero tillage;
- Changes from crop production to livestock production.

## Protection Measure

- Chance row orientation with respect to slopes;
- Apply soil amendments e.g. farm yard manure;
- Increase fertilizer application three days prior to sowing.

## Household Livelihood

- Undertake non-farm economic activities;
- Avoid selling remaining food stocks;
- Reduce expenditure;
- Ration food;
- Migrate.

## Education and Finance

- Access to extension facilities;
- Access to credit facilities;
- Government policies and on increasing the capacity of extension agents and farmers, where needed.
- Farmers' concerns and those of other actors as they address both.

It is well established that climate change poses a serious threat to agricultural production, particularly on rain-fed agriculture. For instance, climate change results in stunted growth and low yield of food crops, late fruiting of tree crops, easy spread of pests and diseases attacks on crops and livestock. Recommended adaptation or coping strategies to climate change include, inter alia water conservation techniques (precision farming); planting of different crops/varieties, mixed farming and early planting. Adding the perspective of climate change to the existing program is an attractive option given the existing focus on developing adaptation strategies and training of stakeholders to add seasonal climate variability as part of their decision making process. Dissemination and promotion of mitigation strategies should also be included, especially strategies that increase the efficiency of inputs, improve soil quality, and may allow the participation of producers in the carbon trading market.

The first step towards developing a climate change Extension program should be to undertake initial or ex ante assessment to understand farmers' perceptions, attitudes, long-term goals, and other cognitive and decision-making information through participatory methods. Once enough data has been elicited and analyzed, we may start to developing adaptation and mitigation strategies targeting and end-goal of economic and ecological sustainability.

## References

Bast, J. , & Taylor, J. M. (2007). Scientific consensus on global warming—Results of an international survey of climate scientists. Chicago, IL: The Heartland Institute. Retrieved from.

Brechin, S. R., & Bhandari, M. (2011). Perceptions of climate change worldwide. Interdisciplinary Reviews: Climate Change (WIRE's), 2(6): 871-885.

Breuer, N. E., Fraisse, C. W., & Cabrera, V. E. (2010). The Cooperative Extension Service as a boundary organization for diffusion of climate forecasts: a 5-year study. Journal of Extension [On-line], 48(4) Article 4RIB7.

Burnett, R. E., Vuola, A. J., Megalos, M. A., Adams, D. C., & Monroe, M. C. (2014). North Carolina Cooperative Extension professionals' climate change perceptions, willingness, and perceived barriers to programming: An educational needs assessment. Journal of Extension [On-line], 52(1) Article 1RIB1. Available at: http://www.joe.org/joe/2014february/rb1.php

Estill, L., Wechsler, D., Schwerin, T., & Tacular, B. (2013). Farming strategies in today's changing climate. Abundance Foundation, Pittsboro, NC. Florida Oceans and Coastal Council. (2010). Climate change and sea-level rise in Florida. Retrieved from: http://www.floridaoceanscouncil.org/reports/Climate_Change_and_Sea_Level_Rise.pdf

Fraisse, C. W., Breuer, N. E., Zierden, D., & Ingram, K. T. (2009). From climate variability to climate change: Challenges and opportunities to Extension. Journal of Extension [On-line], 47(2) Article 2FEA9.

Grotta, A. T., Creighton, J. H., Schnepf, C., & Kantor, S. (2013). Family forest owners and climate change: Understanding, attitudes, and educational needs. Journal of Forestry, 111(2): 87-93.

James, A., Estwick, N., & Bryant, A. (2014). Climate change impacts on agriculture and their effective communication by Extension agents. Journal of Extension [On-line], 52(1) Article 1COM2. Available at: http://www.joe.org/joe/2014february/comm2.php

Kirilenko, A. P., & Sedjo, R. A. (2007). Climate change impacts on forestry. Proceedings of the National Academy of Sciences of the United States of America, 104(50), 19697-19702. doi:10.1073/pnas.0701424104

Lachapelle, E., Borick, C. P., & Rabe, B. (2012). Public attitudes toward climate science and climate policy in federal systems: Canada and the United States compared. Review of Policy Research, 29(3): 334-357.

Layman, C., Doll, J., & Peters, C. (2013). Using stakeholder needs assessments and deliberative dialogue to inform climate change outreach efforts. *Journal of Extension* [On-line], 51(3) Article 3FEA3. Available at: http://www.joe.org/joe/2013june/a3.php

Leiserowitz, A., Maibach, E., Roser-Renouf, C., Feinberg, G., & Howe, P. (2012). Climate change in the American mind: Americans' global warming beliefs and attitudes in September 2012 Yale University and George Mason University. New Haven, CT: Yale Project on Climate Change Communication.

Leiserowitz, A., Maibach, E., & Roser-Renouf, C. (2007). Global warming's "Six Americas": An audience segmentation. Yale Project on Climate Change; George Mason University Center for Climate Change Communication.

Monroe, M. C., & Adams, D. C. (2012). Increasing response rates to Web-based surveys. Journal of Extension [On-line], 50(6) Article 6TOT7. Available at: http://www.joe.org/joe/2012december/tt7.php

Monroe, M. C., Plate, R. R., Adams, D. C., & Wojcik, D. J. (In press). Harnessing homophily to improve climate change education. Environmental Education Research.

National Climate Assessment and Development Advisory Committee. (2013). National climate assessment report: Draft for public comment. U.S. Global Change Research Program.

Oreskes, N. (2007). The scientific consensus on climate change: How do we know we're not wrong? In J. F. C. DiMento, & P. Doughman (Eds.), Climate change: what it means for us, our children, and our grand children (pp. 65-99) MIT Press.

Pathak, T. B., Bernadt, T., & Umphlett, N. (2014). Climate Masters of Nebraska: An innovative action-based approach for climate change education. Journal of Extension [On-line], 52(1) Article 1IAW1.

Roser-Renouf, C. (2013). Communicating with global warming's six Americas: Strategies for building public understanding and issue engagement. 2013 Pacific Northwest Climate Science Conference, Portland, Oregon. Retrieved from: https://www.youtube.com/watch?v=IlgqCeSrcv4

Smith, J. W., Anderson, D. H., & Moore, R. L. (2012). Social capital, place meanings, and perceived resilience to climate change. Rural Sociology, 77(3), 380-407.

Tubiello, F. N., & Fischer, G. (2007). Reducing climate change impacts on agriculture: Global and regional effects of mitigation, 2000–2080. Technological Forecasting and Social Change, 74(7): 1030-1056. doi: 10.1016/j.techfore.2006.05.027

Tyson, R. (2014). The merits of separating global warming from Extension education sustainability programs. Journal of Extension [On-line], 52(1) Article 1COM3. Available at: http://www.joe.org/joe/2014february/comm3.php.

Wojcik, D. J., Monroe, M. C., Adams, D. C., & Plate, R. R. (2014). Message in a bottleneck? Attitudes and perceptions of climate change in the Cooperative Extension Service in the southeast United States. Journal of Human Sciences and Extension, 2(1), 51-70. Available online at: http://www.jhseonline.com/#!current-issue/c227d.

□□□

# 22

# Role of Information and Communication Technologies on Climate Dependant Agriculture

**K.K. Chaturvedi, Anil Rai, Mohd. Samir Farooqi, S.B. Lal Vinod Kumar and U.B. Angadi**

## Introduction

ICT is an acronym for Information and Communication Technology. The information refers to accessibility of digital data in an easy way to retrieve relevant facts. The communication refers to the medium or the way to transmit the information from one place to another place or one media to another media. The advancement in these areas refers to technology. The collectively ICT covers any product that will store, retrieve, manipulate, transmit or receive information electronically in a digital form. For example, personal computers, digital television, email, robots.

Climate change may refer to a change in average weather conditions, or in the time variation of weather around longer-term average conditions (i.e., more or fewer extreme weather events). The change lasts for an extended period of time i.e., decades to millions of years has also been considered as climate change. The causing factors of climate change are biotic processes, variations in solar radiation received by Earth, plate tectonics, and volcanic eruptions. The affecting factors seems to be irreversible, it has become the today's necessity to develop improved next generation technologies which processes the current and future shifts in the climate system. In particular, the Agriculture will be most affected domain due to climate over the next few decades (Brown and Funk, 2008). The development of crop modeling tools pointed out that climate change is likely to decrease agricultural production.

The IPCC is the leading international body for the assessment of climate change and established by the United Nations Environment

Programme (UNEP) and the World Meteorological Organization (WMO). It does not conduct any research nor does it monitor climate related data or parameters but reviewed the essential part of the IPCC process as thousands of scientists from approximately 195 countries over the world contribute to the work of the IPCC on a voluntary basis.

The IPCC collects and reviews the most important scientific contributions in every five years and compiled the evidences for diffusion of climate fluctuations to identify more severe and frequent droughts and floods, which already causes the short-term fluctuations in food production in semi-arid and sub-humid areas. The study reflects the investments in communications as an effective mean to address future climate change (Schmidhuber and Tubiello, 2007).

It is difficult to identify/select the best use of information and communication technologies enabled initiatives that the farmers need in order to meet the challenges faced due to climate change. Long term policy need to be developed and reviewed frequently to tackle with the climate change. The climate change is affecting the people of every country due to large impact and shift in changing climatic conditions. The climate change will severely affects the poor or small landholding farmers because their livelihood mainly depends on agriculture. The majority of Asian and African farmers do not have access to the scientific and technological advances that support agricultural decision-making because of the lack of reliable communication networks in these countries. The major challenges to the current days research is to study the information needs of masses, policy makers, and how effectively data, models and report will be developed and disseminated these results to the end users using an ICT enabled technologies. ICT and climate change learning curve is shown in Figure 1.

## IPCC Initiative

The formation of IPCC have made an significant impact to tackle with the climate change. The historical initiatives taken by the IPCC are mentioned in Table 1.

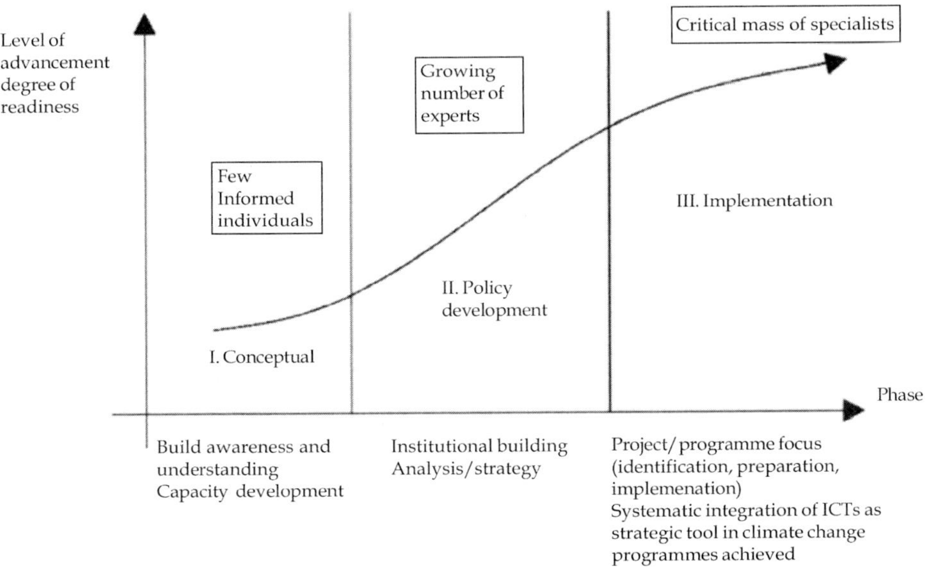


**Figure 1.** ICT and Climate Change Learning Curve

**Table 1.** Historical Initiatives of IPCC

| Time line | Highlights of IPCC history |
|---|---|
| 1988 | The United Nations Environment Programme (UNEP) and the World Meteorological Organization (WMO) establish the Intergovernmental Panel on Climate Change. |
| 1988 | The United Nations General Assembly endorses the action of UNEP and the WMO in setting up the IPCC. |
| 1990 | The IPCC publishes its First Assessment Report Working Group I - Climate Change: The IPCC Scientific AssessmentWorking Group II - Climate Change: The IPCC Impacts AssessmentWorking Group III - Climate Change: The IPCC Response Strategies |
| 1990 | The UN General Assembly notes the report findings and decides to initiate negotiations for a framework convention on climate change. |
| 1992 | The IPCC publishes Supplementary Reports Working Group I - Climate Change 1992: The Supplementary Report to the IPCC Scientific Assessment; Working Group II - Climate Change 1992: The Supplementary Report to the IPCC Impacts Assessment; Climate Change: The IPCC 1990 and 1992 Assessments |
| 1992 | The United Nations Framework Convention on Climate Change (UNFCCC) opens for signature at the UN Conference on Environment and Development in Rio de Janeiro. |

| | |
|---|---|
| 1995 | The IPCC publishes its Second Assessment Report Working Group I – Climate Change 1995: The Science of Climate Change; Working Group II – Climate Change 1995: Impacts, Adaptations and Mitigation of Climate Change: Scientific-Technical Analyses; Working Group III – Climate Change 1995: Economic and Social Dimensions of Climate Change; IPCC Second Assessment: Climate Change 1995 (includes Synthesis Report). |
| 1996 | The IPCC issues the Revised 1996 IPCC Guidelines for National Greenhouse Gas Inventories. |
| 1997 | The UNFCCC's Kyoto Protocol is adopted. It comes into force in 2005. |
| 1998 | The IPCC sets up the Task Force on National Greenhouse Gas Inventories (TFI) to oversee the National Greenhouse Gas Inventories Programme. Since 1999 the Task Force has been supported by the Government of Japan. |
| 2000 | The IPCC issues the Good Practice Guidance and Uncertainty Management in National Greenhouse Gas Inventories. |
| 2001 | The IPCC publishes its Third Assessment Report Working Group I – Climate Change 2001: The Scientific Basis; Working Group II – Climate Change 2001: Impacts, Adaptation, and Vulnerability; Working Group III – Climate Change 2001: Mitigation; Climate Change 2001: Synthesis Report). |
| 2003 | The IPCC issues the Good Practice Guidance for Land Use, Land. |
| 2006 | The IPCC issues the 2006 Guidelines for National Greenhouse Gas Inventories. |
| 2007 | The IPCC publishes its Fourth Assessment Report (AR4) Working Group I – Climate Change 2007: The Physical Science Basis; Working Group II – Climate Change 2007: Impacts, Adaptation and Vulnerability; Working Group III – Climate Change 2007: Mitigation of Climate Change; Climate Change 2007: Synthesis Report. |
| 2007 | The IPCC shares the Nobel Peace Prize which is awarded for its "efforts to build up and disseminate greater knowledge of man. |
| 2009 | The IPCC approves the outlines of the three Working Group contributions to the Fifth Assessment Report (AR5), due to be finalized in 2013 and 2014. |
| 2010 | The three Working Groups complete the selection of the 831 authors for the AR5 and work on the assessment starts. |
| 2010 | The IPCC starts a review of its processes and procedures, completed in 2012,based on recommendations from the Inter Academy Council. |

| | |
|---|---|
| 2011 | The IPCC approves the Special Report on Renewable Energy Sources and Climate Change Mitigation (SRREN), prepared by Working Group III. |
| 2011 | The IPCC approves the Special Report on Managing the Risks of Extreme Events and Disasters to Advance Climate Change Adaptation (SREX), prepared by Working Group II and Working Group I. |
| 2013 | The IPCC approves Climate Change 2013: The Physical Science Basis, the Working Group I contribution to AR5. |
| 2014 | The IPCC approves Climate Change 2014: Impacts Adaptation and Vulnerability and Climate Change 2014: Mitigation of Climate Change, the Working Group II and Working Group III contributions to AR5. The Fifth Assessment Report was completed in November 2014 with the Synthesis Report. |

[*Source:* http://www.ipcc.ch/news_and_events/docs/factsheets/FS_timeline.pdf]

Latest initiative taken by IPCC is approval of climate change policy 2014 which focused on impact adaptation, vulnerability, mitigation of climate change and come out with important recommendations. These recommendations are documented in Fifth assessment report. The emission of green house gases has been documented in the figure 2.

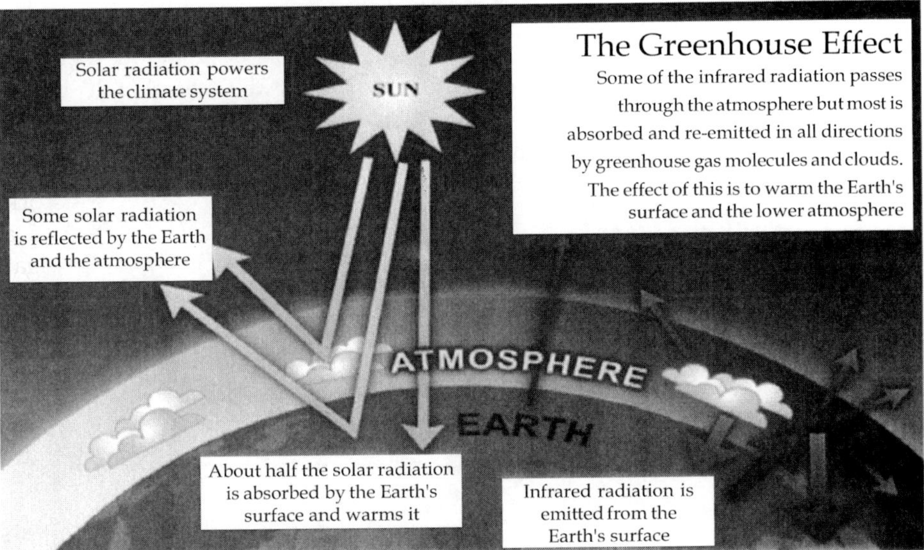


**Figure 2.** The Greenhouse Effect (*Source:* Intergovernmental Panel on Climate Change Working Group I Climate Change 2007: The Physical Science Basis Geneva: IPCC, 2007)

## Climate Change: An Indian Initiative

To meet the challenges apparent from the climate change, Indian government is also taking step forward to tackle this situation. The major organization involved in studying the climate change in India are working under the umbrella of Ministry of Environment and Forests, Indian Institute of Tropical Meteorology, National Initiative on Climatic Resilient in Agriculture (NICRA) National Institute of Abiotic Stress Management (NIASM) from Indian Council of Agricultural Research, Ministry of Agriculture, The Energy and Resources Institute, Indian Institute of Oceanography, Indian Institute of Technology, Indian Institute of Sciences, Indian Institute of Management and other national/state/district level research institutes and non-governmental organizations (NGO). The organizations are working towards the data collection, data decryption, data modeling and demonstration of effects of climate change in Indian conditions. The data scientists are working towards the deciphering of the large historical databases and will be building the effective decision support systems, expert systems and mobile based content delivery and dissemination systems. The Indian government initiatives have been summarized in table 2.

**Table 2.** Indian Government Initiative

| Area | Initiative/Event | Contribution |
|---|---|---|
| **Science & Research** | Indian Network for Climate Change Assessment (INCCA) | Network of 120 research institutions and 250 scientists launched; major conferences planned in May and November 2010. |
| | Himalayan Glaciers Monitoring Programme | Comprehensive programme to scientifically monitor the Himalayan glaciers - Phase I completed; Phase II launched; Discussion Paper on State of Himalayan Glaciers released. |
| | Launch of Indian Satellite to Monitor Greenhouse Gases | ISRO to launch a micro-satellite in 2010 to study aerosols (soot particles), followed by a comprehensive satellite in 2011 to monitor GHG gases; India to join elite club of countries to do so. |
| | India's Forest and Tree Cover as a Carbon Sink | Research estimates the value of India's forests as a carbon sink - assessment shows that they neutralize 11% of India's annual GHG emissions Science & Research. |
| | India's GHG Emissions Profile | India's GHG Emission Pathways until 2030 under different assumptions made |

| | | |
|---|---|---|
| | | public; shows India will remain a minor per capita emitter even in 2030. |
| **Policy Develop-ment** | Expert Group on Low Carbon Economy | Planning Commission-led Group set up to develop strategy for India as a low carbon economy; to feed into twelfth plan process. |
| | State Action Plans on Climate Change | Delhi becomes first State to release Climate Change Action Plan; other States finalizing their Plans Policy Development. |
| | National Policy on Biofuels | National Policy on Bio-fuels approved by Cabinet to promote cultivation, production and use of Bio-fuels for transport and in other applications. |
| **Policy Imple-mentation** | National Missions under National Action Plan on Climate Change | National Missions on Solar Energy, Energy Efficiency and Strategic Knowledge approved; other Missions in final stages of preparation. |
| | First National Conference on Green Building- Materials and Technologies | Conference to stimulate green building sector; to set an example the government proposes that all its new buildings will be GRIHA 4* compliant subject to site conditions. |
| | 30 "Solar Cities" | In-principle approval given to 30 'Solar Cities' with aim of 10% deduction in projected demand of conventional energy through a combination of energy efficiency and renewable. |
| | Energy Efficiency Standards for Appliances | Energy efficiency ratings made mandatory for 4 key appliances — refrigerators, air conditioners, tube lights and transformers from January 7, 2010; more to follow through 2010. |
| | Fuel Efficiency Norms | Plan for fuel economy norms for vehicles announced; to be made operational in two years Policy Implementation. |
| | CDM Program | India assessed as Best CDM Country; Indian projects to neutralize 10% of emissions by 2012. |
| **Interna-tional Co-operation** | India to host 'Rio+20' | 20' India to host 11th COP of Convention on Biodiversity (CBD) in 2012, mark 20th anniversary of Rio. |
| | UN Climate Technology Conference | India successfully hosts global Conference on technology, Delhi Statement adopted. |

| | | |
|---|---|---|
| | SAARC Environment Ministers Conference | India successfully hosts SAARC Ministers Conference and agrees joint actions on Climate Change; 2010 SAARC Summit to be on the theme of Climate Change International Cooperation. |
| | India's Submissions to UNFCCC | Report documenting India's 12 proactive submissions to UNFCCC released. |
| **Forestry** | State of Forests Report 2009 | Latest State of Forest Report released; shows continued rise in India's forest cover. |
| | Launch of CAMPA | Ambitious Rs. 11,700crore (USD2.5Bn) Programme for forest conservation launched. |
| | Green India Mission | New mission under NAPCC to fast-track re-forestation being finalized. |
| | Capacity Building in Forestry Scheme | New Rs. 369 crore (USD 80Mn) scheme for HRD for forest personnel. |
| | Intensification of Forest Management | New Rs. 600 crore (USD 125Mn) scheme to improve forest management, infrastructure, fires, etc. |
| | Inclusion of Forestry within NREGA | Forestry related activities included as part of India's flagship employment guarantee scheme to fast-track reforestation; Pilots being implemented. |

[*Source:* http://www.moef.nic.in/sites/default/files/24_recent_initiatives]

## ICT for climate change in agriculture

The climate change and agriculture are affecting each other because agriculture is the source as well target due to climate change. Many tools have been developed that will guide the farmers in choosing various operations and inputs to survive the climate based crop conditions. The same agronomic practices are not applicable in every regions even though conditions looks pretty change. The tools should allow the analysis of different processes of the agricultural sector, from local crop modeling under climate change conditions to the management of economic impacts of climate change on the agriculture sector (soil value variations, demand and supply, production, *etc.*).

The tools also allowed to simulate the growth of specific crops, validating the variations under different climate change scenarios in sophisticated controlled environment in labs. Generally, these tools are site-specific, but may be useful at national and/or regional level through the use of appropriate Geographic Information System (GIS). The tools have been summarized in following steps:

- The application have been initiated from the definition of basic conditions that include data on crop calendar, soil status, etc., input climate parameters and data namely temperature (Minimum and maximum), rainfall, humidity, precipitations, wind speed, sun shine hours, global radiation, soil moisture, air humidity, water flows etc. Some of the tools also include also data related to crop management conditions/practices.
- The stage of development of growth simulation models to assess the potential crop production e.g. with a certain fixed amount of water resources and nitrogen production for different management options and for a chosen climate change scenario, through the link to ad hoc expert system/decision support system.
- The general output of these simulation tools is the assessment of crop production under given scenarios, facilitating decision making at farm level up to a whole crop system.

Examples of few of these tools are:

- WOFOST, developed by the Centre for World Food Studies, CFWS, in cooperation with the Dutch University of Wageningen. This tools is applicable in barley, field bean, maize, potato, rice, soybean, sunflower, wheat, etc. crops.
- GOSSYM/COMAX, developed by the Universities of Clemson and Mississipi and the Agriculture Department of United States. This tools is useful to simulate cotton growth, with COMAX (CrOp Management eXpert, an expert system) to study the effects of climate change on cotton production.
- APSIM (Agricultural Production Systems SIMulator) developed by a consortium of universities and departments of the Australian state of Queensland named Agricultural Production Systems Research Unit (APSRU). This tool can be applied on more than twenty crops and plants, namely alfalfa, barley, chickpea, cotton, eucalyptus, lupin, maize, peanuts, sugarcane, sunflower, tomato, wheat, etc.

In addition to the above specific set of tools, few generic tools have also been developed at a higher scale i.e., at regional level with a broader perspective. These systems are focused on variety of factors and influence climate change and related responses. These factors may be either exogenous (e.g. government policies, economy, etc.) or endogenous (e.g. location, scale, etc.) with reference to a specific farming conditions.

Mainly on international efforts have been made in this direction and few of them are: DSSAT (Decision Support System for Agrotechnology Transfer), developed by the International Consortium for Agricultural Systems Applications (ICASA); CENTURY, developed by the Natural Resource Ecology Laboratory of Colorado University (NREL); and MAACV (Model of Agricultural Adaptation to Climatic Variation), developed by the Canadian Universities of Guelph and Carleton.

## Key enabling points towards ICT and Climate Change

The key enabling points in relation to ICT in climate change has been summarized into following three steps. The ICT for development has been summarized in figure 3.

The first step indicates the points related to ICT and climate change. The second step towards the development facts on climate change and last step focusses about the linkages between ICT and climate change.

### a. ICTs and climate change

- Climate change is not a new phenomenon but amplifies and magnifies existing development challenges to reduce suffering and alleviate poverty.
- Climate change is a social justice issue.
- Strategically integrated ICTs, such as community radios, mobile phones, knowledge centers and interactive media, are enabling tools that help to reduce climate change vulnerability and risk.
- ICTs contribute tangibly to climate change mitigation/adaptation strategies through providing access to relevant information, raising awareness at the grassroots level, and facilitating learning and practical knowledge.
- Current approaches that integrate ICTs as a strategic tool into development programmes (e.g., education, health, governance) can be directly applied to climate change strategies.
- A multi-stakeholder approach is central to ICT climate change mitigation and adaptation interventions.

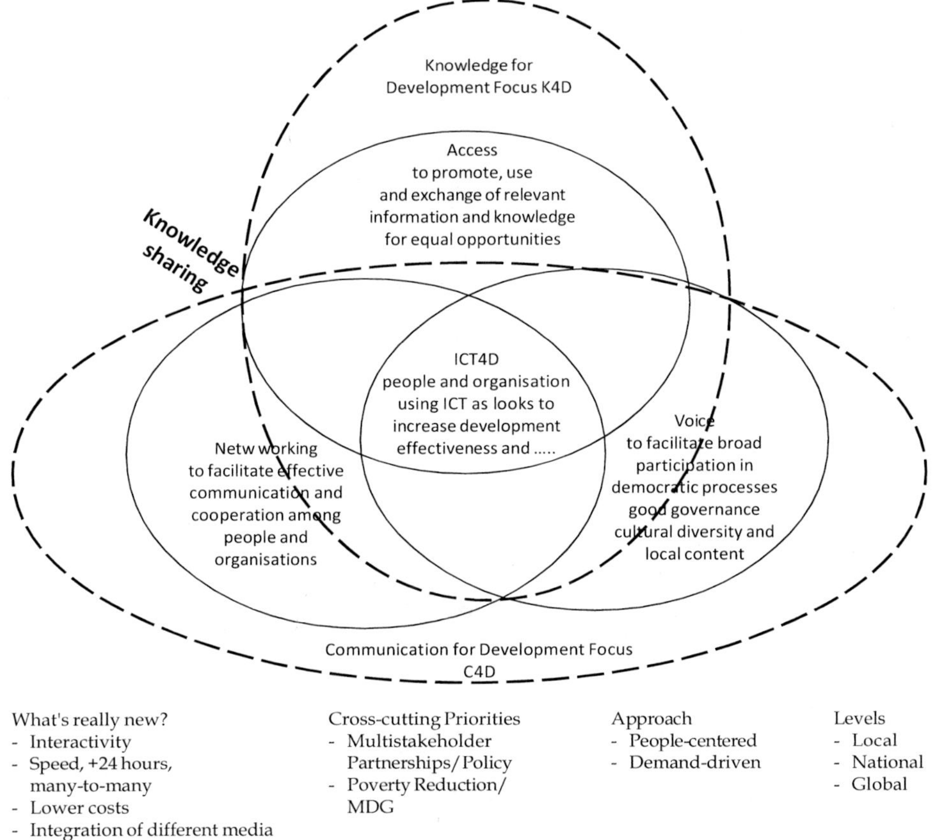


**Figure 3.** ICT for Development (*Source*: Weigel & Waldburger, 2003)

## b. Developmental facts about climate change

- The poor and marginalized are least informed about the potential impact of climate change on their livelihoods.
- Climate change fundamentally changes vulnerability patterns for the poor and marginalized.
- Coping solutions and adaptation strategies need to be localized and decentralized with grassroots interventions to be initiated.
- The voices of those most affected by climate change are not sufficiently included in the policy debate with the subsequent search for solutions and informed decision making.
- Vulnerability and risk can be substantially reduced by enabling access to and the sharing of information and knowledge.

### c. Linkages about ICTs and climate change

- ICTs for mitigation

  i. Production and use of more energy- and CO2-efficient ICTs (www.climatesaverscomputing.org), and the promotion of less energy-consuming technology transfer to developing countries, thereby contributing to the de-carbonization of the economy. Role for development cooperation, regulators and private sector.

  ii. New technologies such as video conferencing and voice over internet protocol (VoIP) technologies which have the potential to reduce travel and subsequent emissions.

  iii. Better waste management and recycling (http://ewasteguide.info).

- ICTs for natural disaster prevention, preparedness and risk management: ICTs offer tools relevant for data analysis, satellite imaging and vulnerability assessment (instedd.org), coordination of emergency efforts, and dissemination of locally specific and relevant information (e.g., early-warning systems, meteorological information for preparedness disseminated through tele-centers or mobile phones).

- ICTs as information and communication and empowerment tools (including community radio, knowledge centers, mobile phones, internet, internet-based media) can be used for both mitigation and adaptation efforts in order to:

  i. Inform and raise awareness (e.g., media campaigns) at all levels of society – including the poor – about the effects of climate change.

  ii. Raise the voices of grassroots communities and those most affected by climate change – again, mostly the poor in developing countries – at the local, national and international level, and carry them to the decision makers to demand actions from their leaders and political accountability (advocacy function/vertical linkages).

  iii. Facilitate networking and building of coalitions and help define positions (including for mitigation and adaptation strategies).

  iv. Capacity building through e-learning as vertical learning and knowledge sharing as horizontal, peer-to-peer learning.

## Conclusion

At present, the majority of applications and systems on climate change with reference to agriculture sector are related to scenario development, simulation studies, impact assessment and adaptation planning. These systems are the result of single Research & Development efforts, rather than collaborative programmes even though the initiatives have been taken up by the IPCC. The Open Source approach is opening the venues for creation of a collaborative community based environment, as it happened within spatial technology.

In addition, the gap still exists between global and local applications for promoting the development of an integrated framework for information sciences, agro-environmental sciences and communication at different levels. The information is vital to tackle climate change effects and a shift is needed in the agriculture to disseminate appropriate knowledge at the right time to the farmers, in both developed and developing countries. At the same time, the information *per se* is not enough, but appropriate communication systems are needed to ensure that information come to farmers in an effective, accurate and clear way. This means that the information provided to farmers must have the following properties:

1. Timing: Farmers need to the right information on right time, especially in production strategy;
2. Quality: Information must be correct and comprehensive in order to be transparent with the recipient;
3. Content: The effective way of creation/development of updated information and its delivery.
4. Reality: The created and processed information must be taking into account the recipient' problems by adapting the content delivery mechanism.
5. Media: The effective media of communication must be utilized such as community radio, mobile based SMS system and any other.

Thus, any knowledge should be transferred to take into account farmers' point of view, with the aim of building on their knowledge and capitalize it. The climate change is a global problem with local impacts, thus information technology jointly with communication sciences, can play a big role in handling different perspectives.

## References

Brown, ME, and Funk, CC. 2008. Food Security Under Climate Change. Science, 319(5863): 580–581.

David B. Lobell, DB, Burke, MB, Tebaldi, C, Mastrandrea, MD, Falcon, WP, and Naylor, RL. (2008). Prioritizing Climate Change Adaptation Needs for Food Security in 2030. Science, 319(5863): 607-610.

Bruinsma, J. (editor). (2003). World Agriculture: Towards 2015/2030 – An FAO Perspective. Earthscan Publications Ltd, London.

Schmidhuber, J, and Tubiello, FN. (2007). Global food security under climate change. Proceedings of the National Academy of Science, 104: 19703-19708.

Weigel, G and Waldburger, D. (eds.). (2004). ICT4D – Connecting People for a Better World: Lessons, Innovations and Perspectives of Information and Communication Technologies for Development (Berne: SDC and GKP, 2004) www.globalknowledge.org/ ict4d.

❑❑❑

# 23

# Role of Extension in Climate Change Mitigation and Adaptation

B.L. Dhaka

## Introduction

In order to define climate it is important to distinguish it from weather. The weather that we experience on a day-to-day basis is a momentary atmospheric state characterized by temperature, precipitation, wind, and so on, and seems to vary in an irregular way, not following any particular pattern. Climate is a very general term that has a variety of closely related meanings. Usually, climate refers to the average weather conditions observed over a long period of time for a given area. In contrast to the instantaneous conditions described by weather, the timescale upon which climate statistics are calculated is typically thirty years. There can be variations in climate from year to year, or one decade to another, one century to another, or any longer time scale. The climate of a region depends on many factors including the amount of sunlight it receives, its height above sea level, the shape of the land, distance from the equator and how close it is to oceans. Generally speaking, the climate remains stable over long periods of time if the various elements within the system remain stable. However, if one or more of the components of the system is altered, the stability of the whole system is compromised and can lead to uncharacteristic behaviour and give rise to weather, which is outside the usual range of expectations. This situation can be described as climate change.

## Climate Change and Agriculture

Agriculture is the mainstay of Indian economy. In addition to providing us with much of our food, the crops, livestock, it contribute significantly to the economy each year. Agriculture is highly dependent on specific climate conditions. Solar radiation, temperature, and precipitation are the main drivers of crop growth; therefore agriculture has always been highly dependent on climate patterns and variations. Since the industrial revolution, humans have been changing the global climate by emitting high amounts of greenhouse gases into the atmosphere, resulting in higher global temperatures, affecting hydrological regimes and increasing climatic variability. Climate change is projected to have significant impacts on agricultural conditions, food supply, and food security. Overall, climate change could result in a variety of impacts on agriculture. Some of these effects are biophysical, some are ecological, and some are economic. Climate change affects agriculture in a number of ways, including through changes in average temperatures, rainfall, and climate extremes; changes in pests and diseases; changes in atmospheric carbon dioxide and ground-level ozone concentrations; changes in the nutritional quality of some foods; and changes in sea level. Changes in the frequency and severity of droughts and floods could pose challenges for farmers.

Overall, climate change could make it more difficult to grow crops, raise animals, and catch fish in the same ways and same places. The effects of climate change also need to be considered along with other evolving factors that affect agricultural production, such as changes in farming practices and technology. Climate change will probably increase the risk of food insecurity for resource poor farmers.

## Role of Extension

Extension traditionally has played a pivotal role in promoting new ways of managing crops and farms. Extension must now go beyond conventional approach to increase resilience to climate change and mitigate. Today's farmers will need to be able to quickly respond to climate change and adeptly manage risk. This will be especially challenging for extension in terms of knowledge and information systems. Farmers need to have access to this kind of information—be it climatic information, forecasts, adaptive technology innovations, or markets—through extension and information systems. Extension agents can introduce locally appropriate technologies and management techniques that enable farmers to adapt to climate change by, for example, developing and disseminating local cultivars of drought-resistant crop varieties with information about the crops' advantages and disadvantages. Additionally, extension staff can share with farmers their knowledge of cropping and management systems that are resilient to changing climate conditions such as agroforestry, intercropping, sequential cropping, and no-till agriculture. Some of these practices have the added advantage of improved natural resource management. Tree planting can also help to improve soil, prevent soil erosion, and increase biodiversity. It is important to provide farmers with information about how the various options will potentially increase income and yields, protect household food security, improve soils, enhance sustainability, and generally help to alleviate the effects of climate change. At the same time, extension staff can play an important role in transferring indigenous technical knowledge to help farmers. A core challenge for extension in the future is to shift from providing packages of technological and management advice to, instead, supporting farmers with the skills they need to choose the best option to deal with the climate uncertainty and variability. Some farmers will also need access to new technologies and management options in those areas where climate change renders their current farming systems inviable.

Mitigation efforts will require information, education, and technology transfer. Agricultural extension services, both public and private, thus have a major role to play in providing farmers with information, technologies, and education on how to cope with climate change and ways to contribute to mitigation. This support is especially important for resource poor smallholders, who contribute little to climate change and yet will be among the most affected. There are several ways that extension systems can help farmers deal with climate change. Extension can help farmers prepare for greater climate variability and uncertainty, create contingency measures to deal with exponentially increasing risk, and alleviate the consequences of climate change by providing advice on how to deal with droughts, floods, and so forth. This assistance may include providing links to new markets, information about new regulatory structures, and new government priorities and policies. Discussed below are the ways in which extension can help with climate change mitigation and adaptation.

## Awareness Creation

One of the greatest challenges with climate change is its intangibility. By focusing farmers on climate change, extension agencies are better able to raise their awareness, engage their attention and inspire individual and community action. Education is an essential element of the response to climate change. It helps farmers understand and address the impact of global warming, encourages changes in their attitudes and behaviour and helps them adapt to climate change-related trends. It is important function of extension to make climate change education a more central and visible part of the response to climate change. It does this by strengthening the capacity of farmers to provide quality climate change education; encouraging innovative teaching approaches to integrate climate change education in school and by raising awareness about climate change as well as enhancing non-formal education through media, networking and partnerships. Awareness raising can cover a large number of activities: anything that involves people understanding, learning or doing something new; visioning the future; working out how to change something in their lives; or talking to someone else about what they've done – all are part of the process of raising awareness about the climate change. One key to successful awareness raising is understanding the community. Understanding the process of change is a key tool in designing effective awareness-raising activities. It is also important to note that different people learn in different ways. Some people learn and absorb information mentally, through exposure to ideas, concepts and written or spoken information. Some respond more on an emotional level, through being engaged by what they care about; what they fear or hope for. Other people need to learn in a more physical way, by making something, through acting out scenarios. If extension agents are able to create events that include something that will appeal to all these different learning types, it is far more likely to attain the depth of awareness you are looking for.

## Advisory Services

Advisory services today are viewed from a broad systems perspective, which focuses on the roles and capacities needed at individual, organizational and system levels to address current challenges. In addition to the traditional role of promoting agricultural innovation and technology adoption, these services now must deal with myriad issues, including risk and disasters, climate change adaptation and rebuilding after emergencies. These issues present additional challenges not only to the extension workers but also especially to the farmers themselves. The capacity of advisory services to provide preventive measures or coping mechanisms to address these issues is a critical component of extension services. If these challenges can be overcome, advisory services may be able to aid in enhancing the resilience of farmers in several ways by providing services that are more relevant. Due to its potential access to timely information, the system can indentify relevant actors with whom to work to ensure that intervention strategies are harmonized, relevant, effective and timely. Advisory services could enhance farmers' resilience by providing information and knowledge regarding weather and climate change so that farmers can make informed decisions.

## Human Resources Development

Development of human resources in the form of education and training for farmers could be critical for climate change mitigation and adaptation. The education and training of farmers and extension workers are at times inadequate and need intensification. One of extension's major activities over time has been adult and non-formal education. This role continues today and is even more important in light of climate change. Recent innovative extension activities include farmer field schools and experiential learning approaches could be utilized on issues of climate change.

The capacity of farmers to scope with such different forms of risk will become ever more crucial, and extension efforts must pay special attention to educating farmers about their options to enhance resilience and response capacity. Education must thus move beyond technical training to enhance farmers' abilities for planning, problem solving, critical thinking, prioritizing, negotiating, building consensus and leadership skills, working with multiple stakeholders, and, finally, being proactive.

Capacity development is important within extension as well. Extension agents should have ample time to learn a technology before they can effectively teach others. If this is not realized, extension work would be like the blind leading a blind. Extension personnel may need more training than our farmers to garner credibility. Extension agents have traditionally been trained only in technical expertise and often lack soft skills such as communication, development of farmer groups, systems thinking, knowledge management, and networking. To improve outcomes in rural development, farmers and extension agents need new skills that will require agricultural education and extension curriculums to include valuing and understanding the knowledge and experiences of rural people and co-learning. The extension agencies should focus on extension systems and programmes that incorporate a good understanding of what practices and skills are needed to best promote activities that help in the climate change effort and on increasing the capacity of extension agents and farmers, where needed.

## Technology Transfer

The role of technology and technology transfer has come to the forefront in the climate change mitigation and adaptation strategies. Technology transfer plays a critical role in an effective response to the climate change challenge. Effective and timely transfer of appropriate technology to farmers is crucial for a concerted action towards tackling the challenges posed by climate change. It is now well recognized that technology can play an important role in climate change mitigation and adaptation. While some existing technologies if diffused properly can mitigate the effects of climate change, there is potential for development of new technologies that can help further which can help to reduce the impacts of climate change. An effective technology transfer system requires a carefully thought out plan, clearly communicable ideas, and a cooperative effort. It should be capable of dealing with a wide spectrum of ground realities. Technologies compatible with felt needs, economic and socio-cultural factors, and government policies are amenable to effective transfer. Technology transfer includes a broad set of activities that make development accessible to a wider range of users in the context of climate change. These activities may include the transfer of tangible objects as well as the transfer of knowledge and capacity building.

Technology transfer is a complex type of communication. It requires skilled personnel, appropriate resources, organization structures and formal recognition. In many models five distinct phases of the technology transfer process are recognized. The knowledge phase describes when the technology is known to exist and is relevant to an organization's problems or opportunities. Key stakeholders are persuaded to examine the technology during the persuasive phase. During the decision phase stakeholders decide whether to use the innovation before embarking on the implementation phase. When the technology has been successfully adopted, the confirmation phase is complete. Clear communication is important effective transfer of technology. Understanding the needs of farmers is vital to the technology transfer process. The most effective way to assess needs is through carefully planned direct contact with the farmers at the appropriate level. It is important that the farmers are highly-motivated and ambitious. Trust and previous experience with the farmers is important in elucidating their needs. Systematic and regular forms of communication are recommended in many technology transfer strategies.

## Social Capital Development

The self-organizational capacity of small-scale farmers is increasingly seen as a way of enhancing their concerted efforts to access appropriate services critical for their agricultural and developmental needs. However, farmer organizations in rural areas are generally weak, and are unable to negotiate and mobilize services for their members. This approach could used to strengthen farmer organizations to address challenges faced by farmers at community level with the aim of enhancing their self organizational capacity to better articulate their demands for services, co-ordinate services in the community, create strong linkages with service providers and engage in innovation processes that address their local technical problems. Overall, strengthening local organizations is important for creating partnerships between farmers and external service providers for client orientated service delivery. In this way, service providers are made more accountable to the farmers.

Farmer organizations can play a key role in agricultural innovation once they have the capacity to pool, aggregate and disseminate agricultural knowledge and information. Overall, strengthening of farmer organizations should lead to the development of local innovations for solving farmers' technical problems. Strong farmer groups are capable of providing basic services to their members. However, in addition to organizing producer groups, it should be noted that organizing rural youth groups is an effective, long-term strategy of building human and social capital within rural communities and continues to receive top priority in a few public extension systems.

## Strengthening Institutional Linkages

Another role of extension, which will be critical for climate change issues, is that of acting as an honest facilitator, bringing together different actors within an umbrella. Because of the many agencies providing extension services, there is a need to realign objectives and clearly identify each agency's role in the climate change mitigation and adaptation and sharing resources so as to limit duplication of efforts and minimizing losses to extension resources.

With climate change, it will be increasingly important for the extension system to link farmers and other people directly private and public institutions that disseminate mitigation technologies, and funding programmes for adaptation. Extension also has an enormous challenge in bringing together farmers' concerns and those of other actors as they address climatic uncertainties. Extension has the chance to make a significant contribution to overcoming this gap through enhanced farmer decision making. Extension agents may also play a role not only in brokering, but also in assisting farmers in implementing policies and programmes that deal with climate change mitigation.

## Conclusion

Extension has proven itself to be a cost-effective means of bringing about greater economic returns for farmers with significant and positive effects on knowledge, adoption, and productivity. Extension is thus an effective tool that can play an important role in dealing with climate change while at the same time helping to increase productivity and reduce poverty. To capture this potential role, extension efforts could be channelized to deliver new technologies, information, and education about climate change mitigation adaptation. All of these roles can be exploited in a cost-effective way to help resource-poor smallholders deal with the issues of climate change that will so radically affect their livelihoods. Perhaps the most important purpose for extension today is to bring about the empowerment of farmers, so that their voices can be heard and they can play a major role in deciding how they will mitigate and adapt to climate change.

❏❏❏

# Other Publications on Climate Change and Environment Sciences

| S.No. | Title | Author | ISBN | Year |
|---|---|---|---|---|
| 1 | Advances in Protected Cultivation | Singh, Brahma, B. Singh | 9789383305179 | 2014 |
| 2 | Agriculture and Waste Management for Sustainable Future | Sannigrahi, Asoke Kumar | 9789380235530 | 2011 |
| 3 | Agri-Horticultural Biodiverstiy of Temperate and Cold Arid Regions | Zeerak, Nazir Ahmad | 9789381450086 | 2012 |
| 4 | Agrobiodiversity and Sustainable Rural Development | Soam, S.K. & M.Balakrishnan | 9789383305896 | 2015 |
| 5 | Agroforestry for Increased Production and Livelihood Security | Sushil Kumar Gupta, Pankaj Panwar & Rajesh Kaushal | 9789385516764 | 2017 |
| 6 | Agrometeorology: A Simplified Textbook | Kathiresan, G. | 9789383305568 | 2015 |
| 7 | Agronomy: Principles and Practices | E. Somasundaram & M. M. Amanullah | 9789385516740 | 2017 |
| 8 | Air Pollution and It's Impacts on Plant Growth | Shyam, S. & H.N.Verma: et al | 9788189422103 | 2006 |
| 9 | Basics of Environmental Science and Engineering | Sivashanmugam, P. | 9788189422271 | 2007 |
| 10 | Climate Change and Agricultural Food Production | Kibria, Golam et.al. | 9789381450512 | 2013 |
| 11 | Climate Change and Chemicals: Environmental & Biological Aspects | Kibria, Golam et.al. | 9789380235301 | 2010 |
| 12 | Climate Change and Environment: Concepts and Strategies to Mitigate Impacts | Devesh Sharma & K C Saha | 9789385516375 | 2016 |
| 13 | Climate Change and Food Security | Datta, M.& N.P.Singh | 9788189422387 | 2008 |
| 14 | Climate Change and Natural Resources Management | Lenka,S. & N.K.Lenka | 9789381450673 | 2013 |
| 15 | Climate Change and Plantations in the Humid Tropics | GSHLV Prasada Rao and C.S. Gopakumar | 9789385516368 | 2016 |
| 16 | Climate Change and Water Security: Impacts, Future Scenarios, Adaptations and Mitigations | Golam Kibria *et al.* | 9789385516269 | 2016 |
| 17 | Climate Mitigation and Carbon Finance: Global Initiatives & Challenges | Sahoo, A.K. | 9789381450024 | 2012 |
| 18 | Climate Resilient Crops for the Future | Peter, K.V. | 9789383305599 | 2015 |
| 19 | Climatic Variability: Impacts on Agriculture and Allied Sectors | Datta, M.:Ed. | 9789381450949 | 2014 |
| 20 | Community Ecology of Tropical Birds | Sivaperruman,C. & E.A.Jayson | 9789380235165 | 2010 |
| 21 | Computers in Agriculture | M. Kumar Sharma & Anil Bhat | 9789385516160 | 2017 |
| 22 | Conservation Agriculture for Carbon Sequestration and Sustaining Soil Health | Somasundaram, J., R.S.Chaudhary: et.al. | 9789383305322 | 2014 |
| 23 | Current Trends in Global Environment | Bhatia, A.L. | 9788189422004 | 2005 |
| 24 | Detritus and Decomposition in Ecosystems | Reshi, Zafar | 9788189422158 | 2007 |
| 25 | Environment,People and Development: Experiences from Desert Ecosystems | Gaur Mahesh Kumar & P.Moharana | 9789381450796 | 2014 |

| S.No. | Title | Author | ISBN | Year |
|---|---|---|---|---|
| 26 | Environmental Changes and Natural Disasters | Babar, Md. | 9788189422752 | 2007 |
| 27 | Forestry Science: Fundamentals and Terms | Sharad Nema | 9789385516382 | 2016 |
| 28 | Geomatics in Applied Geomorphology | Ramasamy, SM | 9789385516429 | 2016 |
| 29 | Managing Soil Health for Maximising Crop Productivity | Obeng, F.k, Avornyo | 9789385516788 | 2017 |
| 30 | Microbes for Restoration of Degraded Ecosystems | D. Joseph Bagyaraj & Jamaluddin | 9789385516665 | 2017 |
| 31 | Microbial Diversity and Its Applications | Barbudde, S.B. R. Ramesh & N.P.Singh | 9789381450666 | 2013 |
| 32 | Physical Climatology | Sellers, Williams | 9789383305575 | 2014 |
| 33 | Practical Agricultural Meteorology | Srivastava, A.K. | 9789380235776 | 2011 |
| 34 | Protected Cultivation of Horticultural Crops | Singh, D.K & K.V.Peter | 9789383305155 | 2014 |
| 35 | Recycling of Industrial Effluents | Manivanan, R. | 9788189422127 | 2006 |
| 36 | Sustainable Environmental Science | Sahu, D.D. | 9789381450208 | 2012 |
| 37 | Technologies for Sustainable Green Environment | Davamani, V. | 9789381450420 | 2012 |
| 38 | Water Quality Modeling: Rivers,Streams and Estuaries | Manivanan, R. | 9788189422936 | 2008 |